ROCKVILLE CAMPUS LIBRARY

W9-DCJ-718

WITHDRAWN

SATELLITE COMMUNICATIONS

2ND EDITION

STAN PRENTISS

AAU6607

SECOND EDITION
FIRST PRINTING

Copyright © 1987 by TAB BOOKS Inc.
Printed in the United States of America

Reproduction or publication of the content in any manner, without express permission of the publisher, is prohibited. No liability is assumed with respect to the use of the information herein.

First edition copyright © 1983 by TAB BOOKS Inc.

Library of Congress Cataloging in Publication Data

Prentiss, Stan.
Satellite communications.

Includes index.
1. Artificial satellites in telecommunication.
1. Title.
TK5104.P73 1987 621.38′0422 86-23163
ISBN 0-8306-0192-9
IBSN 0-8306-2792-8 (pbk.)

Cover photograph courtesy of GTE Spacenet Corporation.

Contents

Acknowledgments

I WISH TO EXPRESS MY SINCERE GRATITUDE TO ALL THOSE INdividuals and the companies and corporations they represent for their technical and editorial contributions to the second edition of *Satellite Communications.*

Dr. Mark F. Medress, Paul Fisher, Randy Raybon, and Tom Winslow of M/A-COM; John Williamson of RCA Americom; Jim Byrnes of AT&T Communications; Bill Johnson, Ron Mohar, and Jay LaBarge of Microwave Filter Co.; Gil Hodges of Pico Products, Inc.; Neil Le Saint, Mike Brubaker, and Rich Renken of R.L. Drake; John Klecker and Mark Fehlig of Harris Corp.; Hal Sorenson and Jerry Von Behren of Winegard Co.; John T Hanley of Andrew; Mark Sheldon of Chaparral; Steven Lett, Roger Herbstritt, Joe Harcarufka and George Fehlner of the FCC; Paul Miller of Belden Electronic Wire and Cable; Frank Finn of CommTek; Wallace Burton of Wiltron; Bill Stark and Tom Thompson of Uniden; Dave Waddington and Harry Greenberg of Channel Master; Mark Sellinger of Avantek; Frank Bremer and Robert Maniaci of Boman Industries; Len Garrett and Terry Burchman of Tektronix; Wendell Bailey of National Cable TV Assn.; Allan Schlosser, Thomas Friel and Tom Mock of the EIA; and, as always, a very special note of thanks to George Bell and Jim Nelson, two *excellent* electronic engineers wherever they are.

Introduction

IN THIS UPDATED SECOND EDITION OF *SATELLITE COMMUNI-cations* I have retained the material concerning the general satellite industry and all the useful equations. I have added the very latest material available from the satellite industry. Instead of a few word and punctuation changes, you will find entirely new sections devoted to satellite TVRO antennas, receivers, scrambling (both VideoCipher® and B-MAC), fresh and timely information on Ku-band and RCA's K-birds, DBS, terrestrial interference and troubleshooting, quick methods of nomograph satellite location from *any* lat/lon position; a discussion of both fiberglass and perforated metal reflectors; new type Ku- and C-band combined feeds; and further descriptions of satellites now flying, along with variation and deviation charts for the entire U.S.

If you thought the first volume was useful, try this updated version for all its new material, and a combination of the two should amply fill most of your needs for general satellite communications, especially TV receive-only terminals (TVRO).

Frankly, we're really on the threshold of this great geosynchronous satellite undertaking for both data and excellent audio/video reproduction. You really haven't heard stereo and seen excellent color until you've received them on an outstanding TVRO system that supplies the appropriate audio and video (L/R) outputs for your TV/monitor. Musicians, photographers, engineers, and artists take particular note: this new medium as seen and heard on good equipment offers noteworthy instruction and special enjoyment. Many others may find these suggestions equally true, particularly those who are eager to learn and appreciate good music and living color.

Chapter 1

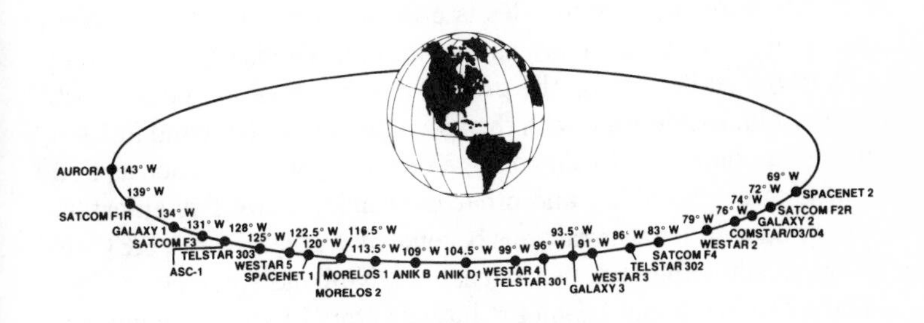

Satellites in Orbit and Approved for Construction

ORBITING GEOSYNCHRONOUS SATELLITES AND FIBEROPTIC cables are fast becoming our principal means of terrestrial communications. Ships at sea, electronic newsprint and photographs, entertainment, religion, business transactions, aircraft, telephone, voice, video music, and messages of any and all descriptions are moving in and out of space at frequencies between 3.7 and 12 GHz right now with applications up to 30 GHz pending. Broadband transmissions are less expensive and highly reliable, well-designed receiver systems and transmitters are producing almost remarkable results without appreciable noise or other side effects, and cable television (CATV) has increased programming immensely as it leans more heavily on satellite reception. Even the three major TV networks are now using SBS, TELSTARs 301/302, and K-2 satellites for a large portion of their feeder and national traffic with excellent results, especially when programming is viewed on video/audio baseband monitors that will also reproduce stereo sound in all its glory.

Conservative estimates indicate numbers of foreign countries directly involved in satellite transmissions/receptions at 110, with tens of thousands of foreign earth stations in active operation in the Americas, Europe, and Asia, as well as others we may not even know about. As you may imagine, electronics and hardware production has increased dramatically over the past several years and will continue to expand with a few bumps and growing pains as programmers, owners, entrepreneurs and 8-10 year lifetime satellites come and go. And while M/A-COM's VideoCipher® and other scrambling devices have depressed the home satellite industry temporarily, a resurgence in sales is now underway by the survivors and there may be as many as two million television receive-only (TVRO) installations on the ground and in operation before 1987 is very old. In addition, with the advent of a *lusty* Ku-band in 1986, these numbers might well go higher as more services such as Federal Express' ZapMail and other data links as well as entertainment and additional attractions become available. Even AT&T, the phone giant, expects to use at least two Ku-band channels in 1986. The smaller 1.2- and 1.8-meter antennas for Ku are becoming very popular, both for homeowners and commercially because they may be mounted almost anywhere there is line-of-sight to the transmitting satellite.

A growing number of 6- and 8-foot C-band antennas are also available in solid, mesh, and perforated versions from the various consumer manufacturers, and operate at varying degrees of efficiency in their diverse locations. With the coming of 2° spacing as mandated now by the Federal Communications Commission for all geosynchronous satellites, carrier-to-interference problems will most certainly increase as *all* C-band "birds" are either moved into new positions or replaced with updated versions. Late tests have shown that parabolic receptors under 9 feet in diameter and more than 2° beamwidths could be subject to C/I and other related problems. What will happen at Ku we don't know since the few hybrid (4 and 12 GHz) and Ku-only geosynchronous orbiters now flying haven't generated enough electromagnetic clutter yet. But when there are more of them, and cross polarization transmissions are the rule, we'll have a much better idea. With more than 25 parking spaces yet to be filled at Ku, the wait-to-see may take some time.

The direct broadcast service (DBS) uplink at 17 GHz and downlink at 12.2-12.7 GHz is momentarily on-hold due to the sad experience of U.S. Communications, Inc., which operated on Canada's ANIK-C-2 for approximately a year and then went "belly up" for lack of sufficient subscribers and into the Chapter 11 bankruptcy courts after shutting down on March 31, 1985. Prudential and other backers could lose as much as $100 million as well as equipment and subscription losses of the reportedly 9,000-10,000 customers left in the lurch.

The ill-fated USCI venture, however, did prove that large sections of the U.S. could be served by a small, relatively low-powered

Ku-band satellite at least in the upper east and mid-west sectors. Satellite Business Systems with its SBS3 and 20 watts of power has been even more successful, and the K birds of RCA at 45 watts per transponder are already broadcasting loud and clear even during heavy rainfalls. In addition, K-1 may carry some direct-to-earth service to hotels and motels beginning in 1987 if presently announced trials prove satisfactory.

While all this is going on in Ku-band, DBS is *not* yet dead, and may even experience a resurgence, though possibly not in the exact form originally intended. Although CBS, Western Union, and others have backed down, and COMSAT has its two DBS powerful orbiters up for sale, there remain four applicants who have full approval for construction and due diligence—which means they have contracted for construction. These include STC (COMSAT), USSB, Dominion Video Satellite, and Hughes. While still authorized as direct-to-home service, USSB has now an approved request by the Federal Communications Commission to offer other than broadcast services. Actually, the two Satellite TV Corp. satellites have already been built by RCA Astro and may be in orbit as soon as U.S. or Ariane launch facilities are available. Unfortunately, the Challenger Shuttle catastrophe has temporarily shut down and severely delayed commercial launches probably through 1986, or until some nonreusable, single-purpose rockets are again capable of carrying payloads.

As for C band, its 30 orbit slots are virtually filled and only replacement satellites, for the most part, are entering this frequency band. Already two COMSTARs, two SATCOMs, and one WESTAR have been retired, with others ready for substitution shortly. Geosynchronous satellites are usually designed for lifetimes of between 7 and 10 years, but this may be prolonged if Shuttle astronauts or specially trained technicians can refuel their station-keeping chemicals inflight. This is a possibility for the better orbiters since a price tag of $100 million for construction and launch isn't unrealistic. The electronics, apparently, will last considerably longer if there are no obvious breakdowns.

Before continuing, this would be a good place to identify the various satellites and their owner/operators:

SATCOM—RCA American Communications, Inc.

WESTAR—Western Union Telegraph Co.

COMSTAR—Owned by COMSAT, operated by AT&T (now being replaced).

GSTAR—GTE Spacenet Corp.
SBS—Satellite Business Systems (MCI and IBM)
GALAXY—Hughes Communications Galaxy, Inc.
TELSTAR—AT&T
SPACENET—GTE Spacenet Corp.
ASC—American Satellite Co.
USSB—U.S. Satellite Broadcasting Corp.

In a recent buyout, IBM sold SBS 1, 2, and 3 to MCI, while retaining SBS 4 and 5 for itself. The first three will probably handle data and video because voice traffic will go by terrestrial transmission instead. IBM hasn't yet announced its intentions for SBS4 and SBS5.

For the future, higher powered and perhaps somewhat heavier satellites are planned, as well as a great deal of attention devoted to Ku-band and, later, DBS. Techniques and electronics are now well understood and fully applicable, especially by Hughes and RCA. Video/audio scrambling may well peak in 1987 and subside thereafter due to smart, but illegal, black decoder boxes, sagging TVRO owner response, and the eagle eyes of advertisers looking for maximum markets. Thereafter, we would expect both C and Ku-bands to flourish, along with digital television, college educational credit programs, work-at-home formats, and many new market and financial services for both individuals and businesses. In truth, we are now undergoing a veritable video/audio explosion of remarkable proportions. A piece of the action could be profitable.

SATELLITE BUSINESS SYSTEMS

Meanwhile, Satellite Business Systems (SBS) has been using the 14/12 GHz uplink/downlink service in what it calls wideband digital integration (Fig. 1-1). It transmits some video, all forms of business communications, and is designed for working organizations with large volumes of traffic. Smaller receiving dishes make urban reception practical and feasible, with 5.5- to 7.7-meter receptors permitting locations at key company sites. Digital transmission techniques offer considerable security, along with random traffic interleaving from the various sources, in addition to a customer-controlled "bulk encryption capability" which will become available shortly. Communications Network Service transmits between 2400 and "several" megabits per second. Most customers, SBS says, will "use CNS for long distance telephone services."

SBS has also announced plans for service in medium-to-high-speed data communications. The FCC has already been requested to approve applications for data speeds in the kilobit range of 56, 112, 224, 448, and 896, in addition to 1.344 and 1.544 megabits per sec for point-to-point simplex and duplex transmissions, as well as point-to-multipoint broadcast transmissions. DNS earth stations will have rf terminals with 3.6-meter or a few 4.7-meter antennas and TDMA (time-division multiple access) controllers. This will be called the data network service (DNS) and will offer lower volume rates than communications network service (CNS), which is designed for very high capacity earth stations already in service. Digital video transmissions are part of this new program in the Ku-band. SBS tracking and telemetry stations are located at Castle Rock, Colorado, and Clarksburg, Maryland, with headquarters in McLean, Virginia. The system is expected to have some 120 earth stations.

Although traffic growth is considerable even now, when computer-to-computer exchange begins, massive transfers of information are projected. Work-at-home terminals are also under study even as industry looks at the possibility of decentralizing to the more sparsely populated portions of the country such as Idaho, South Dakota, and Montana. In addition to huge data communications programs, there's teleconferencing, plus the prospect of electronic mail at rates of a page per second with costs of only pennies per page. Considering that government alone spends upwards of four billion dollars per year in travel, all this means a complete reorientation in the conduct of both government and private business during the coming decade. The vendor (seller) of the future, we're told, must also be a first-rate consultant with personality to match, and he or she will be paid accordingly. Well-modulated voices, good looks, and sharp wits, apparently, are to continue in ever increasing demand.

COMSAT

The Communications Satellite Corporation, a huge space services and communications conglomerate formed in February 1963, remains an enormous factor in the space industry. Following Congressional passage of the Communications Satellite Act in 1962, COMSAT and then INTELSAT (Fig. 1-2) emerged which, with its six or more active satellites in synchronous 22,240 mile orbits over the Atlantic, Pacific, and Indian Oceans (at the equator) now

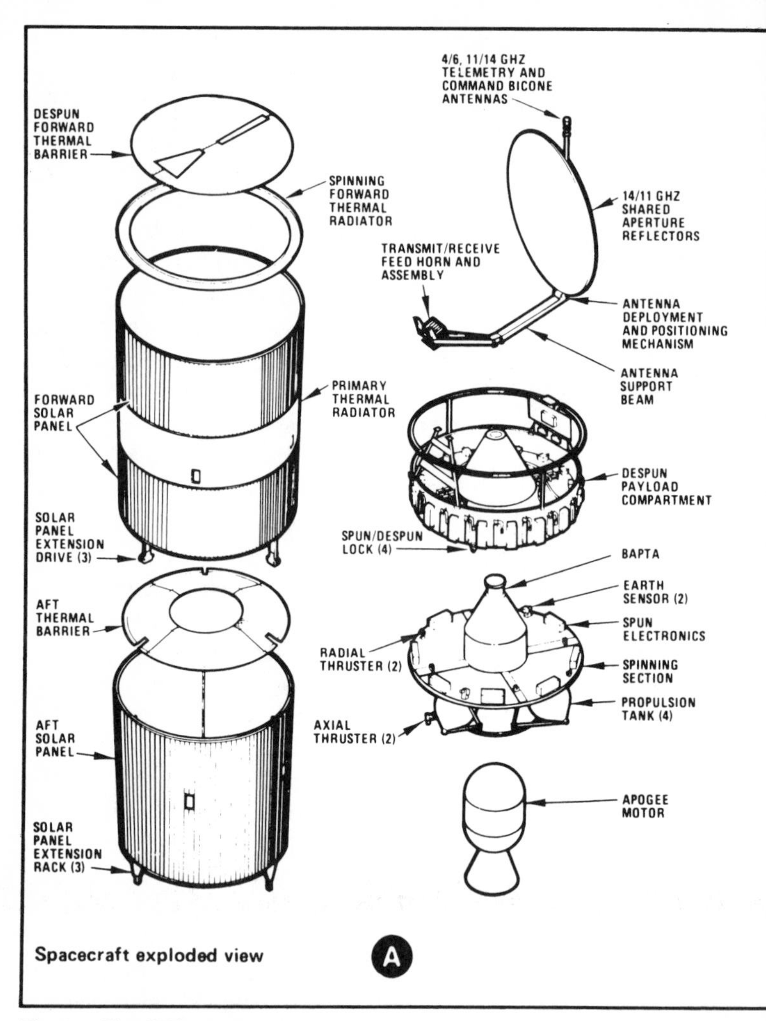

Fig. 1-1. The SBS spacecraft built by Hughes (A) and the launch sequences (B) (courtesy of Hughes Aircraft).

carries about two-thirds of all transoceanic traffic. Under international joint ownership, INTELSAT has a 105-nation owner-equity of some 729 million dollars, with COMSAT as the U.S. representative. INTELSAT maintains and operates all space vehicles and traffic. The earth stations in the various member countries are government owned or privately owned. Most countries except the Russian Warsaw Pact group are members.

MARISAT, as a joint venture of COMSAT General and three

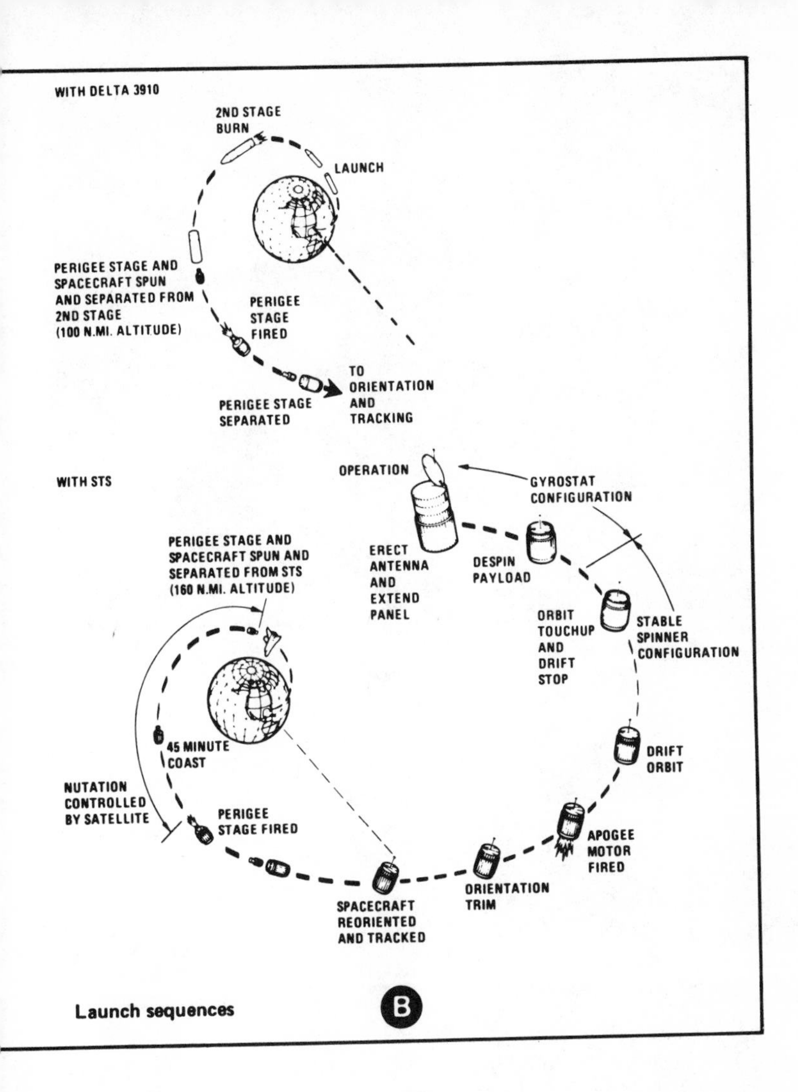

Launch sequences

other international carriers, Fig. 1-3, has three multifrequency satellites serving the Indian, Atlantic, and Pacific Oceans. Its mission provides modern satellite communications to ships at sea, including the U.S. Navy, and commercial shipping. More than 600 commercial ships and offshore drilling rigs are now equipped with MARISAT terminals; COMSAT supplies over 200. Great Britain also leases some capacity in the Atlantic. These satellites have only a 5-year design life, but are expected to continue in service much

Fig. 1-2. The INTELSAT V operating in the 11/14 and 4/6 GHz bands delivering 12,000 circuits and two TV channels (courtesy of COMSAT).

Fig. 1-3. The MARISAT with L/C and C/L band repeaters. The design life is 5 years (courtesy of COMSAT).

longer. Another maritime satellite organization in which COMSAT has a large interest is called INMARSAT, which stands for International Maritime Satellite Organization. Over 30 nations are members of this group which began operations in 1982.

COMSTAR (Fig. 1-4), the final operational satellite grouping under COMSAT (General), has four satellites in operation, all originally leased to American Telephone and Telegraph for its long-distance telephone network. Each has a design life of seven years and can handle 18,000 simultaneous telephone conversations. Seven of the earth stations in this system are owned by AT&T (four) and GTE (three). COMSAT General has monitoring stations at Southbury, Connecticut and Santa Paula, California.

Begun but not yet serviceable (1982), COMSAT has another subsidiary—this time wholly owned—that promised even greater diversity than those already named. The U.S. consumer now becomes the direct beneficiary, as the FCC gives its approval. Satellite Television Corporation applied to the FCC on December 17, 1980 for permission to start up its three-channel pay-TV system with two experimental satellites for the Eastern U.S. time zone (Fig. 1-5), both operational and the extra transponder in backup. Even-

Fig. 1-4. COMSTAR handles 18,000 phone calls and has 24 transponders used by AT&T (courtesy of COMSAT).

Fig. 1-5. Artist's conception of an STC satellite. Power for each channel is 185 watts (courtesy of Satellite TV Corp.).

tually, all of the U.S. mainland and Alaska-Hawaii would be covered by four of these 14 GHz uplink and 12 GHz downlink, relatively high powered satellites spaced 20° apart in geosynchronous orbits. While the FCC continues approval, STC on March 8, 1982 released technical specifications for home TVRO dishes (plus attendant electronics) which are parabolic reflectors of 2.5-feet in diameter, with a low-noise downconverter attached to the reflector. From there, a cable will carry the received intelligence (at about 1 GHz) directly to an FM demodulator, channel selector, and AM video modulator. Nonspecific downlink frequencies are between 12.2 and 12.7 GHz. STC's descrambler units are not included in these specifications.

STC in the meantime, however, let contracts for its first two satellites to RCA Astro Electronics, which won contracts over comepetitors Ford Aerospace, General Electric, and Hughes Aircraft. With FCC conditional approval already secured, pending the outcome of the RARC Regional Administrative Radio Conference in Geneva, set for June 1983. These satellites are to begin working our Eastern time zone, with others to follow covering the rest of the country. STC has also purchased 40 acres of land northwest of Las Vegas for its broadcasting center and contracted for architectural engineering and equipment location setup. RCA Astro Electronics expects to take somewhat more than three years to construct these first two space vehicles, now virtually complete but up for sale to either U.S. or possibly foreign interests.

At the same time, COMSAT laboratories have already tested prototype antennas, LNAs, receivers, and set top converters, preparatory to selecting one or more of the dozen manufacturers bidding for this considerable business.

Other companies given tentative approval on the same RARC basis were: CBS, Inc.; DBS Satellite Corp.; Graphic Scanning Corp.; Western Union Telegraph; RCA Americom; U.S. Satellite Broadcasting Co.; and Video Satellite Systems. For the most part, these companies are doing what STC has done, and are sticking to service for the Eastern time zone. U.S. Satellite Communications, Inc., it was also proposed to use some five transponders on Canada's ANIK-C-2 to commence Boston-Washington-Chicago service by or before 1984, securing a headstart on its competition with DBS-type service in the 11.7-GHz downlink.

There are presently three main commercial satellite types suitable for general television signal transmissions: one is by Hughes, Western Union (WESTAR); the second, A.T.&T. (TELSTAR); and the third by RCA (SATCOM). All operate at C-band in the 4/6 GHz standard ranges with the usual 500 MHz bandwidths. WESTARS have 12 transponders that will handle signals 40 MHz wide, with each transponder separated by 40 MHz. SATCOMs are equipped with 24 transponders, each operating within 40 MHz but only separated by 20 MHz. This is done by vertically polarizing 12 transponders and horizontally polarizing 12 transponders so that each set may carry different signals without crosstalk, aided by a 20 MHz frequency offset. Hughes also has 24 transponders.

Most fixed satellite communications now operate at 6 MHz on the uplink (transmitter) and 4 GHz on the downlink (receiver). Gain over temperature (G/T) characterizes the receiver, while EIRP (effective isotropic radiated power) is usually specified around the saturation figure for a transponder amplifier. Typical domestic satellite EIRP amounts to between 30 and 40 dBW in prime coverage surroundings. However, picture or transmission quality depends very much on carrier-to-noise, which is the ratio of carrier-to-noise power for some specified bandwidth. The greater this ratio, the better the signals are produced by (and through) the receiver. Other additional definitions and explanations of transmit and receive electronics will be covered in the following chapters.

Fixed and direct broadcast services (DBS), in one way or another, will be reaching an estimated 91.6 million U.S. TV homes by the year 1990, either directly, by broadcast, or CATV. Both pay TV and subscription TV will have their respective shares of the market.

Geosynchronous satellites orbiting 22,300 miles above the U.S. are assigned to the following three frequency bands:

C-Band (Fixed)

Uplink 5.925-6.425 GHz *Downlink* 3.7-4.2 GHz

Original Orbit Separation 4°, but now reduced to 2°

Ku-Band (Communications)

Uplink 14-14.5 GHz *Downlink* 11.7-12.2 GHz

Orbit Separation 2°

K-Band (DBS)

Uplink 17.3-17.8 GHz *Downlink* 12.2-12.7 GHz

At the moment, Satellite Business Systems (SBS) RCA, and Spacenet use the Ku-band because of C-band crowding and the 12 GHz downlink, but others are certain to follow since they are already successful in their operations. As these C- and K-band station assignments are filled, additional requests will be filed with the FCC for extra slots at even higher frequencies. So don't be surprised if 20 GHz and then 30 GHz becomes occupied in the next few years with even greater technology and higher frequency satellites.

SATCOM

Resembling a coffee grinder with two paddles (Fig. 1-6), SATCOMs are going into orbit as fast as launch dates permit. Built first for RCA and then for its wholly owned subsidiary (RCA Americom) by RCA Astro-Electronics, Princeton, New Jersey, these satellites, variously known as SATCOMs I through IV (D & E), and the Advanced SATCOM series F through H, occupy equatorial posi-

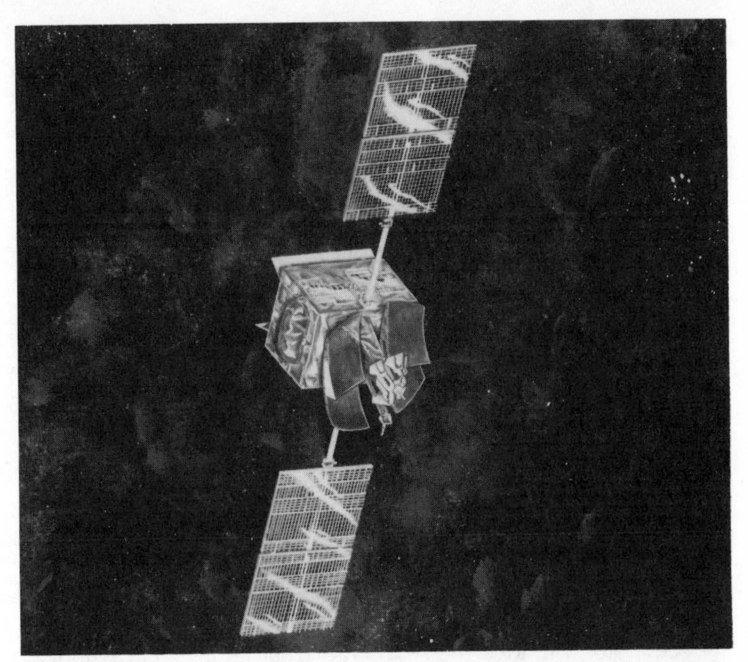

Fig. 1-6. The RCA SATCOM D and E series is built primarily for TV (courtesy RCA Americom).

tions of 136° WL (SATCOM I); 119° WL (SATCOM II); 131° WL (SATCOM III-R); 83° WL (SATCOM IV); 139° WL (SATCOMI-R); and 66° WL (SATCOM II-R). An Alascom satellite is also positioned at 143° West Longitude (WL) for the Alaska-mainland service.

All SATCOMs have 24 channels which can handle 1,400 voice circuits, FM/color TV transmissions or 64 megabits/sec of computer information. They can cover all 48 continental (CONUS) states, and Hawaii/Alaska, in addition to Puerto Rico. Of the 22 million homes served by CATV, almost all see at least one satellite channel program. And as SATCOM IV flies, there will be two satellites devoted to CATV programming with a capacity of more than 1,000 hours each day. Solar arrays and three nickel-cadmium batteries provide power, which will be increased from 5 W to 8.5 W for six transponders as SATCOMs III-R and IV (D & E) enter service, supported by 90 square feet of solar cells instead of the 75 flown by their predecessors. There are also four spare transponders aboard, extra orbit-keeping fuel, better attitude control subsystems, and enhanced ground-control override of automatic satellite station keeping equipment.

Advanced SATCOMs F, G, and H can also shape their antenna beam patterns using large-aperture antenna reflectors with multiple feedhorns and provide 24 channels to Alaska and Hawaii when in favorable orbital slots. They will also have solid-state transponders instead of traveling-wave tube amplifiers (TWTs) which offer improved linearity resulting from better intermodulation suppression between signal carriers. Design improvements, in addition to greater antenna gain, will double the traffic capacity per channel over prior satellites, and deliver at least a 3 dB improvement in the noise figure. Signal-to-third-order intermod distortion between carriers is between 3 and 8 dB better than with the TWTs. Linearity backoff is also 5 dB less than for traveling-wave tube amplifiers. Therefore, an Advanced SATCOM transponder with its greater linearity and phase-equalized elliptic function filters, will be able to handle two high quality simultaneous video transmissions without crosstalk.

With a transfer orbit weight of 2431 lbs., the solid-propellant apogee motor will kick the spacecraft from its inclined, elliptical transfer orbit at launch to an equatorial, geosynchronous orbit at 22,300 miles above the earth's surface. There, a three-axis fixed-pitch momentum wheel and magnetic torquing with thruster backup will keep the satellite steady in north-south position. Yaw and roll problems are handled by inclination control thrusters under microprocessor control. Compandors compressing and expanding dynamic voice signals will increase existing FDM/FM capacities from one- to two-thousand circuits coupled with improvement in signal-to-noise. (FDMA stands for *frequency-division multiple access*, something you'll hear much more about in the next chapter.) There are also echo cancellers on private leased channel services to subtract any part of this problem, so that a satellite channel may operate just as well as a terrestrial channel. Five-meter duplex earth stations are located at customer offices and are remotely monitored and controlled from a master station.

WESTAR

Western Union, the first U.S. firm to place domestic communications satellites in orbit—WESTARs I and II in 1974—launched WESTAR III in 1979, and WESTAR IV and V in 1982 (Fig. 1-7). The first three carried 12 transponders and are used for data, video,

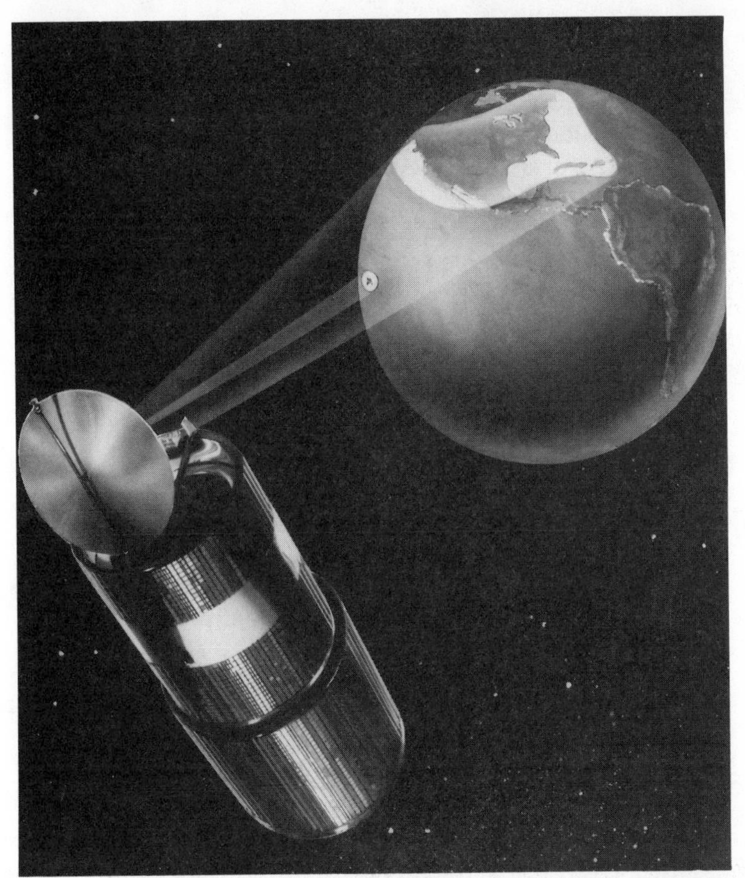
Fig. 1-7. WESTARS IV and V orbited in 1982 (courtesy of Western Union).

voice and facsimile communications traffic. WESTARs IV and V are 24-transponder satellites, each with a design life of 10 years and were lifted into orbit with Thor-Delta rockets. Respective dates of launch and orbits are: WESTAR I, April 13, 1974; WESTAR II, October 10, 1974; WESTAR III, 1979, 91° WL; WESTAR IV, February 25, 1982, 99° WL; and WESTAR V, June 1982, 123° WL.

WESTARs continue to serve Public Broadcasting's 200 television stations via 169 satellite earth stations; Public Radio and its 237 stations; and Mutual Broadcasting's 900 radio stations. The Wall Street Journal is also using satellite distribution to print its Eastern, Midwestern and Western Editions by facsimile. U.S. News & World Report sends material to plants on both the East and West coasts, and the Associated Press distributes its services to newspapers and broadcasters nationwide. Others such as Martin Marietta, Harris, and Digital Communications lease channels on WESTAR, and many additional companies hold their meetings via teleconferencing over this very considerable network (Fig. 1-8).

WESTARs IV and V, now double the weight of the first three WESTARs, have twice the number of transponders and have almost triple the end-of-orbit power. All are built by Hughes Aircraft and have model numbers of HS-333A for the first three, and HS-376G for the other two. These latter satellites have 7,200 one-way voice circuits, or one full color TV signal and audio, or data transmission of up to 64 megabits/sec on each transponder. Frequency bands are 5,925 to 6,405 MHz for uplink and 3,700 to 4,180 downlink, with design life listed as 10 years.

GSTAR

Only recently announced by GTE Satellite Corporation (GSAT) of Stamford, Connecticut, this new series launched in the second and third quarter of 1985, will operate in the 12/14 GHz Ku-band, be body stabilized, and have a mission design life of 10 years. Putting out 20 watts for CONUS (continental U.S.) and 30 watts for CONUS, Hawaii, and Alaska per transponder, two of these satellites will occupy orbital positions of 103° WL and 106° WL and a third retained as a ground spare. Each will have 16 transponders using cross polarization, along with switchable east and west spot beams for better East and West Coast coverage. Hawaii and Alaska are included. Communications will be digitized in time division multiple access and demand multiple access systems (TDMA and DAMA). Service is to begin in the middle of 1984 (Fig. 1-9).

Each of the 16 transponders will have capacities of 60 megabits-per-second, which is the highest satellite capacity proposed to date, according to GTE. Total cost is $100 million. The two operational "birds" will handle 30,000 simultaneous telephone conversations, 300 two-way video conferences, or combinations of both.

A total of seven earth stations will serve this GSTAR and GSAT joint enterprise with AT&T for telephone operations. Prior to the May 1985 launch of GSTAR, GTE will provide 10 leased transponders in January 1983 to POP Satellite, Inc. and Allstar Satellite Network, Inc., operating as United Satellite Television in a multi-program, multi-channel programmed transmission within the U.S. GSAT will also construct satellite earth stations in New York,

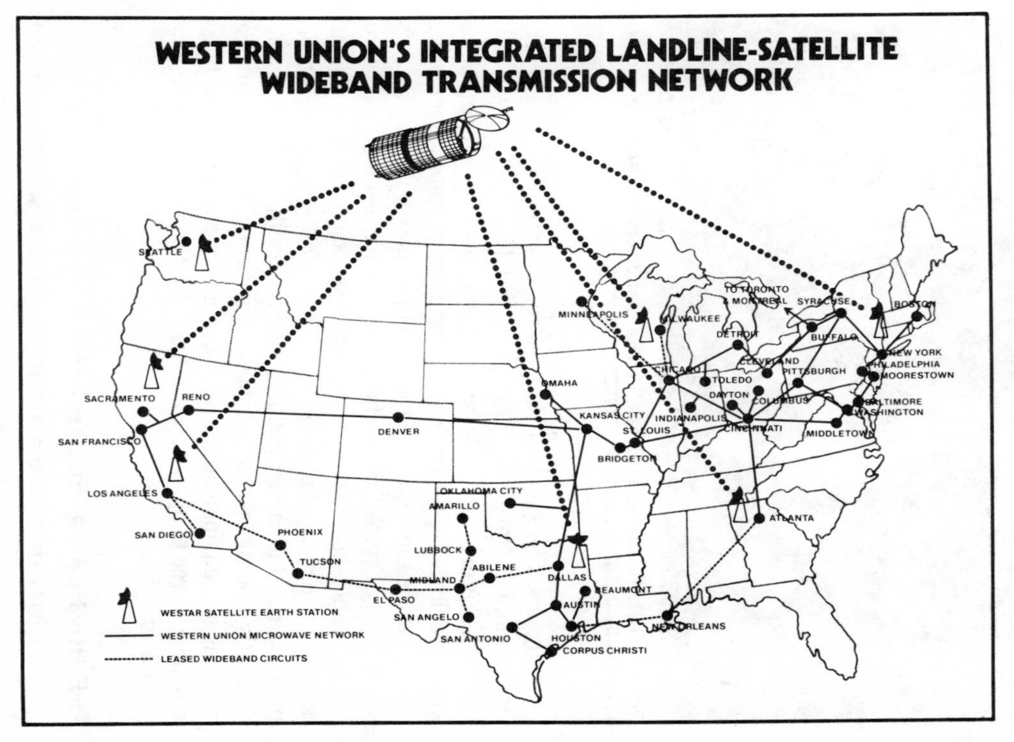

Fig. 1-8. The WESTAR U.S. Satellite network (courtesy of Western Union).

Fig. 1-9. GSTAR, the Ku-band satellites for both telephone and video transmissions fly in 1984 (courtesy of GTE).

Los Angeles, Chicago, and Houston for private line channel service beginning November 1982 at 12/14 GHz.

In the future, there will be larger satellites, even clusters, as well as dedicated space stations. At the moment, those in the 2,000 kgm class are the largest and will occupy about half the Shuttle's payload capacity. A limiting factor to date has been the on-board fuel for station-keeping requirements. When such fuel depletes, the satellite cannot be maintained in 0.1° to 0.05° synchronous orbit and its earth stations are not able to successfully track it and consequently lose signal and control management. Later, as refueling methods and equipment are refined, some of the newer satellites may be restored to operational status if their power and electronics remain functional.

THE HUGHES SATELLITE FAMILY

This has expanded considerably in the past several years and now includes Aussat (Australia Nat. Sat. Comm.); Palapa-A and B (Indonesia); Morelos (Mexico); TELSTAR III (AT&T); Intelsat VI (International Telecommunications); Leasat (Dept. of Defense); Galaxy (Hughes series); SBS (Satellite Business Systems); ANIK-C (Canada); and a brand new group designated as HS 393 intended for SBS F6 Prime, Intelsat 6, and JC satellite. A successor to the highly successful HS 376 series, this Ku-band vehicle may be launched either from the NASA Shuttle or Europe's Ariane. Similar in many respects to the HS 376 Hughes group in payload and electronics, first scheduled to launch should occur December 1987 where it will replace SBS-1 at 99° WL, designed for 10 years of service to CONUS, Alaska and Hawaii. Major differences between this series and the HS 376 series are power and capacity. Each of its 19 transponders transmit 40 W, with more than double the capacity of its forerunners, offering fully switchable voice, high-speed data, fax, and broadband video networks and can focus independent and fully polarized beams over all coverage at bandwidths of 43 MHz.

With the HS 376 satellites now flying and in various stages of completion, the Hughes family of satellites grow rapidly as Delta, Ariane, or Space Shuttles place them in orbit, depending on what's available. Versions will offer C- and K-band facilities including voice, fax, and video, with a design life of 10 years.

Gyroscopically spin-stabilized, the HS 376 measures 85-inches in diameter with a bright silver band around the middle that acts as a thermal radiator. Figure 1-10 shows both stabilizing action and the radiator. With antennas and telescoping cylindrical solar panel in the stowed (launch) position, the spacecraft measures 111 inches high. Pinions driven by three redundant motor and gear box combinations operate the outer solar drum so that control occurs while spinning. After injection into elliptical transfer orbit, the spacecraft weighs some 2,325 lbs. of which 211 lbs. is station-keeping fuel for the position thrusters.

ANIK-C

A somewhat earlier version of this Canadian satellite (C-3) orbited via the Space Shuttle November 11, 1982,and C-2 also attained launch on 18 June 1983. The final satellite in this series, ANIK-C-1, was lofted by the Shuttle April 12, 1985. Their arrays of K7 solar cells generate 19.7 mW/cm^2, producing more than 900 W of dc power initially, with NiCad batteries providing backup during

Fig. 1-10. Hughes HS376 stabilizes itself while spinning (courtesy of Hughes Aircraft).

periods of eclipse. In each of the 16 channels, 15 W TWT amplifiers with multicollectors complement a 1.8-meter shared aperture grid antenna for transmit and receive beams using two reflecting surfaces, with one surface sensitive to vertical and the other to horizontal polarization. Stowed for launch, these 12/14 GHz satellites measure only 9 feet 3 inches high and 7 feet 1 inch in diameter. At launch the solar arrays are telescoped together. When deployed, however, height is increased to 21 feet with an on-station weight of 1,250 lbs. The deployed satellite is shown in Fig. 1-11.

ANIK-D

These advanced Canadian satellites supplant the three ANIK-A satellites during 1982 and 1983. Their mission is to transmit cable and educational TV, and long distance phone service, although D-1 does a good deal of video (much of it scrambled). Spar Aerospace Ltd., is the designated prime contractor, with Hughes as the largest subcontractor, and TELESAT of Canada, the final recipient. While ANIK-D covers all of Canada and much of the U.S., ANIK-C delivers services over Canada's more densely populated southern sector. ANIK-D has 24 channels, each of which will operate 960 one-way voice communications or a color TV program at 36 dBW. Mission lifetimes are 8 years with 10 years of overall operation.

PALAPA B

This is a series of second generation satellites built for Indone-

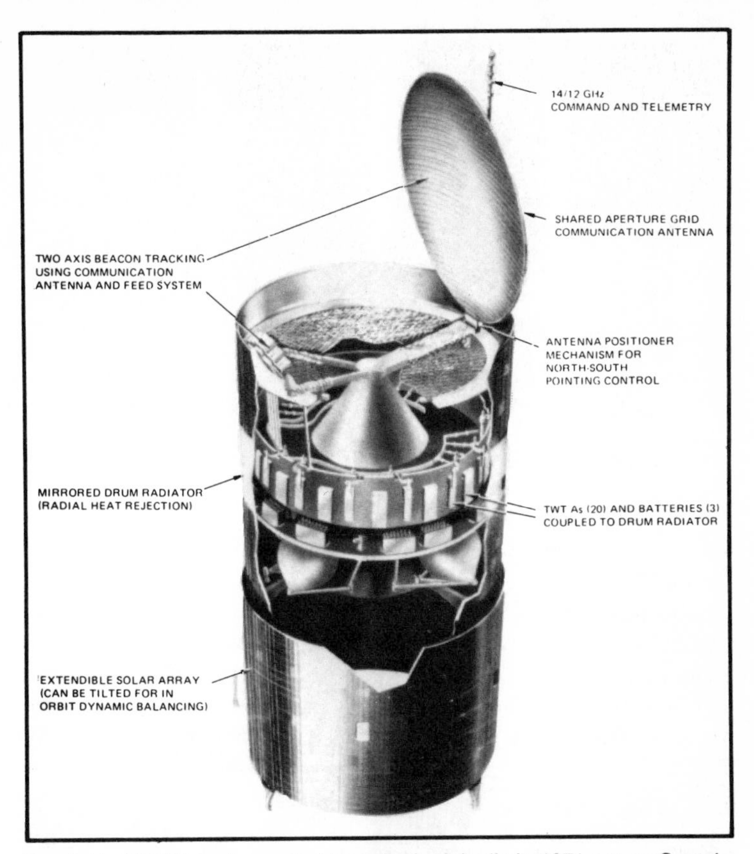

Fig. 1-11. ANIK-C (early version) launched originally in 1971 serves Canada with audio, video, and data communications (courtesy of Telesat Canada).

sia by Hughes Aircraft Space and Communications and operated by the state-owned telecommunications company PERUMTEL. B1 was launched in June of 1983, and B2 during February 1984. Unfortunately, this latter satellite experienced perigee-kick motor failure after leaving the Shuttle, did not achieve proper orbit, and was recovered in November 1984 by the Shuttle crew and returned to earth. A third Palapa B will probably be delivered by Hughes in 1986. (See Fig. 1-12).

These new satellites have twice the capacity, four times the power, and are twice as large as the forerunner A series that preceded them. Their 24 transponders can carry 1,000 one-way voice circuits or color television. Diameters are 7 feet 1 inch, 9 feet 4 inches high in the stowed position for launch. With solar panels extended, these spacecraft are 22 feet 10 inches high and weigh

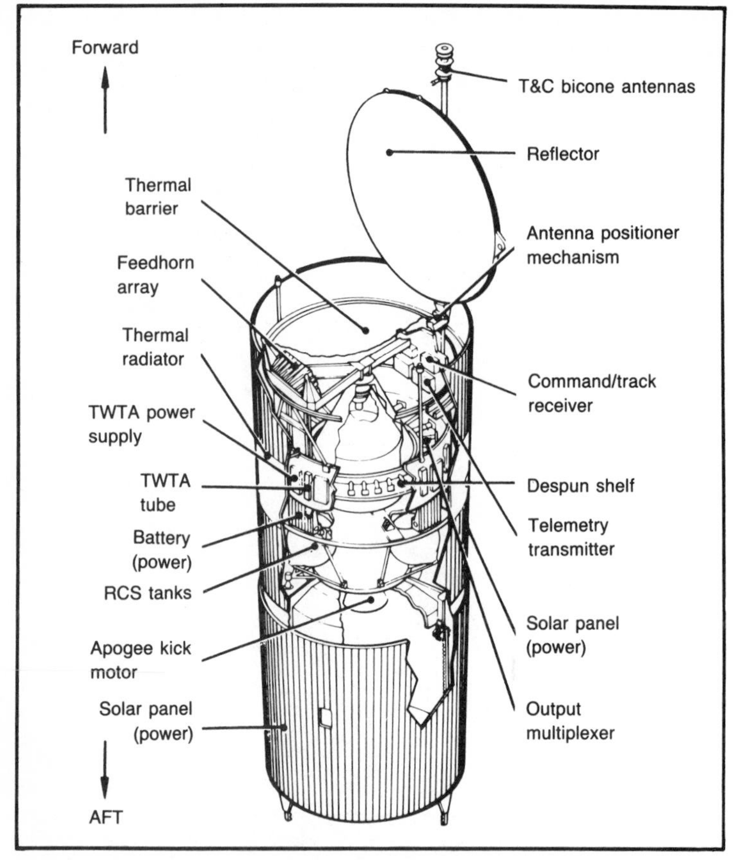

Fig. 1-12. Palapa B, the Indonesian satellite serving Indonesia, Thailand, the Phillipines, Malasia, and Singapore (courtesy of Hughes Aircraft).

1437 pounds at the beginning of orbit. Power consists of 1,062 watts and design life should exceed eight years, with a pointing accuracy of better than 0.05°.

TELSTAR

These Hughes HS 376 spacecraft replace the older COMSTAR group now used by AT&T, and all three have now been boosted into orbit parking positions at 96° WL for TELSTAR 301, 86° WL for TELSTAR 302, and 125° for TELSTAR 303. Launch dates were July 1983, Sept. 1984, and June 1985, respectively, by (1) a Delta rocket and (2) and (3) by the Space Shuttle. All have 10-year design lives, solid-state power amplifiers for 18 of the 30 high-powered transmitters, and are spin-stabilized. Solar cells develop 915 W for each "bird" and measurements are 7 feet 1 inch in diameter and 9 feet high. With antennas and solar panels deployed, they are 22 feet 5 inches tall and have orbit weights of 1438 pounds. They operate in the 4/6 GHz C-band and have 24 transponders that may be switched by ground command for various regional coverages across CONUS, Hawaii, Alaska, and/or Puerto Rico. Both 301 and 303 TELSTARs also provide some video traffic for ABC and CBS, with probably more to come as time progresses. Simultaneous long distance phone capacity is 21,600 in addition to voice and high speed data services. (See Fig. 1-13.)

AT&T's transponder service, called Skynet, offers a full 36 MHz bandwidth for high volume C-band transmissions, including

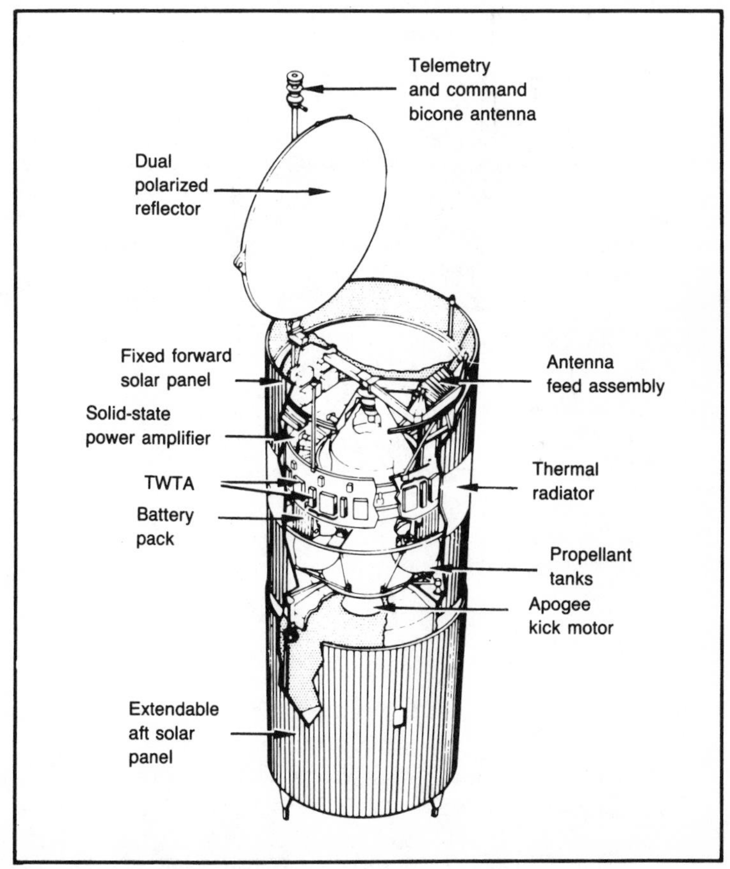

Fig. 1-13. The TELSTAR group of satellites are replacing the COMSTARS (courtesy of Hughes Aircraft).

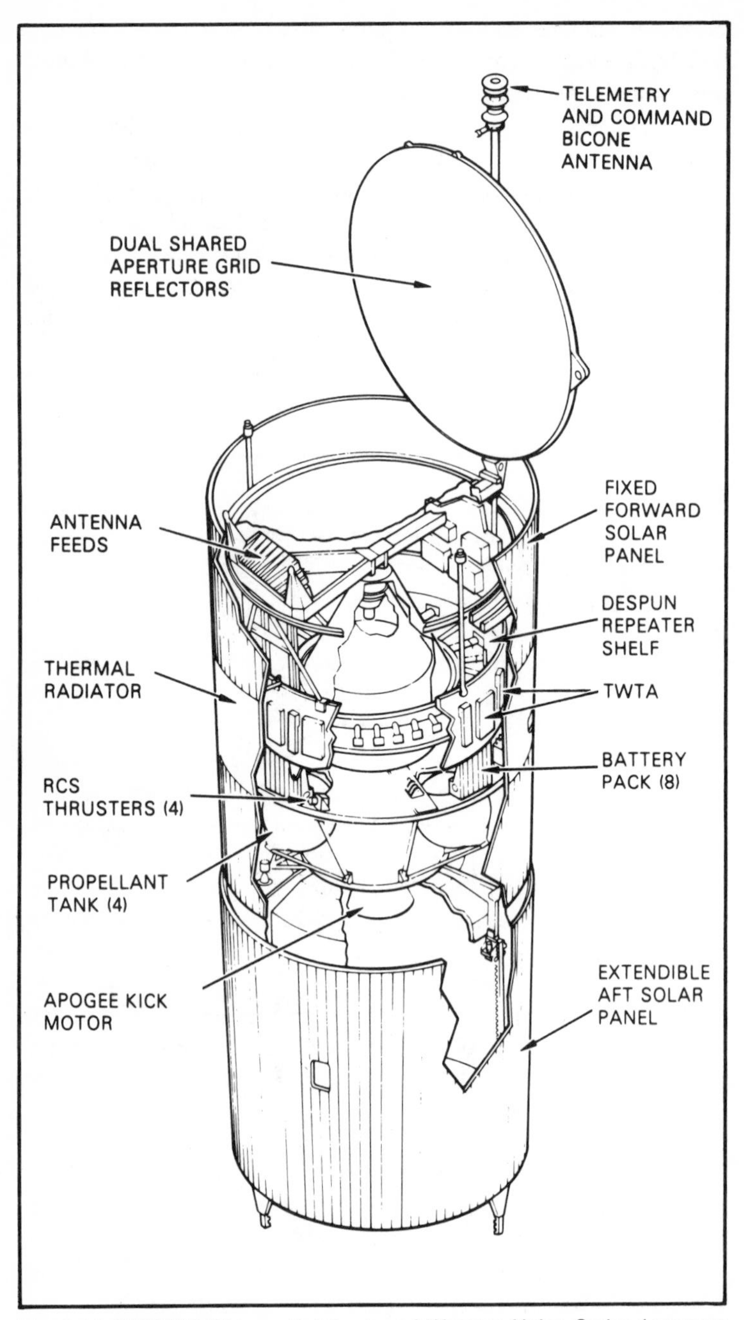

Fig. 1-14. WESTAR IV is part of the broad Western Union Series (courtesy of Hughes Aircraft).

television, and are available on monthly or 5- and 10-year leases at varying levels of "protection", in addition to part time. Shared earth stations are also available.

WESTAR

WESTARs IV and V are also part of the Hughes HS 376 family, which are double the size of the original WESTAR I, II, and III. Some 296 inches high when deployed in space, diameters are 86 inches and weighs 1,290 pounds when first placed in orbit. These satellites have 7.5 W TWT amplifiers and 24 channels, with solar power of 800 watts (Fig. 1-14). Orbit locations are 99° for WESTAR IV and 122.5° for WESTAR V. WESTAR I was retired on 8/83.

GALAXY

Built by Hughes Aircraft and its Hughes Space and Communications Group as "powerful versions" of its HS 376 series, there are three of this C-band series now on station at 134° W, 74° W, and 93.5° W, respectively.

Galaxy 1 (134° W) was launched on June 28, 1983, and distributes cable television video programming and other video signals to CONUS, Alaska, and Hawaii.

Galaxy II (74° W) gained orbit September 22, 1983, and one year later Galaxy III (93.5° W) reached position on September 21, 1984. These two satellites relay video, voice, data, and facsimile to CONUS. Galaxy IV has been assigned orbital position 140° WL. The first three were launched on Delta rockets; at the moment we have no further information on Galaxy IV. We also note there is a Galaxy B approved for orbit positions at 130° W, which will operate at 12/14 GHz in the Ku-band.

The HS 376, said by Hughes to be the world's most "widely purchased commercial communications satellite", has a diameter of 7′ 1″ by 9′ 4″ stowed for launch. After reaching orbit, the solar panel deploys, the antenna opens, and the spacecraft then measures a total height of 21′ 8″.

Its K7 solar cells generate 19.7 mW/cm^2, producing 900 Wdc, charging two nicad batteries that supply power during solar eclipses. Initial orbit weight is 1441 lbs., traveling-wave-tube amplifiers (TWTA) produce 9 W of output power for each of its 24 transponders and there are 6 spares. Design life is 9 years.

Chapter 2

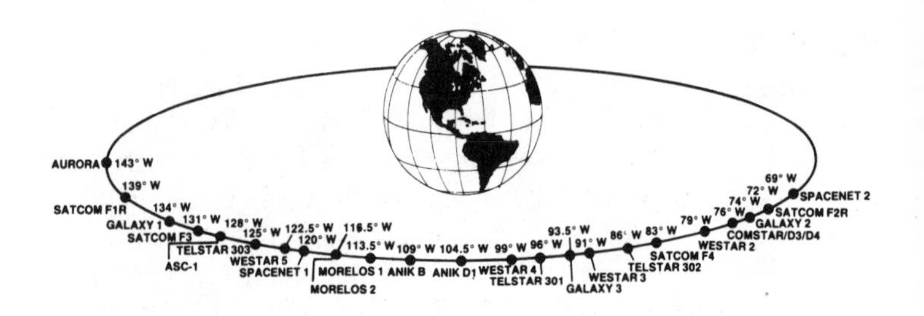

The Satellite Signals

JUST AS IT TAKES TWO TO TANGO, SO SATELLITE COMMUNIcations has to have message origination and message reception. There must be a line-of-sight microwave "uplink" to the relaying satellite, and there must be a "downlink" that accurately reproduces the same information. To avoid undesirable crosstalk between the two, the uplink is transmitted at a higher frequency than the downlink. The satellite electronics that links the two is called a "transponder," which is nothing more than a repeater-translator with certain amplification, some of it large and some small, depending on both the area covered and designed reception. You may have a low-powered transponder and a large receiving dish, or a high-powered transponder with a small receiving dish. At certain frequencies the satellite can produce considerable power with the TVRO (television receive only) dish being small. Naturally, this could be true for voice, facsimile, Teletext, radio or other similar communications as well. But many receive-only satellite ground stations are privately owned and used wholly for video-audio reception, although this is expected to change to some extent with multi-service offerings that are due in the late 1980s and early 1990s.

In the meantime, the fixed satellite service at 6 GHz uplink and 4 GHz downlink will continue to prosper, as will the 14 GHz uplink of the Ku-band system and its 12 GHz downlink that's now in operation with SBS and RCA's K-1, K-2 and eventually K-3 series. GHz, of course, stands for gigahertz, which is 1,000 megahertz (MHz), or 1,000,000,000 Hz (hertz). Heinrich Hertz was a German physicist who proved that electromagnetic waves alternate at various frequencies even before such things as oscilloscopes were invented. Such alterations are now called sine waves, for ± sinusoidal swings between zero (0 °) and (360 °).

Simple alternating intelligence is mostly used in AM or amplitude modulation systems where information actually modulates the carrier. More advanced systems make use of frequency modulation (FM), which deviates a constant amplitude carrier. A third analog type of modulation is phase modulation (PM), and this responds to changes in intelligence phase shifts rather than simply amplitude variations or frequency deviation. However, because of noise and distance problems which are always more prevalent in analog rather than digital systems, the language of microprocessors and computers will soon be applied to audio and video satellite transmissions as this rapidly advancing technique becomes just a little more practical and cost effective. The accuracy of such transmissions, when they are instituted, should also become much easier to evaluate via digital and signature analysis—two quick and accurate methods developed to handle both microprocessors and logic systems in general. Satellite Business Systems (SBS) is already doing some of this work at 12 GHz, and is planning more as its service expands. For now, however, most of its video is transmitted analog.

As you might imagine, there are several different techniques involved in getting transmittable intelligence to the satellite and its subsequent recovery at the receiving earth stations. Therefore, to acquaint you with the general systems prior to discussions of specific transmitters, we'll outline broad system applications for both *up* and *downlink* considerations, all of which are carried on the "bird."

MULTIPLE ACCESS

There are three usual methods of uplink transmissions, two of which are prime feeds, and the other worth honorable mention. These are: *Frequency-Division Multiple Access* (FDMA); *Time-Division Multiple Access* (TDMA); and *Code-Division Multiple Access* (CDMA). There are probably others under development, but these are the main ones at the moment—especially the first two—so we'll stick with the three.

FDMA

Frequency time-division multiple access (Fig. 2-1), a most common system at this writing, was begun back in the 1960s, and is still in prime use. Each station in some particular grouping transmits a select and unique series of frequencies. The transmitters multiplex this information and deliver it to the satellite and its transponders as composite intelligence for frequency translation and rebroadcast to earth receiving stations.

In the course of such activity, however, many different frequencies in any amplifier, which is even slightly nonlinear, can result in intermodulation distortion because sum and difference products are generated that are difference frequencies from the originals. Therefore, satellite electronics must not be overdriven so they produce such intermod distortion, and the way to prevent this is by applying power backup from the transmitters. If the satellite is *not* operated at full power, then such distortions do not occur, but the satellite will then transmit at reduced power, say 3 dB, and this reduces power output to half its rated potential. Consequently, many transponders actually rated 5 and 10 watts or more, are really only putting out 2.5 to 5 watts when working FDMA. However, at least they're operating, and good earth station dish and electronics design will pull in the signals even during heavy squalls of rain, which

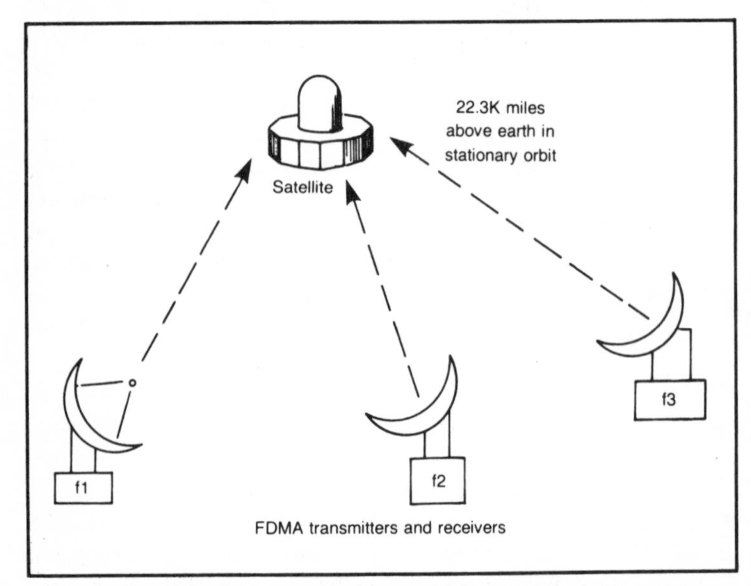

Fig. 2-1. FDMA (frequency time-division multiple access) combines frequencies which then reach the satellite as composite signals.

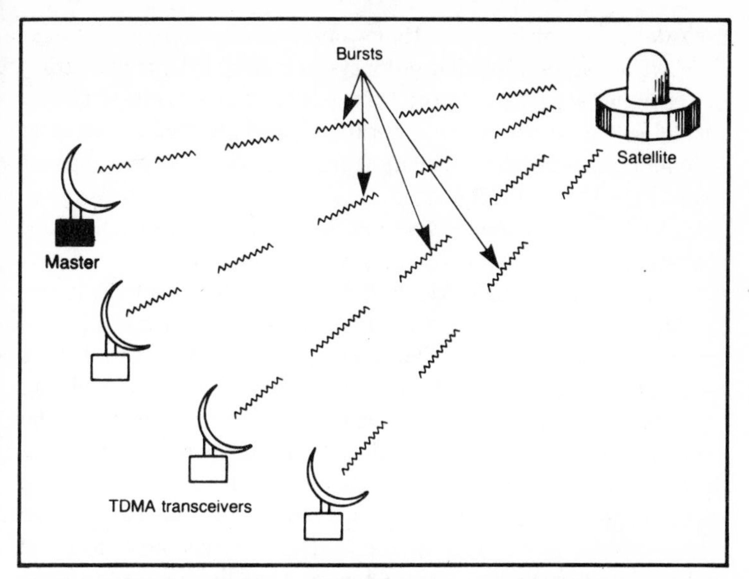

Fig. 2-2. TDMA (time-division multiple access) delivers satellite information in rapid bursts with only one signal reaching the satellite at any one instant.

interfere with earth-bound transmissions several dB more at the higher GHz frequencies.

Earth stations sort out these multiplexed signals by tuning to whatever frequency they wish to receive and process available information accordingly. Naturally, exceptional electronics and dish gain with better than average sidelobes offer best reception possibilities—a subject we will discuss at length in the chapter on TVRO Design Considerations. Remember, especially, the term "sidelobes," for these will be dealt with very specifically since it is now recognized that antenna gain delivers only part of an interference-free signal. The rest depends on rejection of as many adjacent frequencies as possible in addition to the gain/temperature performance of low-noise amplifiers, downconverters, receivers and any remodulators.

TDMA

Time-division multiple access (Fig. 2-2) constitutes the second method in satellite uplink transmissions and is rather different from FDMA in that modulation on the various carriers arrives at the satellite from transmitters in bursts. But as opposed to frequency-division multiple access, only one package of information reaches the satellite at any instant of time, consequently, there is no power division or intermodulation products to concern the transmitting or receiving stations, and the transponder can operate at its full saturated power output without distortion. This, then, is actually a doubling of available capacity from the satellite if such output is compared with FDMA.

The way this system works is that each station has a buffer that stores its information output. With control from some designated master station, individual buffers of each transmitter in the system deliver all stored signals at a very high bit-rate via the uplink and into the satellite. The satellite actually sees these bits as virtually continuous transmissions, even though some transmitting stations are farther from the satellite than others. This has to be taken into account in transmit calculations and propagation timing adjustments. Nonetheless, different burst rates for the individual transmitters are permitted if within design limits, as long as these bursts can be received by the satellite and processed before the next transmitter unloads its intelligence. Usually the slowest burst rate reaches the data frame first, and synchronization begins in this mode and works up. The modems (modulator/demodulator) units can sort out the incoming intelligence either at continuous or specially selected bit rates *if they are so designed*. Larger receive stations are usually capable of receiving high bit rates, while the smaller stations have capacities only for slower collections. Separate networks may also be devised for large and small stations so they can receive all their traffic at more or less the same bit rates.

A variation of this scheme, called Satellite Switched TDMA, is also possible where the satellite serves several different areas. Uplinks and downlinks may be connected to one another by a switch matrix. Traffic is then permuted and directed by very fast switching diodes from uplink to downlink in whatever pattern has been programmed both by the uplink and the satellite transponder itself. Permutations may be changed and reprogrammed from earth according to needs of the program originator and the system.

ALOHA—or random access TDMA—is another method within the major main TDMA discipline (Fig. 2-3) which simply fills a transmit data buffer and then sends its contents randomly. Occasionally, as you might well imagine, some loaded buffer contents collide with others. The receive station simply waits a certain period until this same information is retransmitted and then, hopefully, has a clear enough channel to pick up the intended transmission. This, of course, is the problem with ALOHA. If there are too many

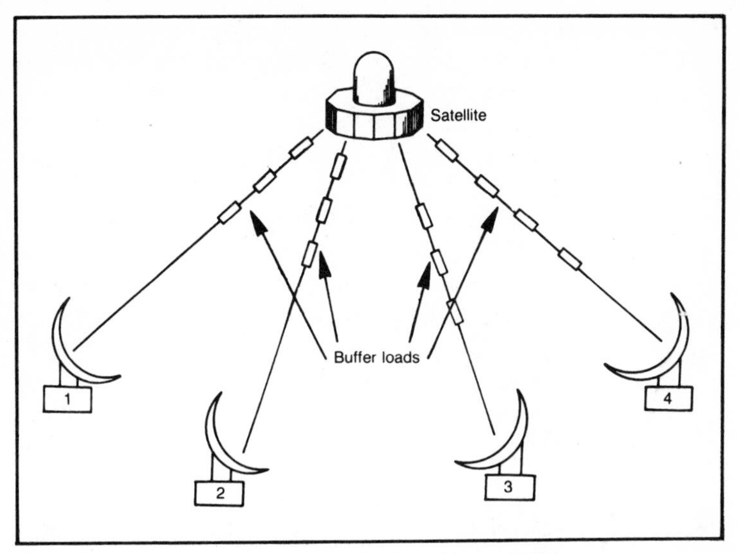

Fig. 2-3. The ALOHA system sends filled buffers randomly.

retransmissions a great deal more interference occurs and soon transponder efficiency decreases to an unacceptable level. Slotted ALOHA *does* permit emptying the buffers from transmitters only at specific times, but this adds complexity to a relatively simple system that works extremely well when it can be controlled and not overloaded. Additions to the basic system, of course, increase costs and reduce the initial attractions considerably.

CDMA

Code-division multiple access (Fig. 2-4) becomes relatively more sophisticated than those concepts just described and is *not* used to any extent in the commercial world, but has found a secure home with the military. You'll see why as the explanation develops.

In CDMA, all stations transmit on the same frequency and at the same time. Fortunately, and by design, each transmission has its own unique code. This can be a pseudo-noise code, for instance, that cannot be decoded until the receiver has suitable equipment to detect and reproduce the original intelligence. And when interfering information is received, it still has no effect since it is not within the limits of a specific code. Naturally, each receiving terminal has its own code and all others, including either intentional (jamming) or random interference is rejected. There's also frequency hopping, where one and then another frequency carries the

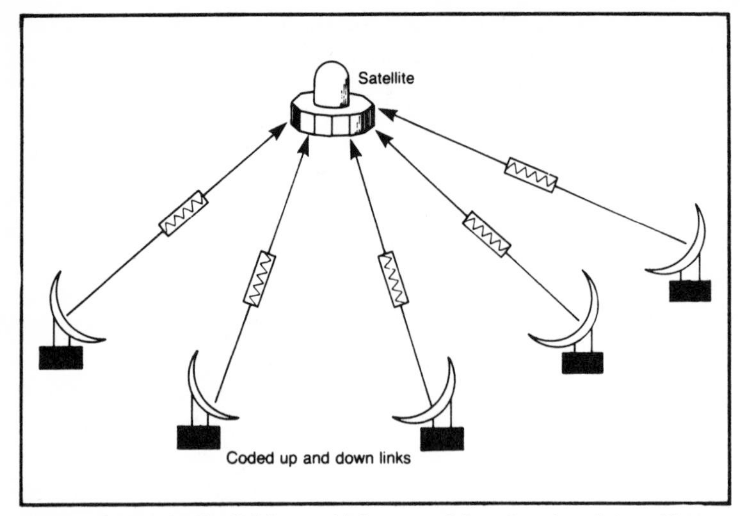

Fig. 2-4 CDMA (code-division multiple access) is a favorite of the military.

intelligence. These, obviously, are expensive systems, and do suit the military need for secure communications much better than commercial operations where a few dollars saved means the difference between profit and bankruptcy. Apparently, both code and decode methods are the prime cost factors in such systems. We're told that even a spectrum analyzer doesn't provide much information other than blobs of activity during transmissions, so imagine what the demodulator/receiver must do in the decoding process. Spread spectrum is not quite the same but is often used as another designation for CDMA.

DAMA

Demand assign multiple access permits reception of information when the receiver requires it. Otherwise, such information goes to a pool where others may also have access when needed. Any of the three previous systems covered can be used with DAMA. However, even though test systems have been built, no one is yet thought to be using DAMA with CDMA. The most popular form in use now is called *Single Channel Per Carrier* (SCPCMA). This separates the output of a particular transponder into many possible channel assignments and these are available as a pool for many stations to acquire on demand. Here, a pair of frequencies are assigned upon request of the receiving station, establishing a full duplex link for the duration of the call. When the call is completed, these same two frequencies are returned to the pool for others to use in the meantime. Of course too many stations trying to access the pool will sometimes receive a busy signal. So just like the telephone, you hang up and try again until the link becomes completed.

If the system handles voice transmissions only, then the talker can key on the transmitter when he is speaking and turn it off when listening. This offers a power saving of some 4 dB, salvaging what some other systems throw away in lost power.

THE SPACECRAFT ITSELF

The spacecraft proper is usually a sealed unit having either spin stabilized or tri-axis stabilized motion restriction with rechargeable batteries to carry it through periods of eclipse with the earth which occur twice a year during spring and autumn for some 72 minutes each.

There also must be communications between the "bird" and ground stations to tell if there is sufficient power, the transponder is operating as it should, batteries are charging, etc. So there is continuous monitoring of satellite systems to see that all's well up above and that the satellite can respond to commands and has both incoming and stored energy to do so. This is all done by telemetry modulating an oscillator called a *beacon* that appears to anyone else as just another signal. It not only carries the necessary telemetry but also provides a path for earth station tracking so all can keep their antennas fixed on the satellite. For smaller stations this is not critical, but for larger receptors it is critical.

In spin-stabilized units, the satellite is actually spinning and its outside is covered with solar cells. More recent satellites have a body stabilized system where three momentum wheels inside its body do the spinning, with the satellite remaining perfectly still and solar cells extend north and south. Prime failure modes in these satellites are the power systems and traveling-wave amplifiers (TWTs).

ORBITS

Usually, satellites are placed in orbits that are almost circular and the angle between the plane of the equator and that of the orbit is called the inclination angle. If it is in the exact plane of the equator and has a precise 24-hour earth rotation period, it is called a synchronous satellite. However, it's next to impossible to keep such a precise path and, therefore, stationkeeping becomes necessary to return to the equatorial plane and keep the inclination angle as low as possible. If you were to watch the satellite from the ground continuously for 24 hours, you would see it describe an actual figure eight, since for 12 hours it would appear above the equator and then 12 hours below the equator from the point on which you were standing. Now, if the beamwidth of the antenna is less than twice that of the satellite's orbit, then the satellite will move outside the antenna's range or capture area and you can't communicate. Small earth stations have enough beamwidth ordinarily to track the "birds" through their figure eights, but for larger arrays the problem is entirely different, since the larger the antenna the narrower the beamwidth but the greater the gain.

Satellite stationkeeping is another problem, and earth tracking also depends on whether your satellite system is designed to keep the bird and its transponder within 0.1° or less, and this is directly related to how much fuel you want to expend from time to time during the life of the satellite. Often, fuel becomes the critical factor in how many days or years these geosynchronous orbiters remain active. Sometimes you have to balance your designs between on-board fuel availability and total electronics, which could amount to a rather difficult choice.

Earth coverage from these satellite antennas has a spread of about 18°. If plotted as a function of the elevation angle (local horizontal) versus the circular cross section you have contours, and these describe the maximum area of coverage.

Another kind of orbit is the polar orbit and this is worth considerable attention in terms of mapping the entire surface of the earth. There is also an inclined planet-type, essentially nonsynchronous orbit used by the Soviets in their Molniya series launched in 1965. It has a highly elliptical arrangement with the satellite hanging in the upper section and moving very fast in the lower part. With an inclination angle of some 63.5°, several of these satellites could continuously cover the northern hemisphere. Perturbing forces that act on most satellites actually balance out under these conditions, and no power is used during speedup times. The Molniyas have now been replaced with three Statsionar geo-sync satellites located over the Indian Ocean and downlink operating at 3.4 to 3.9 GHz.

Overall, the Soviets have at least 23 satellites listed as being in orbit, with perhaps more we don't know about. An ellipse (for elliptical orbit), is defined as a curved line with the sum of distances from any two fixed points on it being constant.

U.S. FIXED SATELLITE SYSTEMS

It is time now to begin describing our own special commercial satellites that are carrying all sorts of video, telecommunications, Teletext, phone, stereo, and other information whose volume, it seems, actually increases hourly. GTE and the SATCOMs are built by RCA, and many of the rest by Hughes. Nonetheless, they differ considerably in electronics and applications, and so a fairly rigorous description of each will be forthcoming depending, of course, on what system owners and designers have supplied in the way of information. There are some slight restrictions, but on the whole you should be able to fully understand each according to mission and electronic execution. Available and pertinent illustrations have been included. At the moment, all currently operational fixed satellite commercial systems are working C-band, except Satellite Business Systems, RCA in Ku GSTARS and the hybrids. Later, many multi-frequency transmissions will be transported in space, especially at C and Ku, as sophistication and satellite numbers take up assigned stations. Obviously the rush is on.

Therefore, the remainder of the chapter is a mix of space electronics now on station and that which has been government solicited or approved and is due to be rocket-launched in the next several years. Figure 2-5 is courtesy of *Communications News* and shows available and forthcoming commercial "birds" for the mid 1980s. As you can see, 15 are already in use and 23 more coming as rapidly as construction and launch vehicles can put them up. International and specialized satellites have already been described in Chapter 1. In Fig. 2-6 is an updated listing of these and other satellites, adding just a little more information to Fig. 2-5.

AT&T's TELSTARS

The American Telephone and Telegraph Company has already petitioned the Federal Communications Commission for three new (replacement) satellites operating uplink between 5.925 and 6.425 and downlink from 3.7 to 4.2 GHz. They approximately assume the 87°, 95°, and 128° slots formerly occupied by the original COMSTAR series, which are Hughes HS303 units. All use analog and digital single access, frequency-division multiple access (FDMA), and time-division multiple access systems (TDMA). Launches by Delta 3920, the Shuttle, or a European Ariane rocket have been scheduled and these new Hughes HS376 communications satellites

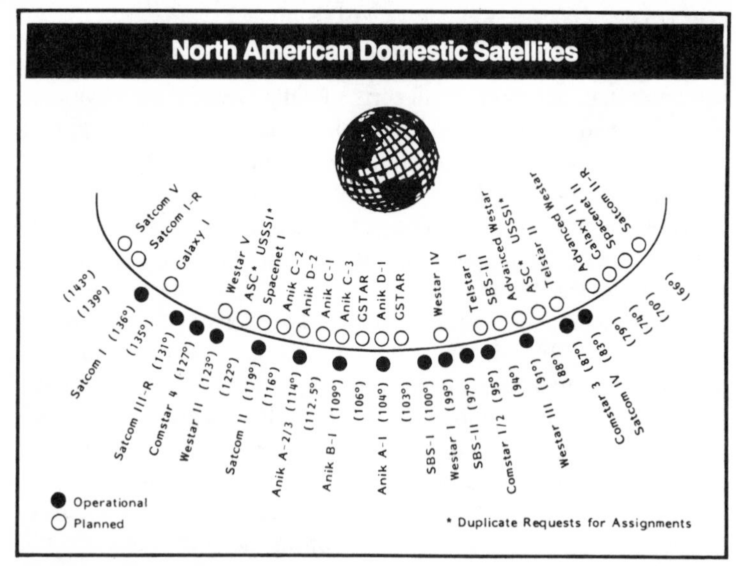

Fig. 2-5. North American commercial satellites arrayed across the equator at assigned west longitudes (courtesy of Communications News).

entered service by 1985. We are also told that the Bell System began single sideband operation in 1983, permitting 6000 one-way voice channels per COMSTAR transponder initially, increasing to 7800 one-way channels with TELSTAR IIIs as these enter service. Coverage includes the continental U.S., Alaska, Hawaii, Puerto Rico, and the U.S. Virgin Islands.

Each of the TELSTAR IIIs weigh (with perigee motor and mounting equipment for shuttle launch) about 10,000 lbs., with apogee propulsion unit weight of some 1400 lbs. and propellant loading of 1300 lbs. Payload is to be mounted vertically in the shuttle bay. All units will contain 24 transponders divided in vertical/horizontal polarizations having bandpasses of 36 MHz.

One 500 MHz, 6-GHz wideband receiver common to all 12 transponder channels for either polarization is included, and to increase reliability, four receivers have transfer switches for 4 × 2 redundancy with receivers 1, 2, or 3 usable for horizontal channels and receivers 4, 3, or 2 for vertical channels. Excellent linearity minimizes crosstalk among channels in the common amplifier, along with low noise for best G/T (gain over temperature) which is expected to result in a G/T of −5 dB/K at beam edge. Frequency translation from 6 to 4 GHz takes place in a single conversion before channel separation, with local oscillator stability of at least 1 ppm/month and better than 10 ppm/satellite lifetime.

Channels

All 24 channels have an input bandpass filter for selectivity, a delay equalizer for envelope delay distortion, a driver and power amplifier, and an output bandpass filter which reduces out-of-band radiation and allows several channel connections to one waveguide. There is also an 8-step gain attenuator for each channel that's transmitter adjustable.

Power amplification consists of 24 primary TWT tubes and 12 spare solid-gold GaAsFET amplifiers in 3 × 2 redundancy. Six amplifiers in each subgroup have outputs combined into a single waveguide by a multiplexer and all but odd vertical channels connect to output multiplexers by T switches (or other equivalent networks).

Antennas for the TELSTARS are orthogonally polarized, offset-fed, parabolic reflectors, including multihorn feed networks having receive/phasing networks, and frequency diplexers separating receive and transmit frequencies. Representative antenna coverage included figures of 24.5 dB for receive gain, 27 dB horizontal for transmit gain (edge coverage) and 26.5 dB for vertical, and cross-polarization isolation of 34 dB.

Satellites are to be stationkept within a ± 0.1° accuracy for both latitude and longitude, and nominal pointing accuracy of pitch and roll has been specified as ± 0.2° or better, with yaw at less than ± 0.6°. System design projects an operational life of 10 years, and redundancy is to be built in power, attitude control subsystems, and communications subsystem. New positions are: T301, 96°; T302, 86°; T303, 125° (all WL).

Telemetry

Control facilities are shown in Fig. 2-7, with stations in Pennsylvania and California, and launch support centers at Kennedy, Johnson, and Guiana.

Uplink signals are received by omnidirectional antennas, passed to two independent receivers, including initial demodulation, commands are crosstrapped to dual command units for further demodulation, processing, decoding, and command execution.

Downlink telemetry transfers data from sensors, transducers and other spacecraft status indicators via two independent and redundant beacon channels. Some analog information is FM'd

Satellite	Orbit Locations West Longitude	Frequency Band (GHz)	Date Launched	# of Xpdrs/ BW (MHz)
SATCOM V	143°	4/6	Oct. 1982	24/36
SATCOM I-R	139°	4/6	Apr. 1983	24/36
GALAXY I	134°	4/6	June 1983	24/36
SATCOM III-R	131°	4/6	Nov. 1981	24/36
ASC-1	128°	4/6; 12/14	Aug. 1985	12/36 & 6/72 6/72
TELSTAR 303	125°	4/6	June 1985	24/36
SPACENET I	120°	4/6; 12/14	May 1984	12/36 & 6/72; 6/72
WESTAR V	122.5°	4/6	June 1982	24/36
GSTAR II	105°	12/14	March 1986	16/54
GSTAR I	103°	12/14	May 1985	16/54
SBS IV	101° (temporary)	12/14	Sept. 1984	10/43
SBS I	99°	12/14	Nov. 1980	10/43
WESTAR IV	99°	4/6	Feb. 1982	24/36
SBS II	97°	12/14	Oct. 1981	10/43
TELSTAR 301	96°	4/6	July 1983	24/36
SBS III	95°	12/14	Nov. 1982	10/43
GALAXY III	93.5°	4/6	Sept. 1984	24/36
WESTAR III	91°	4/6	Aug. 1979	12/36
TELSTAR 302	86°	4/6	Sept. 1984	24/36
SATCOM Ku-1	85°	12/14	Jan. 1986	16/54
SATCOM IV	84°	4/6	Jan. 1982	24/36
SATCOM Ku-2	81°	12/14	Nov. 1985	16/54
WESTAR II	79°	4/6	June 1974	12/36

COMSTAR D_2 & D_4	76°	4/6	Sept. 1976 & Feb. 1981	24/36
GALAXY II	74°	4/6	Sept. 1983	24/36
SATCOM II-R	72°	4/6	Sept. 1983	24/36
SPACENET II	69°	4/6; 12/14	Nov. 1984	12/36 & 6/72; 6/72
WESTAR I	retired (8/83)	4/6	April 1974	12/36
SATCOM I	retired (5/84)	4/6	Dec. 1975	24/36
COMSTAR D_1	retired (9/84)	4/6	July 1976	24/36
COMSTAR D_3	retired (8/25)	4/6	Sept. 1978	24/36
SATCOM II	retired (2/85)	4/6	March 1976	24/36

In Orbit as of April 3, 1986

SATCOM - RCA American Communications, Inc.
WESTAR - Western Union Telegraph Company
SBS - Satellite Business Systems
GALAXY - Hughes Communications Galaxy, Inc.
COMSTAR - owned - Comsat General Corporation
- operated - AT&T Co.
TELSTAR - AT&T Co.
SPACENET - GTE Spacenet Corporation
GSTAR - GTE Satellite Corporation
ASC - American Satellite Company

Planned Launches in 1986

Shuttle	**Ariane**
Westar IV-S	Spacenet F3
GSTAR 3	SBS-5

Fig. 2-6. United States domestic satellite system.

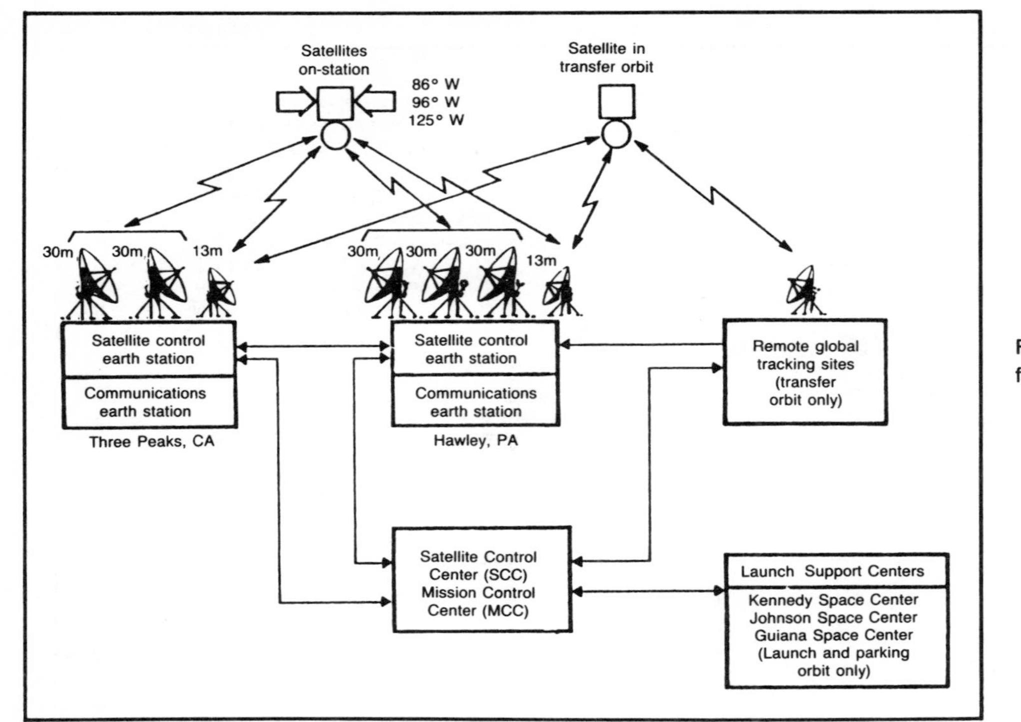

Fig. 2-7. AT&T satellite control facilities (courtesy of AT&T).

directly on to a subcarrier, and other analog data is first digitized as PCM then modulated on a subcarrier for phase-shift keying. Either subcarriers or ranging tones from uplink receivers are then phase modulated on the microwave downlink carriers. Transmission takes place with the dual omnidirectional antenna system when the satellite is in transfer orbit and with directional communication antennas during the time the spacecraft is on station.

Redundant components will be switched by ground command, and redundancy is also to be included in the satellite control center. Satellites may be positioned once following launch, if necessary, and will operate satisfactorily within 4 ° of any other satellite. Telemetry, tracking, and command signals are to be relayed by ground control and the control center will monitor all functions and coordinate all test and launch facilities. A block diagram of such a control station is illustrated in Fig. 2-8.

THE GSTAR SERIES

GTE Satellite Corporation now joins SATCOM, WESTAR, TELSTAR, COMSTAR and the rest with its own satellite network, scheduled for initial launch in the second quarter of 1985. Built by RCA, the network consists of two Ku-band units operating at 12/14 GHz, with a third for backup to be launched later.

Fourteen transponders, with four in reserve will provide 20 watts each for the continental U.S. (CONUS) only and two deliver 30 watts for CONUS, Hawaii, and Alaska. Covering all the 50 states with both vertical and horizontal polarization, each satellite (Fig. 2-9) is body stabilized, has a mission design life of 10 years, and delivers nominal EIRP per transponder of 42 dBW for CONUS, 39 dBW for Hawaii, and 40 dBW for Alaska. Transponder bandwidths are 54 MHz with polarization isolation of 33 dB minimum. Launch vehicles can be the Space Shuttle, Delta 3920, or Ariane 3 expendable rockets, and parking spaces assigned are 103 ° and 105 ° WL. Switchable east and west spot beams also offer increased EIRP over both East and West coasts.

Ground segments have two tracking, telemetry and control earth stations, and shared-use. Dedicated earth stations are to be added, according to customer response. TDMA will be used as the communications medium, which requires precise transmit timing for assigned bursts and time slots without overlapping. Ordinary acquisition and sync methods are used with the broad CONUS coverage, but spot beams with satellite-switched TDMA required

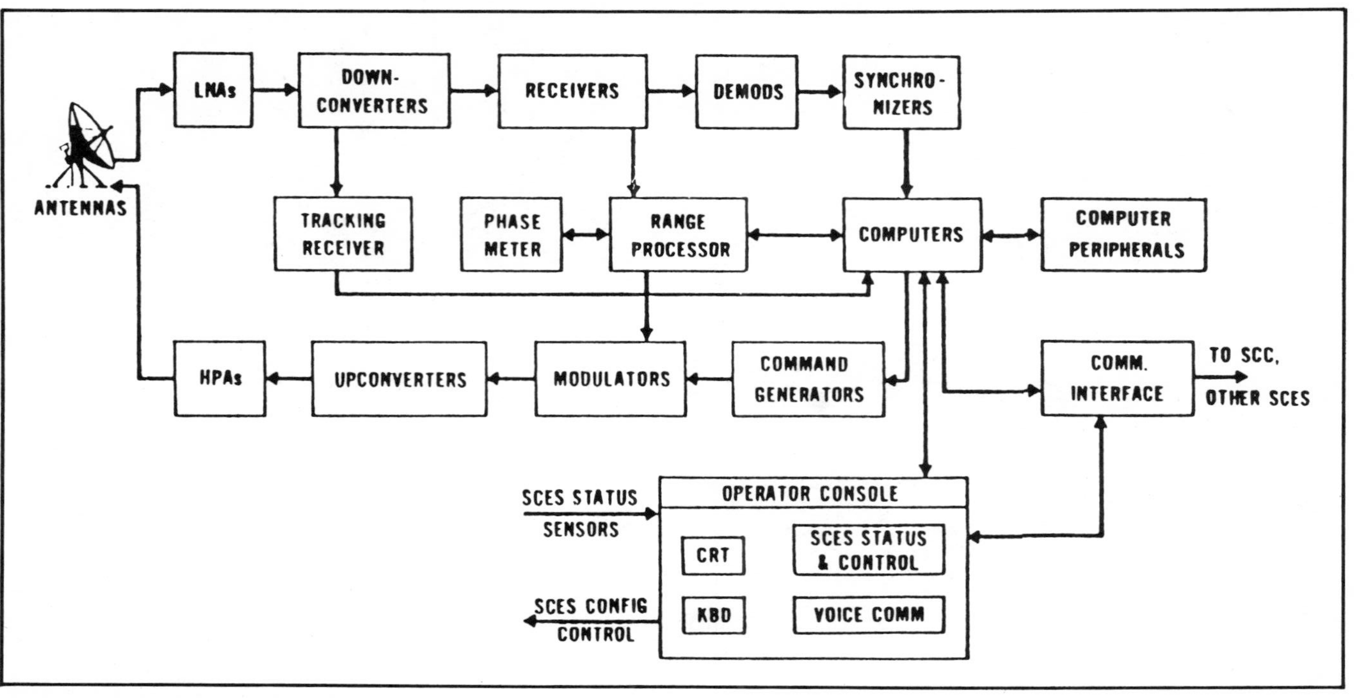

Fig. 2-8. AT&T satellite earth station control (courtesy of AT&T).

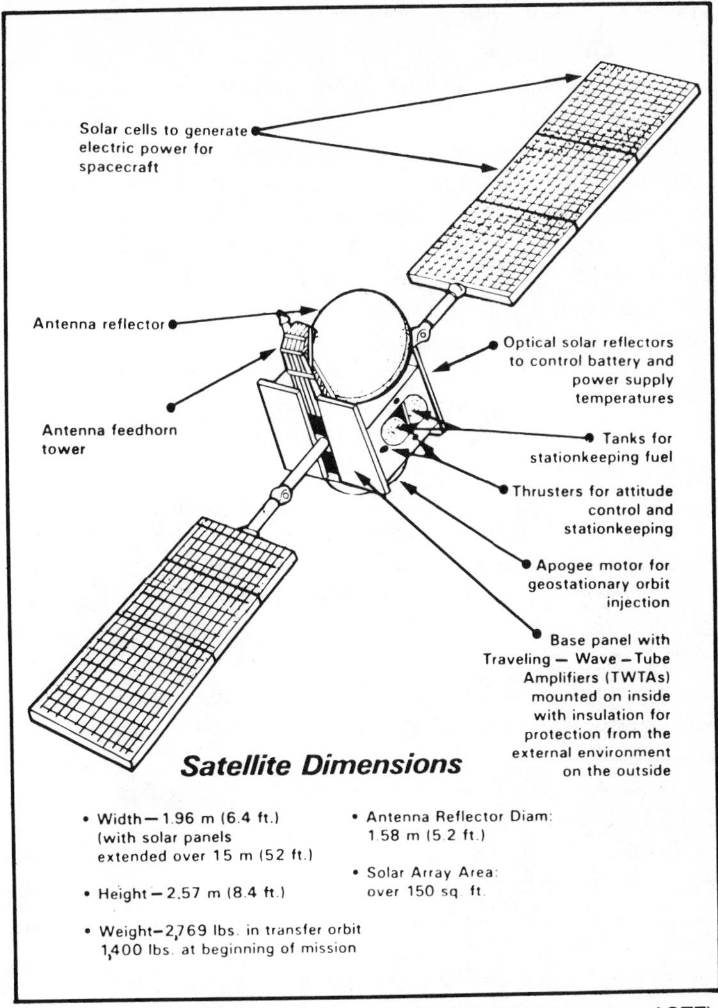

Fig. 2-9. The GTE GSTAR series now on station in 1985-86 (courtesy of GTE).

additional developments. Therefore, small portions of a combined beam transponder on the satellite are used for network reception and sync, while all timing burst transmissions to other transponders are derived from combined beam sync and burst gating. The satellite's vital statistics are: 8.4 feet H, 6.4 feet W, and 1,400 lbs. at mission start. The solar array covers over 150 square feet. Attitude determination and control subsystems are improved bias momentum, 3-axis design, but with the addition of diurnal error compensation and momentum wheel pivoting.

Traffic Considerations

For first generation GSTAR satellite systems, an East/West/CONUS combined beam configuration has been designed for multiple traffic needs. CONUS beam coverage is able to offer 60 Mb/s service, spot beam coverage 90 Mb/s, and transponder beam configuration may be altered by ground command.

The combined beam transponder, providing transmit and receive coverage for CONUS, Hawaii, and Alaska, has the designation of common timing reference transponder. Here all TDMA acquisition and sync timing takes place, with one TDMA frame constituting both a common and redundant timing reference burst, sync time slots, and traffic bursts. A common timing reference burst consists of a preamble, ID code, sync control word, and processor signaling. The preamble has a section for carrier recovery, symbol timing recovery, and burst code word detection.

A control station delivers ID to all earth stations within network control, while sync control provides burst transmit timing for any designated earth station. The processor signaling channel then selects transponder assignment and burst position for the particular earth station receiver. Acquisition and sync control is directed to each earth station at normal TDMA frame rate to update burst transmit timing. Thereafter, the principal control station, a redundant control station, ranging stations, and network control take over.

Based on satellite range information, the principal control station calculates distances from satellite to traffic stations, and sets up timing needs in each traffic station for acquisition and sync. It also monitors acquisition and sync burst positions and offers correct timing to all traffic stations. The redundant station does likewise, but is ignored if the system is normal.

Ranging stations make measurements of satellite positions for TDMA transmissions. Three of these stations in separate locations can measure the satellite's range. The principal control station then calculates any time corrections and passes this information on to the various traffic stations which demodulate for common timing.

The network control center, thereafter, is responsible for network traffic management. It can transmit and load the network plan to TDMA stations via the principal control stations, as well as deciding TDMA capacities. For hitless traffic reconfiguration, burst positions and burst lengths have to be synchronized for all affected stations. Burst overlap in the satellite transponder is thereby avoided.

Output multiplexers in the satellite sum all transponder outputs and drive harmonic filters in each feed system. Alternate channels 1-7, 2-8, as an example, are summed by way of manifolds having the required output filters, which eliminates severe restrictions on output filter design. Channel-to-channel isolation in dual mode networks appears for each feed system after harmonic filtering at the multiplexer output.

SPACENET SATELLITES

Manufactured by RCA Astro Electronics with a life expectancy of 10 years, the three hybrid satellites in this series have twelve 36-MHz transponders with 8.5-W solid-state amplifiers and six 72-MHz 16-W traveling-wave-tube (TWT) transponders, all operating at C-band. At Ku-band there are six 72-MHz transponders with 16-W TWT amplifiers. Spacenet I covers all 50 states with C-band, and CONUS with Ku. Spacenet II, covers CONUS and Puerto Rico with C and CONUS with Ku. Spacenet III has been assigned CONUS and Puerto Rico along with east and west Ku spot beams. Launch dates were: /May 22, 1984 for Spacenet I and November 9, 1984 for Spacenet II, Spacenet III was lost during its aborted Ariane Guiana launch. It is expected to be eventually replaced. The first two now operate at 120° WL and 69° WL, respectively.

These satellites are tri-axis stabilized, dual-band spacecraft with frequency reuse of C-band. Twenty four transponders offer almost 1500 MHz of usable bandwidth with seven power amplifiers for each of the four groups of six transponders. They are to be stationkept within ±0.05° of assigned longitudes and the equatorial plane. Transponder frequency differences for C-band uplinks and downlinks amounts to 2225 MHz, and 2300 MHz for Ku-band (without spectrum inversion). C-band crosspolar isolation is specified at 30 dB or more with possible exceptions in Alaska and Hawaii since EIRPs there normally range between 28 and 31 dBW for the 36-MHz transponders. This coverage with the 72-MHz transponders increases to 38 dB in central CONUS and 44 dB with 72-MHz Ku-band EIRP.

A block diagram for the C-band portion of these satellites is shown for both horizontal and vertical channels, including amplifiers, circulators, isolators, multiplexers, and filter equalizers. A legend on the left identifies the various components (Fig. 2-10).

SATCOM

One of the very major factors in all geosynchronous communications was and is the SATCOM series by RCA American Communications Incorporated (Americom) whose parent corporation launched SATCOM I December 12, 1975, after leasing channels on Canada's ANIK II satellite in December 1973. SATCOM II followed on March 26, 1976, then SATCOM III, which was subsequently lost during its December 19, 1979 launch. Formed in 1976 as a wholly owned subsidiary of RCA Corporation to own and operate RCA's domestic communications satellite system, RCA American Communications then put into orbit SATCOM III-R on November 18, 1981, followed by SATCOM IV which was launched in January 1982 and became operational in April of the same year. These were followed by a satellite for Alaskan service in October 1982, and a replacement for SATCOM I, in April 1983 (Fig. 2-11).

SATCOMs III-R and IV have a design life of 10 years at continuous full power, and with the launch of SATCOM IV, RCA Americom will have two satellites dedicated exclusively to cable TV traffic, offering a programming capacity of more than 1,000 hours each day. The Alaska satellite, by the way, is to be purchased by Alascom Services, a subsidiary of Pacific Power and Light, and will provide internal satellite communications for the state of Alaska as well as between Alaska and the remainder of the U.S. RCA Americom is to operate this satellite as a joint licensee with Alascom.

SATCOMs have an average launch weight of some 2,385 lbs., with attitude control thrusters, thermal control, propulsion, in addition to ranging and telemetry equipment. All are controlled by tracking and telemetry stations operated by Americom at Vernon Valley, New Jersey, and South Mountain, California. Initially, maximum output was 770 watts at launch to 550 watts at the end of their 8 to 10-year design life. Spacepower is derived from two solar array panels and three nickel-cadmium batteries. Each is built by RCA Astro-Electronics, Princeton, New Jersey.

All RCA SATCOMs have 24 cross-polarized, four-reflector antenna systems and offer 24 channels versus 12 for many other competing satellites. Twelve channels accept only horizontally polarized signals and 12 vertically polarized ones. Wire grids embedded in the four antenna reflectors aid in preventing interference between channels.

These satellites are three-axis stabilized versus spin stabilized,

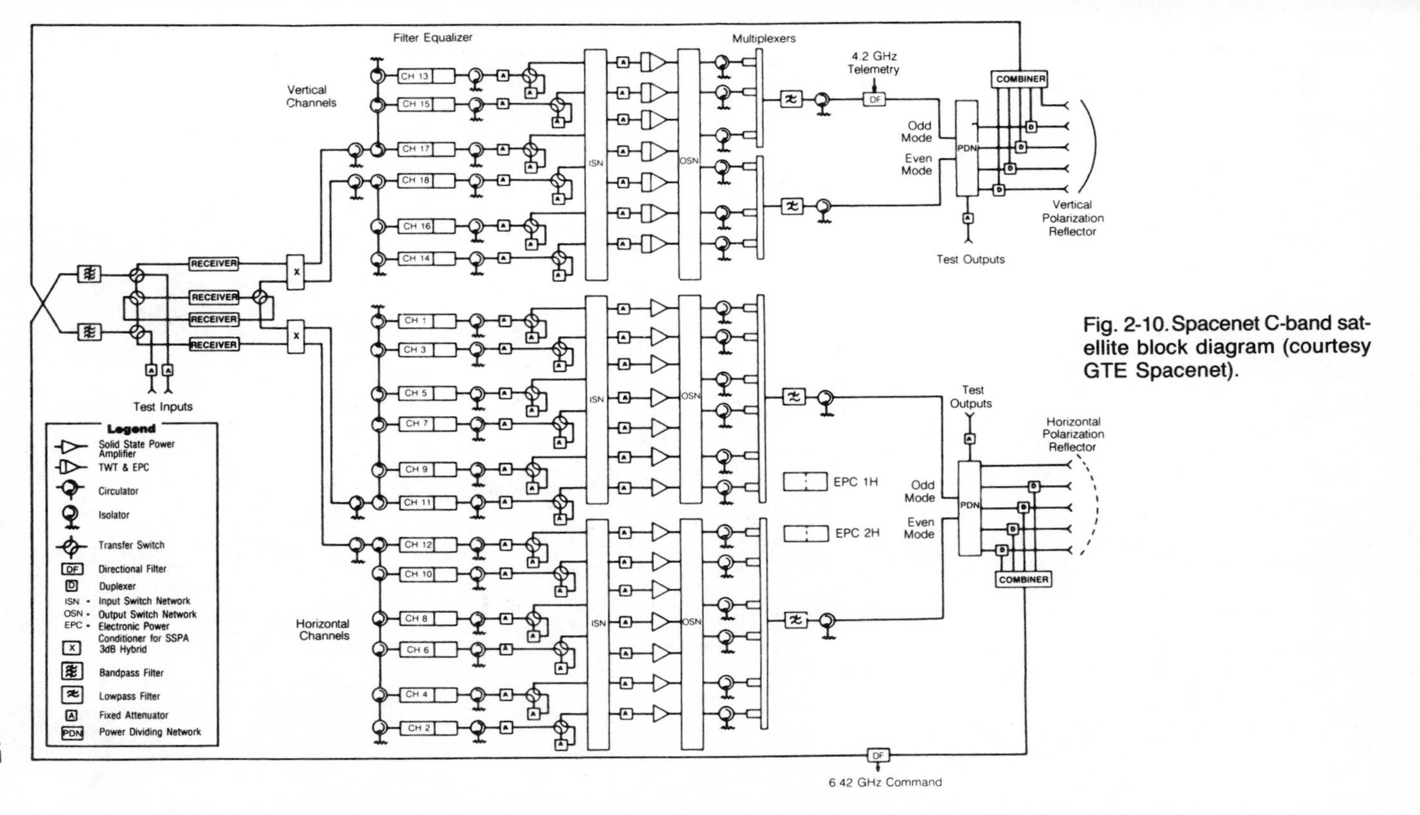

Fig. 2-10. Spacenet C-band satellite block diagram (courtesy GTE Spacenet).

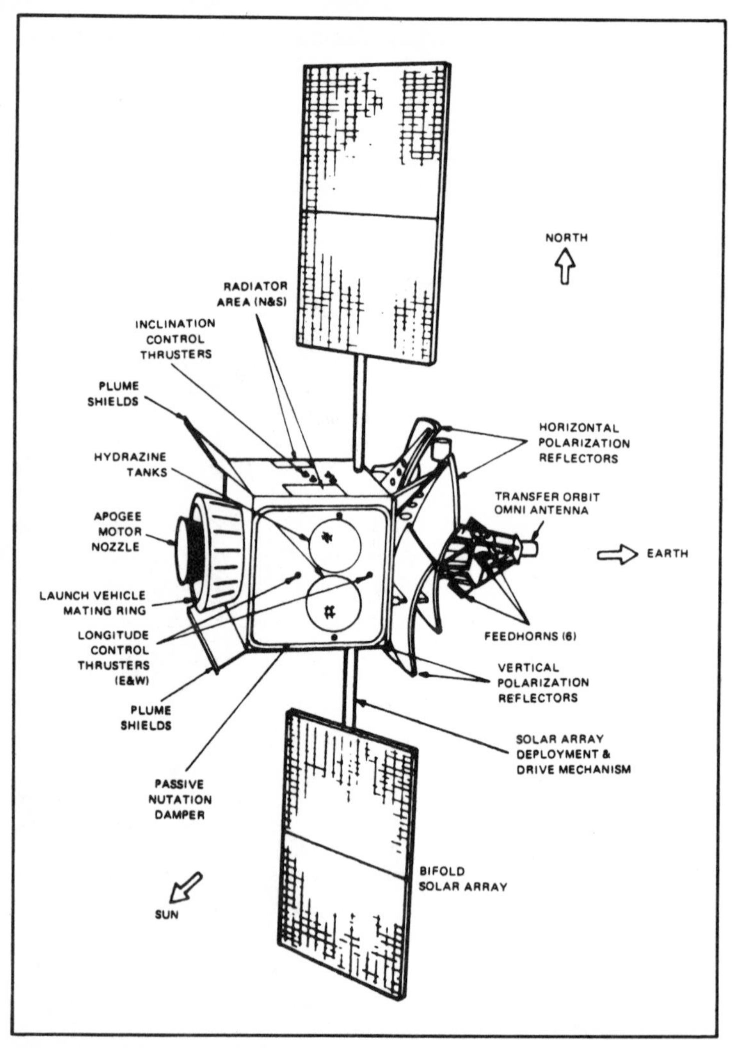

Fig. 2-11. The RCA SATCOM series (courtesy of RCA Americom).

and have extended flat solar cell panels which are always pointed at the sun, and continually collect maximum solar energy. Lightweight materials such as graphite-fiber-epoxy composite materials for communications input and output multiplexers, antenna horns, waveguide and support towers, all help reduce weight. The fixed parabolic reflectors of the antenna systems are made of Kevlar in overlapping, honeycomb structures embedded with printed circuit orthogonal polarization grids. Other honeycomb equipment panels, plus extensive use of aluminum, magnesium, and beryllium also aid in reducing weight. Even so, on-orbit weights have increased from 1,021 lbs. for SATCOMs I and II to 1,288 lbs. for Advanced SATCOM, which will be described shortly.

SATCOMs III-R and IV, although externally similar to their forerunners, have better receiver noise performance, greater bandwidth, higher powered TWT transponders, and an increased design lifetime of 10 years. They also have four spare transponders on board, one for each group of six that are operational. The added fuel supply will extend stationkeeping life from 8 to 10 years, plus more advanced attitude control subsystems, and more flexibility for ground control override of normally automatic on-board stationkeeping equipment.

Advanced SATCOM

As electronic needs and additional traffic expansion is required, so do efficiencies and flexibilities of the newer geosynchronous satellites. RCA's Advanced SATCOM satellite is a good example. It will become the first commercial communications spacecraft to use all solid-state power amplifiers, containing 28 C-band amplifiers having individual power outputs of 8.5 watts. With four in redundant configuration, this allows 24 operating channels in cross-polarization configuration. Its shaped beam antenna will maximize EIRP performance in CONUS, and also offer 24-channel Hawaiian and Alaskan coverage, depending on selected positions in orbital arc. Advanced SATCOM is also to have six electronic power conditioners (EPC) containing two dc/dc power converters for three of the four solid-state power amplifiers (SSPA) on common 8.5 V, 3.5 V, and 3 V buses. Redundancy, according to RCA, includes three EPCs for each two required and seven SSPAs for each six required.

Solid-state power amplifiers are said to have better reliability over traveling-wave tubes (TWT), simpler power supply requirements, improved carrier to 3rd order intermodulation ratios, and increased phase linearity characteristics. Advanced SATCOMs are to have a 50% increase in voice/data capacity and have been designed to be compatible with Delta 3910/PAM launchers. There are four of these satellites, with the initial launch scheduled for October 1982 in the Alaska service, followed by the second in April 1983 to replace SATCOM I. SATCOM II-R flew in September 1983, and the fourth will be held in reserve as a ground spare. All are C-band satellites.

The RCA SATCOM geosynchronous positions along the equator are as follows: SATCOM I, 136° W; II, 119° W; III-R, 131° W; IV, 83° W; Alascom, 143° W; I-R, 139° W; II-R, 72°; SATCOM VI, 67° (launch postponed).

SBS (ORIGINALLY A JOINT VENTURE)

Aetna Life & Casualty Co., Comsat General, and IBM all combined to form a business communications services satellite network directed toward business markets of almost $50 billion in 1984, and doubling that figure by 1989. Three satellites, initially costing $20 million each and transceiving in the Ku-band of 14/12 GHz, handle traffic consisting of teleconferencing, facsimile, voice, and data information, all of which are digitized at current rates of 1.5 megabits (with duplex) and, eventually to 3 and 6.3 megabits when customers are willing to pay for the extra service.

Launched November 15, 1980, SBS-1 was positioned at 100° west longitude (WL) on January 5, 1981 to begin immediate operations. This satellite has been augmented by SBS-2 at 97° WL, and SBS-3 at 94° WL, and joined by SBS-4 in 1984 and SBS-5 later. These latter two are to be permanently parked at yet unannounced positions. SBS-4 now at 101° WL.

SBS-1 measures 7-feet, 1 inch in diameter and 21-feet by 9 inches high. It is the highest powered satellite ever launched by a Delta rocket, and its solar cells and rechargeable batteries provide direct and stored power of 900 watts. Spin stabilized, SBS-1 will be kept within 0.03° of assigned station, with antenna pointing of 0.05° or less. There are 10 transponders aboard, delivering 20 watts each, and one or more customer networks are assigned to each transponder, with six extra TWT amplifiers and three spare receivers. Uplink band assignments are from 14 to 14.5 GHz, and downlinks from 11.7 to 12.2 GHz, with coverage including both the U.S. and southern Canada.

Rf terminals have been procured from Hughes aircraft, and Nippon Electric, port adapters and controllers from IBM, and high speed burst modems from Fujitsu. Tracking and telemetry stations are located at Castle Rock, Colorado and Clarksburg, Md.

TDMA

SBS (Fig. 2-12) uses time-division multiple access (TDMA) with demand assignment—a time sharing arrangement where each user has some complete transponder available during specific time

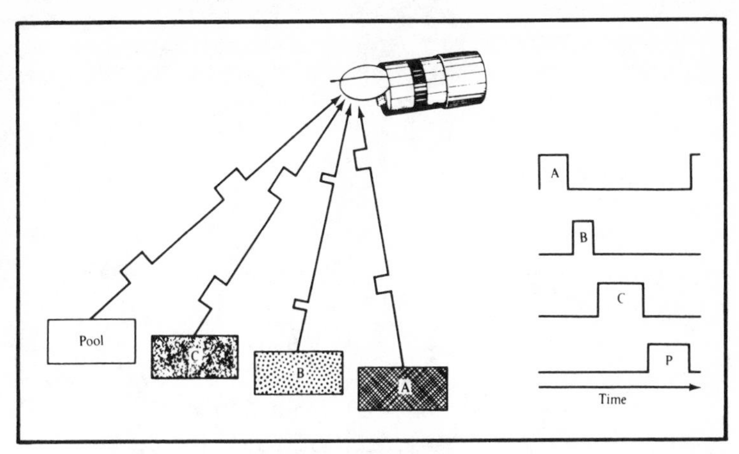

Fig. 2-12. TDMA system for SBS satellites. Note the relative times at the right (courtesy of SBS).

periods. A reference and control station assigns the duration and bit stream position of such special burst information. Then the various earth stations transmit bursts of digitized intelligence with origin, addressee, voice, data, and image traffic. Meanwhile, the reference station synchronizes earth station timing so that reception may be set up accordingly. Each earth station receives all transmissions but decodes only the data specifically addressed.

According to knowledgeable industry sources, TDMA will very likely become the "premier satellite technology of the 1980s." New techniques do accommodate slow scan video teleconferencing, high speed FAX, computer data, etc. Wideband 60- to 64-megabit systems are even now being offered by common carriers and, in the future, for private uses as well. At the moment NTSC color can be digitized at about 90 megabits per second with no visible degradation; at 20 Mb/s without redundancy and only slight loss in video quality; and limited motion at 6 Mb/s. According to SBS, however, improved techniques at 1.5 Mb/s do permit some motion, and freeze frame with some color at 56 kilobits is going on through their network right now. Full color TV, however, is not specified since bandpass requirements are considerably greater. New advances, nonetheless, are on the way.

In the SBS system, a 15-millisecond frame is the basic repetitive unit, which has both control and traffic fields. The control field is used for synchronization, capacity assignments, and to signify the beginning of each frame. Traffic fields contain message intelligence. Every network has a specified piece of the traffic frame

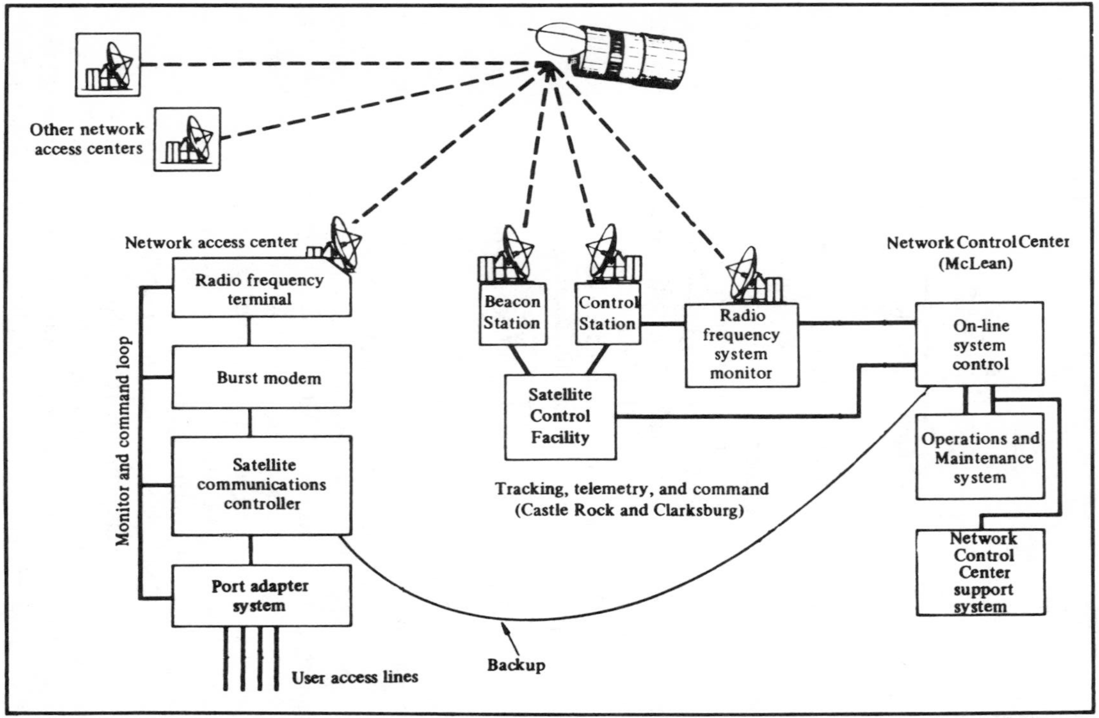

Fig. 2-13. Network control for SBS satellite communications (courtesy of SBS).

which also contains some unassigned time called spare capacity. Networks (Fig. 2-13) requiring additional capacity during peak traffic can draw upon this spare capacity pool.

Voice Compression

During normal telephone conversations, word and sentence pause are commonplace. SBS does not transmit these silent periods but instead inserts "channel noise" to maintain audible continuity. All this is called voice compression and permits reduction of transmission capacity offering more efficient and cost-effective communications. Later, "capacity transfer control" will become available allowing capacity shifting among several stations in a connected network. Under high data rate conditions, this should reduce costs considerably. SBS's switched mode is especially useful for teleconferencing, electronic mail, and high speed data exchange between 56 kilobits and 1.544 megabits per second.

WESTERN UNION'S WESTARS

Also built by Hughes Aircraft, WESTAR I, launched April 13, 1974, was America's first domestic communications satellite and remained in service until 1983. WESTAR II, also launched in 1974, retired in 1986. WESTAR III, launched in Aug., 1979, remains positioned at 91° WL, with WESTARs IV and V at 99° and 122.5°, respectively (Fig. 2-14).

The WESTARs and their seven large earth stations are all combined in Western Unions's 11,000-mile terrestrial microwave network, with tracking and telemetry operations at Glenwood, New Jersey. They relay voice, video, and data communications to the continental U.S. (CONUS), as well as Alaska, Hawaii, and Puerto Rico.

The Corporation for Public Broadcasting and Public Broadcasting Service (PBS) use WESTAR for all its PBS stations in the 50 states, Puerto Rico, and the U.S. Virgin Islands. National Public Radio (NPR) and member stations are also connected by satellite, and Western Union has AM radio programming with full 15 kHz passband signals that makes AM sound like hi-fi FM. Mutual Broadcasting is another radio customer, with over 600 earth stations receiving programming. All these, in addition to Cable News Network, AP Radio Network, Westinghouse Broadcasting, etc., continue to be Western Union customers, plus a large amount of video conferencing.

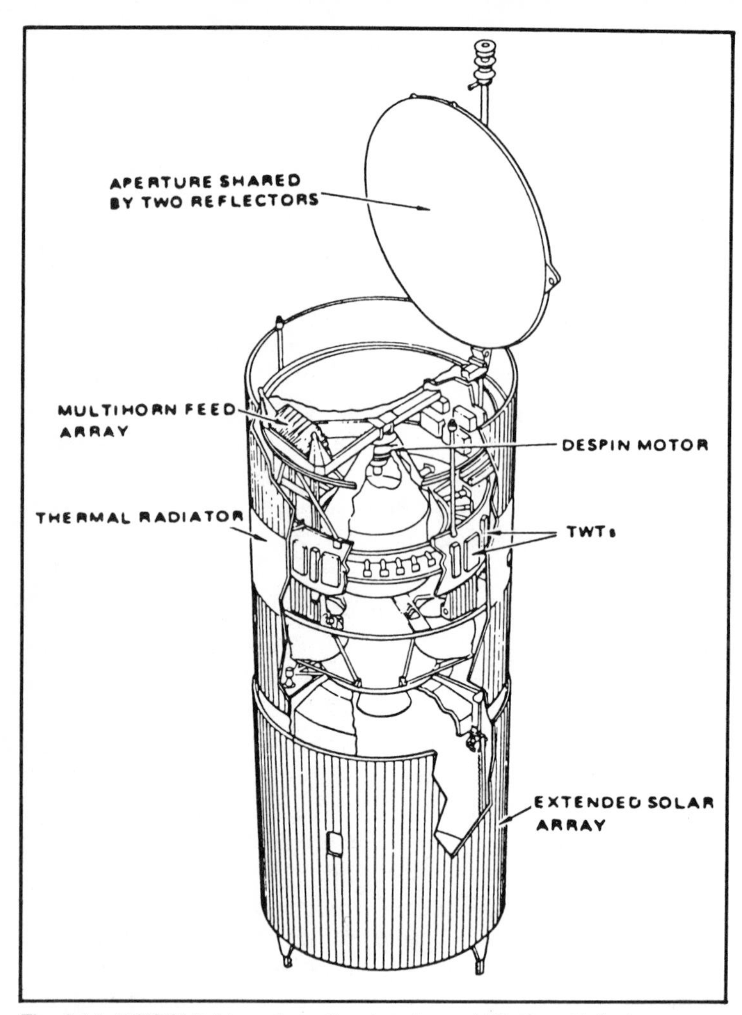

Fig. 2-14. WESTAR IV configuration (courtesy of Western Union).

Information is transmitted from an earth or control station to a satellite transponder which, conventionally, amplifies then down-converts for the transmit downlink (Fig. 2-15). WESTARs I, II, and III have 12 transponders, capable of relaying a single color television signal and its audio or 1,500 one-way voice channels at a nominal 60 megabits per second. WESTARs IV and V, however, have 24 transponders and a design life of 10 years. WESTAR VI, a 24-transponder satellite was launched in February 1984, but did not go into geostationary orbit. It was later recovered by astronauts on the Shuttle and returned to earth. A replacement, WESTAR

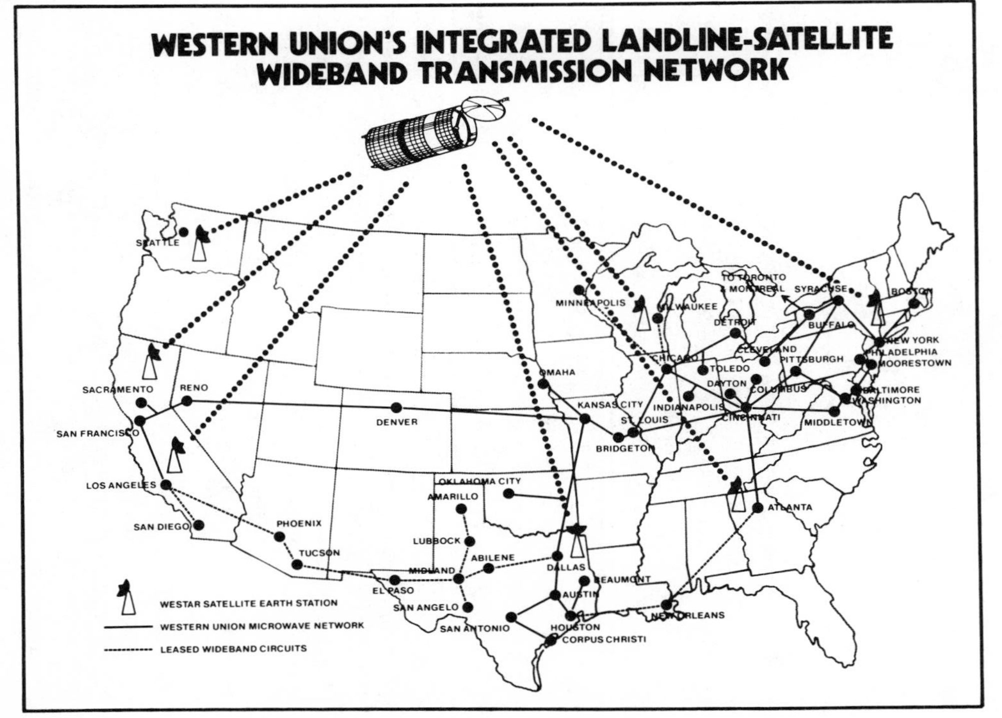

Fig. 2-15. The Westar U.S. Satellite network (courtesy of Western Union).

VI-S, is now ready for launch (1986) but due to the Space Shuttle disaster, will be delayed for some time, possible as late as 1987.

RCA'S K-BIRDS

Undaunted by the USCI experience and thundershower wipeouts on small antennas by 20 W/transponder SBS 3, RCA now has on-station Ku-band satellites K-1 and K-2, with K-3 due for launch in 1987. With their 45-W transponders and 54-MHz transponder bandpasses, these geosynchronous space vehicles are sky-parked at 81° WL, 85° WL, and 67° WL, with launches occurring for K-2, November 28, 1985, and on January 12, 1986 for K-1, by NASA shuttles Atlantis and Columbia, respectively. K-2 carries such programs as NBC network TV, Independent Network News, USFL football, CNN news and TV syndication news. K-1 delivers joint venture HBO/RCA cable programming, SMATV (direct-to-home) services, in addition to private business networking. K-3 will also carry private business networking, TV, and some K-1 programs.

These powerful satellites weight 4,245 lbs. at launch, 2,255 lbs. in orbit, have three-axis, earth-oriented stabilization. Their heights, including antenna towers, measure 98 inches, with an array area of 280 square feet and maximum span (with arrays deployed) of 65 square feet. Each has 16 traveling-wave-tube amplifiers delivering 45 watts, plus six redundants, with each group of eight containing three spare TWTAs. The cross-pole H/V output is linear. Transponders and housekeeping components are mounted about four panels on the north and south sides of the spacecraft. A 2114 pound apogee-kick motor and its four propellant tanks position the satellites and their payload of feed horn arrays, dual reflector antennas, low-noise amplifier/receivers, light weight multiplexers, driver limiter amplifiers, power dividers and high efficiency TWTAs. Dielectric antenna reflectors have orthogonal conducting grids to double channel capacity with frequency re-use and cross-polar isolation. Each channel may serve East or West CONUS or evenly distribute power over the entire U.S. Magnetic torquing, wheel-speed control, and wheel-axis roll trim automatically correct any attitude errors that may arise during their design lifetimes of 10 years.

Power derives from a solar array of eight hinged panels in sets of four, all normal to the plane of orbit. Three parallel-connected, 50-amp-hour nickel-hydrogen batteries provide reserve power and are charged simultaneously but controlled separately during sun-

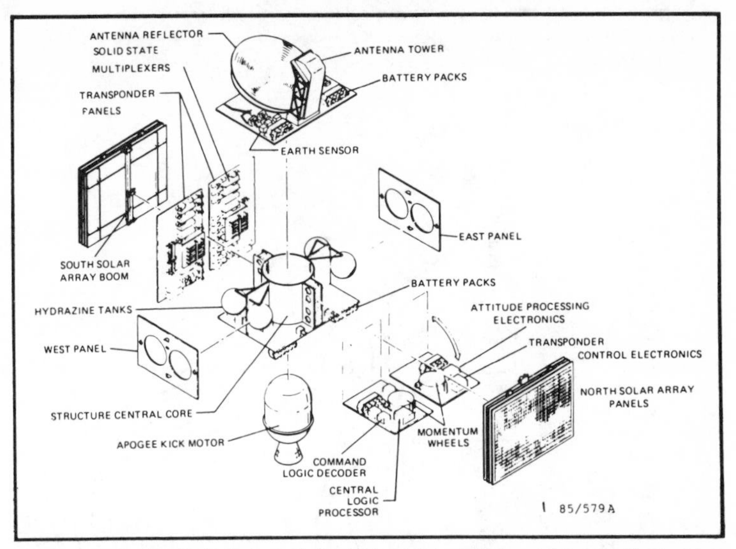

Fig. 2-16. An exploded view of the K-1, K-2 and K-3 satellites—all of which are or soon will be in orbit and fully operational (courtesy RCA Americom).

times and are active during eclipses. Bus voltage varies between 24.5 and 35.5 volts, with dc-dc conversion for specific power requirements. The reaction control system uses conventional blowdown monopropellant hydrazine and has four electrothermal hydrazine thruster engines for north-south stationkeeping. See Figs. 2-16 through 2-18.

Control and telemetry terrestrial guidance for the K-birds originates from Vernon Valley, New Jersey and South Mountain, Somas, California.

IBM RETAINS SOME SBS SATELLITES

Although no longer a Satellite Business Systems owner, IBM retains control and proprietorship of SBS 4 (on station at 94° WL) as well as SBS-5 (due for launch by Ariane 3 in early 1987), and SBS-6, scheduled for the Shuttle in mid 1988 or 1989. All have 43 MHz bandwidth transponders that are capable of at least 20 W output. SBS 5, however, may strap two channels together for 40 W, when needed, and this same satellite can also deliver a 110 MHz bandwidth from four transponders upon extraordinary traffic demands. These three satellites, although owned by IBM, will have *no* ground facilities and depend entirely on customer up/down links for terrestrial signal processing. All are equipped to handle either data or video.

EIRP CONTOURS* FOR FULL-CONUS COVERAGE BEAM

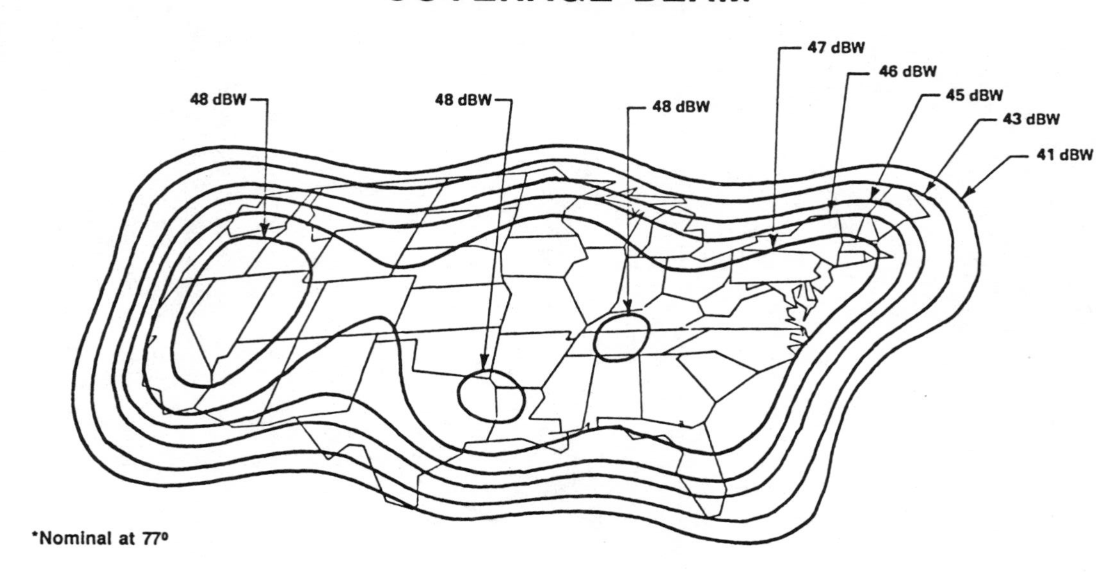

*Nominal at 77°

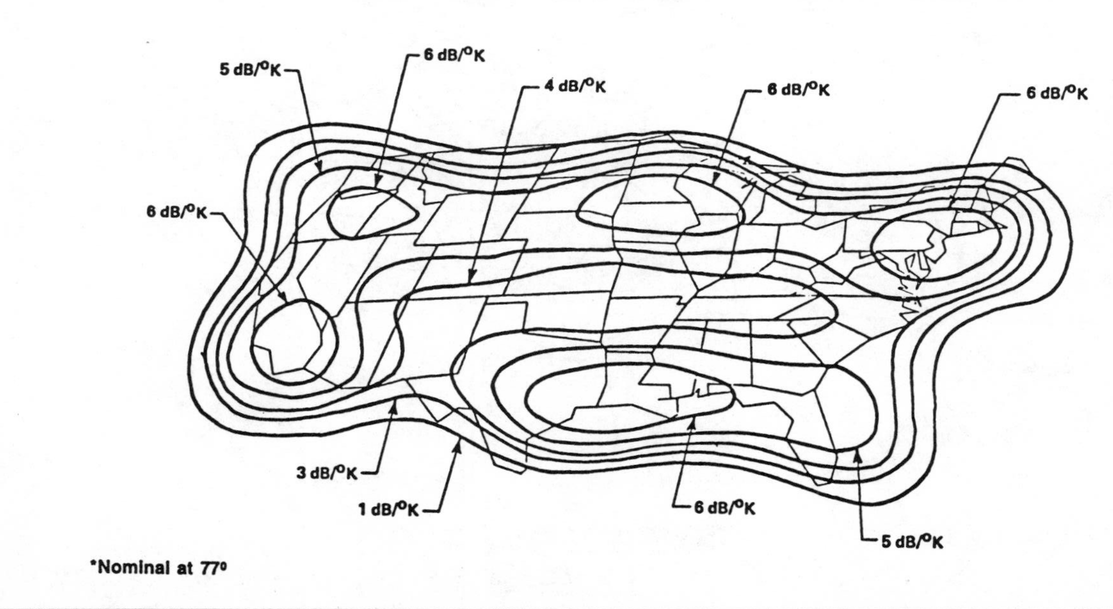

Fig. 2-17. Full CONUS (Continental U.S.) coverage by RCA's K-birds (courtesy RCA Americom).

EIRP CONTOURS* FOR EASTERN HALF-CONUS COVERAGE BEAM

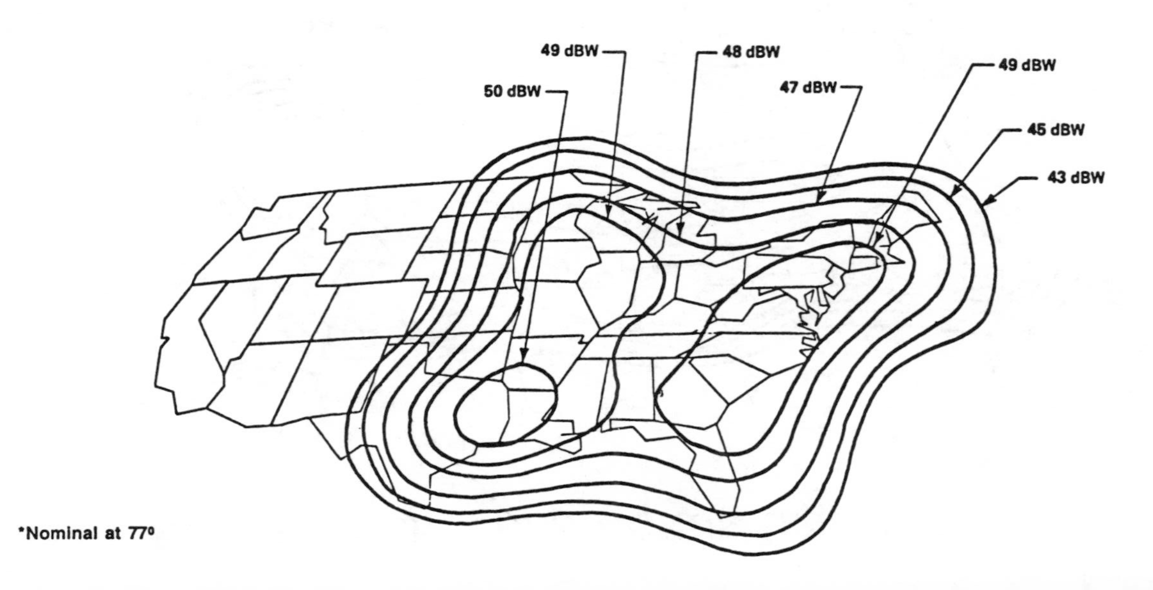

*Nominal at 77°

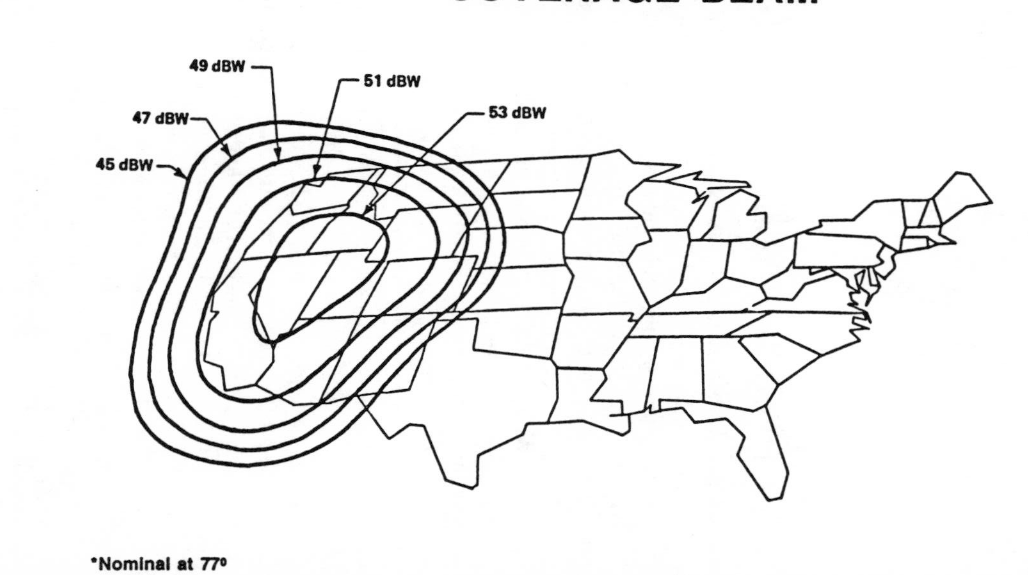

Fig. 2-18. Combined East and West coverage by RCA's Ku-band satellites (courtesy RCA Americom).

Chapter 3

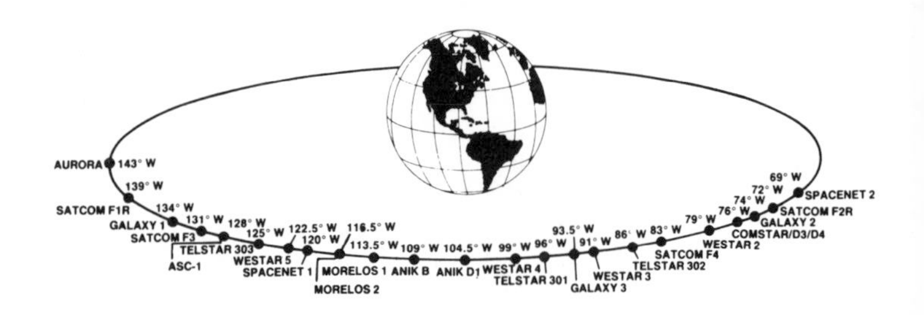

The Fixed Satellite Service

A GREAT DEAL IS NOW GOING ON IN THE SATELLITE COMMUnications world that directly affects downlink/dish considerations, whether there should be broadband electronics for both 4 and 12-GHz downlink services, what the geosynchronous parking separation should be, and just how many space "birds" can be approved by the Federal Communications Commission for each downlink available. We are told that even at this mid 1986 writing, in the 4/6 GHz band there are 30 satellites assigned and eight applications pending, and in the 12/14 GHz band there are another 30 assigned, plus five hybrids. Of the totals there are 16 C-band, 8 Ku-band, and 3 hybrids now on station. Further, theoretical curves between prime focus and Cassegrain feeds suggest there is an approximately 10 dB sidelobe advantage to the former with similar gains when the two are compared at 2° satellite separation.

Since such studies are by no means totally conclusive, we'll have to leave this for future practical tests and applications to verify. Our job at the moment is to call such hypothesis to your attention and let time bear sweet or sour fruit. Certainly many prominent manufacturers will have all sorts of opinions and scientific postulates as soon as they are informed. When all's said and done, however, the FCC has made its ruling and earthlings must conform. But eventually, say those already responding to the FCC's notice of inquiry, a 2° separation is practical. As all this comes to pass a 10-12-foot TVRO parabolic dish will certainly be very reasonable for better gain, less sidelobes, but somewhat harder to aim. Then the need for automatic azimuth settings via microprocessor or other assorted electronics could become more of a necessity than luxury—if you would have most satellite services at your fingertip command. Specifications are even tighter at Ku band.

SYSTEMS AND SUBSYSTEMS

The earth station actually consists of three main sections (Fig. 3-1).

1. Antenna
2. Receiver
3. Transmitter

If the antenna is large and has a very narrow beamwidth there has to be a tracking system which automatically moves drive motors in some sort of pedestal to keep the antenna on target. The antenna (often called a radiator) is designed to transmit as well as receive. Output goes through a diplexer which separates signals transmitted from those received. A well-designed filter is also attached to prevent transmit signals from leaking into the electronics that receive.

UP/DOWNCONVERSIONS

Key to the entire electronics portion of the receiving subsystem is the low-noise amplifier (LNA) or block downconverter (LNB). How the LNA/B operates substantially establishes the entire earth station's performance. Generally, the LNA/B is not operating in the same location with the rest of the earth station's electronics. In addition, you may have a second *inter-facilities amplifier* (IFL) that is placed between the LNA and other electronics to boost any cumulative losses because of extra cabling distances found on many of the huge transceiver dishes.

If this station receives more than one frequency, all are put into a simple power divider usually made up of passive devices and the various energies are then delivered equally to a number of outputs—one of which can be the return signal to keep the antenna tracking correctly as it follows the path of the satellite. We emphasize this is only a divider and not a diplexer since it has no frequency selectivity and passes the entire 500 MHz C-band.

All this goes into a downconverter that amounts, basically, to

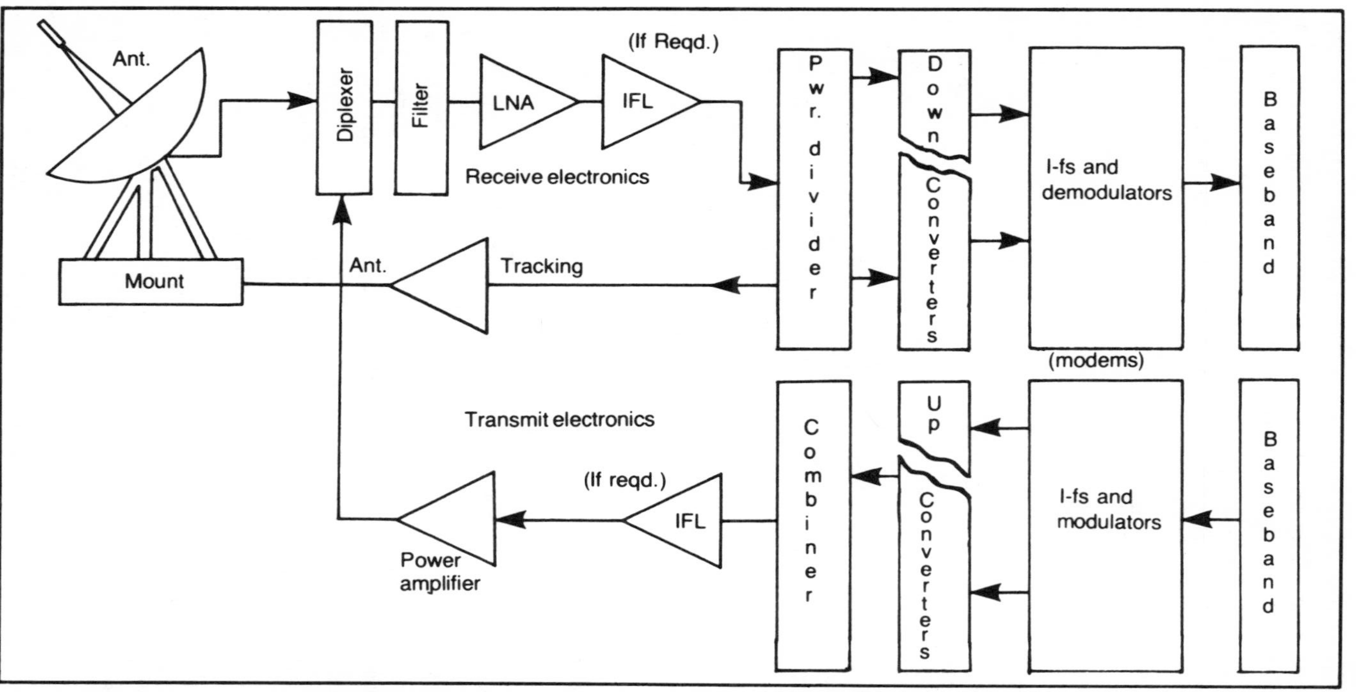

Fig. 3-1. Transmit and receive electronics in a large commercial satellite earth station transceiver.

a high frequency tunable radio. Here all special downlinks destined for this particular earth station are selected and converted. The output of this downconverter—which has also reduced the GHz frequencies often to something like 70 MHz or more by heterodyning—then supplies such lower frequency signals to intermediate frequency amplifiers called i-fs. Then, just like a color television receiver, this new carrier is demodulated and baseband intelligence extracted for use by other electronics.

Returning to transmit, this same 70 MHz carrier and its modulation is once again i-f-amplified and put into upconverters which translate into uplink frequencies assigned to this particular earth station. Depending on the method of transmission there is one or more converters which are once more summed in a power combiner, further IFL amplified, if need be, and put into the power amplifier where discipline is most important, especially in FDMA because there are other contributors offering their inputs also. Power levels, however, are usually determined and adjusted in the upconverters. In baseband, there are normally two main categories: analog and digital. Analog consists of voice and video, which is becoming more digitized as time goes on. (1984, for instance, saw for the first time, digital television receivers.)

MODEMS

In normal video and audio satellite transmissions, this is now almost entirely frequency modulation, as is sound. That's done by a *modem*—a modulator/demodulator—and, as with audio, the information is processed either from a carrier or on a carrier. Digital modems are used for binary transmissions in the same way, although the method of operation is called phase-shift keying. For a digital system there can be error correction. Here a certain number of bit errors can be detected and corrected. And there may also be several digital basebands available. NASA right now has a system running at 300 megabits which is considerably faster than any recognized commercial system at this time. So, even now, a very high bit rate for digital transmission is entirely possible. Using video for standard television pictures, however, now requires either AM baseband or rf for ordinary 525 line amplitude modulation scan. Sound, of course, remains FM and is handled accordingly.

CAPACITY FACTORS

If backoffs are required, then we're not using the maximum

capacity of the satellite. Another factor is rain margin. Its importance varies, depending on the frequency bands involved and the location. The intensity of rain also makes considerable difference. In the tropics, downpours count more than in the arid areas, since their incidence is more frequent and much heavier. To compensate for this, you add enough additional gain or capacity to the system to offset probable losses, and this is called a "rain margin."

In commercial C-band 1 or 2 dB is all that's necessary. But in the military band at 7 and 8 GHz you'll need about 6 dB rain margin (a factor of 4), so you're operating at only 25% of potential, with the remainder waiting to overcome this expected rain, and that's not a very efficient means of operation. In the Ku-band at 12 or 14 GHz, there are differing opinions, but 4 to 12 dB are numbers most people talk about. At 20 or so GHz, the safe margin is approximately 20 dB, which means gross overdesign of the system for anything like adequate compensation.

Other capacity factors which have to be recognized are: free space loss, atmospheric loss, implementation loss, satellite backoff, and the obvious antenna pattern. The higher the frequency, the greater the rain loss, and the more additional overdesign required. Yet these higher portions of the spectrum are, so far, the least used and that's where some of our more useful communications will originate. Therefore, once again, a great deal of attention has to be paid to earth station terminals to be sure whatever signals are available get through and are seen under favorable circumstances.

Most or all of these factors may be calculated and reduced to usable equations, beginning with free space loss (L). Using a latitude of 37° north, 76° west longitude for an earth station looking at a satellite situated at 120° W, the basic math is as follows:

$$L = 20 \log (4\pi\, r/\lambda)$$

with "r" being the transmitter-receiver distance in statute miles and lambda (λ) the wavelength of radiation.

$$L = 185.05 + 10 \log (1\text{-}0.295 \cos H \cos DL) + 20 \log f \text{ (in dB)}$$

"H" stands for the earth station's latitude, "DL" the earth station's and satellite's difference in longitude, and "f" the operating frequency in GHz. The final figures, then, would be:

$$L = 185.05 + 10 \log (1\text{-}0.169) + 20 \log 4 = 185.05 - 0.81 + 12.04 \text{ or } 196.28 \text{ dB}$$

We are told that atmospheric attenuation of rf frequencies is primarily due to oxygen and water, and that clear weather attenuation originates from the earth station's operating frequency and its angle of elevation. Fortunately, for transmissions below 8 GHz, the problem amounts to less than 1/2 dB with dish angles greater than minimums of 5°. This, of course, increases with higher frequencies.

FOOTPRINTS

Each satellite, however, has what is called a "footprint" which is the terrestrial area that it is designed to cover. As this pattern spreads out (Fig. 3-2) the more difficult it becomes for the receiving dish to pick up usable signals. A typical example would be the dBW figures projected in terms of EIRP contours for SATCOM patterns were the particular "bird" to be located at 131° W longitude. Note that as each contour comes closer to the U.S. mainland, the stronger the signal becomes and the better earth terminals can receive it. Cross-polarization (vertical/horizontal) permits up to 24 transponders to be operated from an ordinary 12-transponder satellite using certain additional electronics and frequency separation (offsets). EIRP, of course, stands for effective isotropic radiated power in terms of radiated strength of the satellite signal. RCA's SATCOM series has 24 transponders, of which 21 are usually in use, with others acting as spares or kept out of service to avoid crosstalk.

As you can see, a fairly strong signal from the satellite becomes tiny when it reaches the ground, necessitating an extremely sensitive receiver. At higher frequency bands, the path loss goes up as the square of the frequency, and the square of the distance, as well. And rain not only attenuates the signal, but its temperature also constitutes a loss because of all the moving molecules, which add noise. The main problem with snow is its accumulation on the dishes where heaters may have to be installed to melt it. As you can imagine, there are all sorts of imperfections that contribute to nonlinearity signal handling and loss. Even the antenna has some pointing loss, as well as a polarization problem if the antenna's polarization isn't matched to that of the satellite.

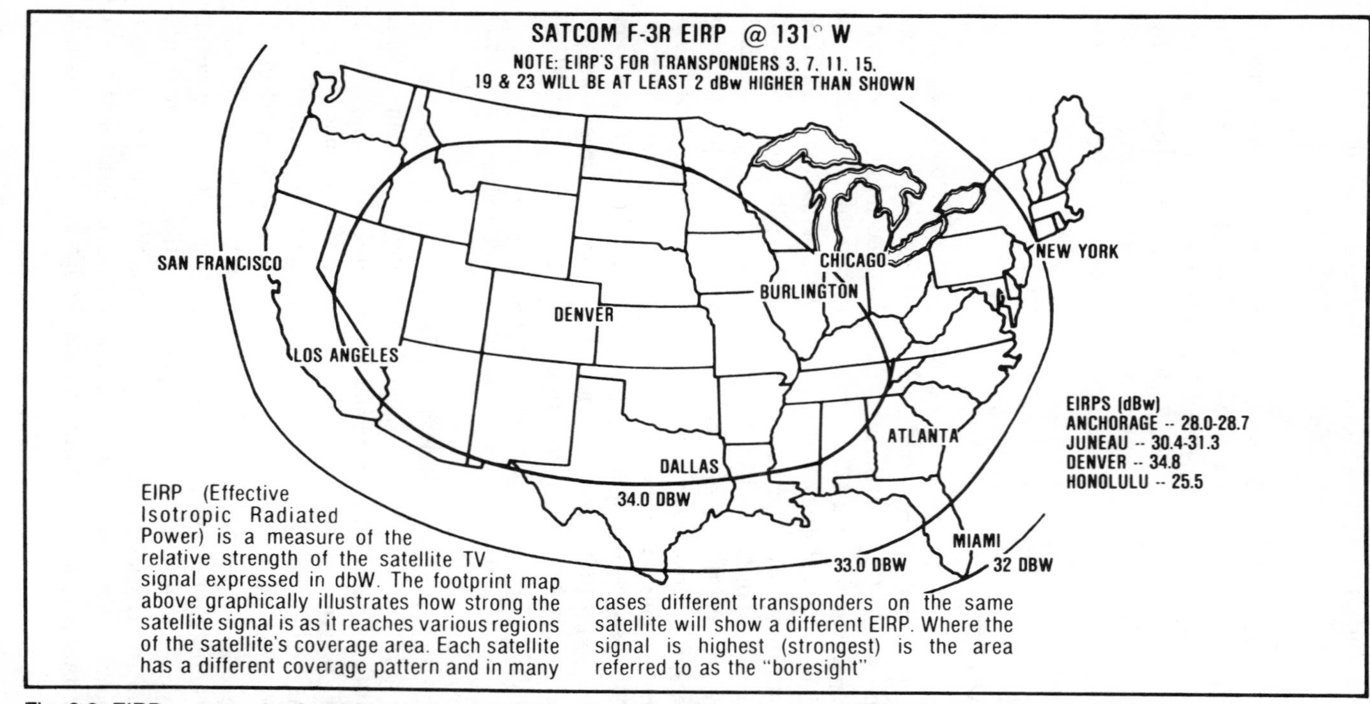

Fig. 3-2. EIRP contours for SATCOM F-3R at 131°W longitude (courtesy of Winegard Satellite Communications Div.).

THE EARTH STATION

You will find this portion of the chapter the focal point of what has both gone before and what's to come. For the receiving equipment actually dominates the design of much uplink and downlink gear since signals must be received and processed before being seen or heard. Here, signal-to-noise (S/N) ratios are highly important since intelligence must be above the noise level sufficiently for recognition and downconversion and/or baseband detection (Fig. 3-3).

The G/T figure of merit immediately indicates the downlink capacity of some earth station receiver for a particular satellite transponder output. This is found by the ratio of capacity (C) over the signal-to-noise density ratio (KT), in an equality involving satellite power times antenna gain times receiver antenna gain over path loss (K) and system noise temperature.

$$C/KT = \frac{\text{Satellite power} \times \text{Ant. gain} \times \text{Receiver ant. gain}}{(\text{Path loss}) \times (\text{K}) \times (\text{System noise temperature})}$$

Therefore: the capacity is directly proportional to G/T where G is the net antenna gain to the LNA, and T stands for resistive loss between the antenna and LNA, antenna temperature, and all

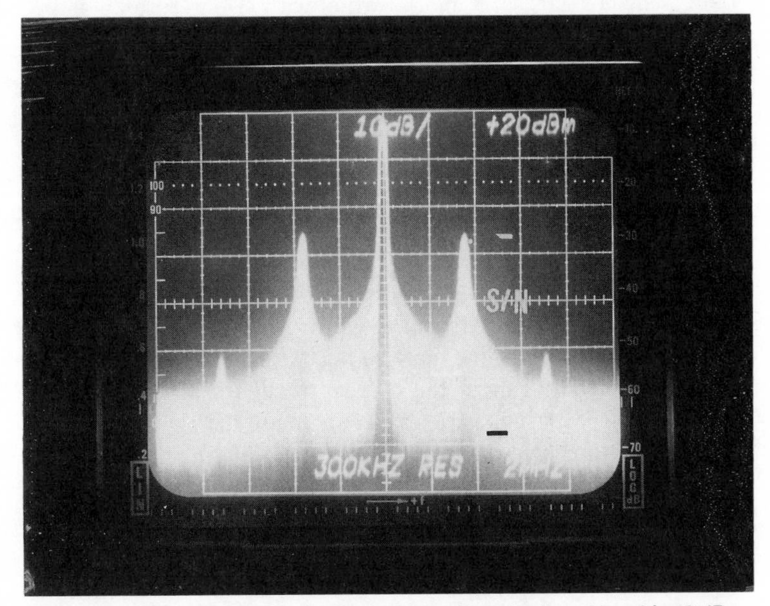

Fig. 3-3. Signal/noise ratios are measured from the noise floor. Here it's 42 dB.

equivalent noise temperature characteristics of the receiving system. As stated, the higher the G/T, the better performance of the system. (K is Boltzmann's constant.)

$G/T = G - (10 \log T)$ dB/°K (for receive side of the earth station)
Backtracking; $T_L = (1 - 1/L)\, T_0$ (in °K)

where L = numeric loss and T_0 the earth temperature at 290 °K and system noise temperature amounts to:

$$T_S = T_a/L_r + T_L + T_e \text{ (in °K)}$$

Note: noise temperature referenced to *input* of receiver. Where T_a is the antenna noise contribution; L_r is the antenna, LNA loss (between the two); T_e becomes the receiving subsystem equivalent noise temperature; and T_L the noise contribution loss, which is a ratio of input to output, not expressed in dB. Finally, system gain (G) derives from:

$$G = G_r - L_r \text{ (in dB)}$$

(bigger dishes can carry more traffic) with G_r interpreted as receive antenna gain; and L_r is the antenna/LNA loss, both in dB. °K must be convertedto log scale by $10 \log_{10} T$. Were you to take an example of G_r being equal to 60 dB, T_a 50 °K; L_r at 0.3 dB; and T_e at 20 °K, then: G/T = 59.7 − 19.33 = 40.37 dB/°K.

Match G/T to capacity for best results, and reference G/T to input of the subsystem since both have to be taken at the same place.

Three main sources in the station constitute G/T. These are the antenna itself and its environment, the antenna subsystem, and any transmission line between the antenna and the subsystem.

This transmission line must be kept as short as possible. The antenna has some gain at the receive frequency and some noise temperature (T_a) which varies as a function of the elevation angle.

The earth is said to be at a constant of 290 °K. So, if the antenna sees only the earth, the noise temperature at the antenna must be as warm as the earth's. The temperature decreases, of course, as some of the antenna sees only part of the earth. Looking directly at 90° vertically, you will see only a few degrees K. Were you to downtilt this antenna to, say, +45°, you have the main beam, also other smaller beams called sidelobes. These sidelobes do look at the earth and therefore attract noise. Fortunately, sidelobe strength is not as great as that of the main beam, but it's important they be as small as possible. As frequency goes up the noise temperature of the antenna increases. By the way, try to avoid pointing your antenna at the sun, it could easily melt the feed, especially if the antenna is shiny. Thereafter the noise temperature could go to thousands of degrees and do more damage along the way.

SATELLITE TRANSMISSIONS

Although the receiver and its dish usually have a major influence on system design, the transmitting satellite plays a part also. And a brief computation of its contribution here could be helpful in understanding the relationship.

EIRP—the effective isotropic radiated power—is a figure of merit of the transmitter and becomes a measure of how the transmitting antenna receives power from the uplink electronics and directs it toward the satellite. All computations are in dB.

$$EIRP = P_a + G_t - L_r \text{ (dBW)}$$

where P_a represents power amplifier output; G_t the antenna transmitting gain; and L_t power amplifier-antenna loss. Given a station with 5 kW output, an antenna gain of 40 dB and an L_t loss of 1 dB, the EIRP of this particular station would amount to:

$$EIRP = 37 + 40 - 1 = 76 \text{ dBW}$$

(The 37 coming from 10 log 5 kW, or 10 log P_2/P_1, where P_1 is consider 1 W).

EARTH STATION ANTENNAS (FOCAL-FED AND CASSEGRAIN)

Initially, this discussion will be based on dish-type parabolic antennas which are normally the common garden variety and are the most numerous among all earth station installations. They are like an ordinary flashlight reflector (Fig. 3-4).

At the focus of the parabola, any ray that strikes the parabola emerges parallel to any other ray. So all lines are at right angles

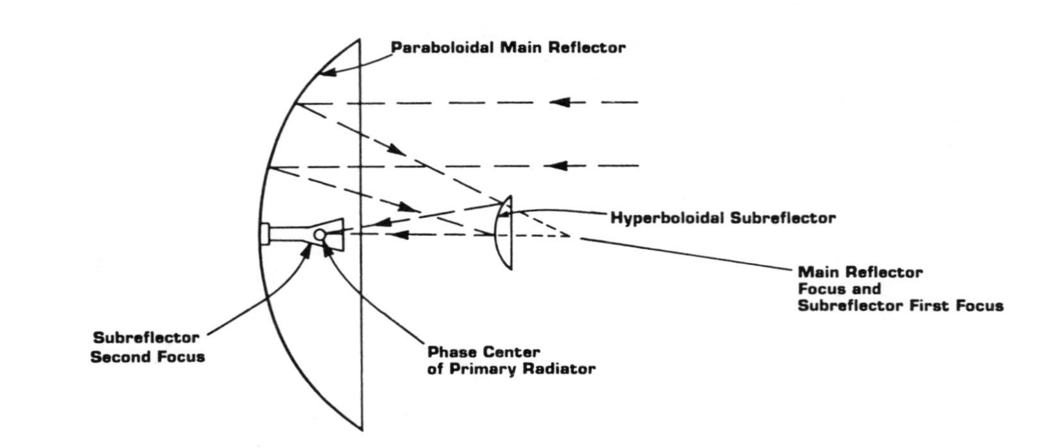

Subreflector blockage is greater than for prime focus types for D<6 meters. Hence, its sidelobes are larger.

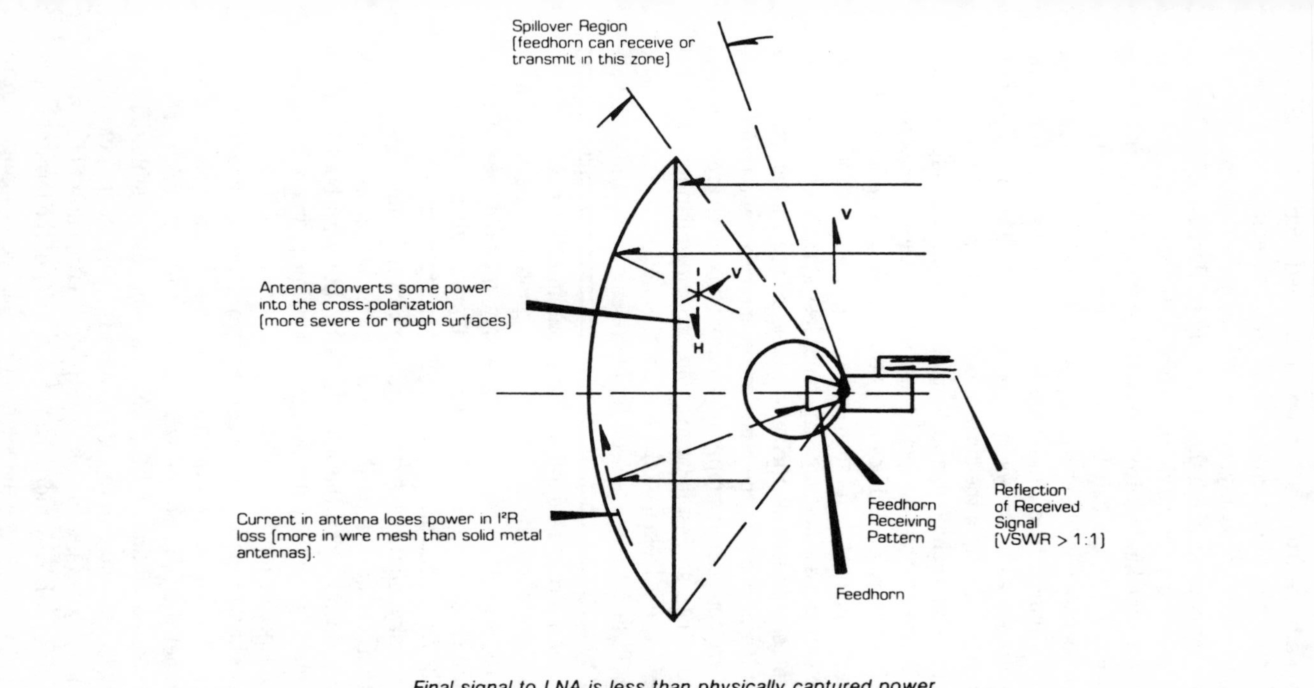

Fig. 3-4. Cassegrain and prime focus fed reflectors—note spillover effects (courtesy Microwave Filter Co.).

to the focus entry plane. This antenna, therefore, has gain since it forms a beam from the focal point in the direction of the satellite. Connecting the antenna and its transmission line is a device called a "feed." Basically, this is nothing more than the open end of a waveguide, although the sides may be flared and it gives the appearance of a small horn. This is the transition piece or impedance-matching device between the waveguide and free space. Its purpose is to illuminate the surface of the reflector so the reflector can focus available energy towards the satellite or receive it.

The simplest feed can be placed at the focus of the parabola and is known as a *focal-fed* parabola antenna. All of the energy should be retained on the surface of the reflector with none spilling over. This technique approximates that of a bandpass filter, which is never wholly perfect in passing only those frequencies for which it was designed. The energy that misses the reflector, then, is termed "spillover," and such energy is lost, consequently decreasing the antenna's efficiency. The taper loss at the antenna's edges, also represents a loss in efficiency. In terms of reciprocity, the same kind of losses occur at the receiver since receive and transmit antennas are considered simply as radiators, and both are equivalent to one another, at least in large commercial installations. Spillover affects G, and is looking at the hot earth, so noise temperature rises, also at T. Focal feed, as seen by many engineers, must be very carefully designed, without much room for taper and spillover losses unless there are considerable reductions in efficiency.

Cassegrain, on the other hand, has another reflector in front of the focal point and that reflects a new focus in the vicinity of the main reflector. Take the same dish, put a subreflector in, and the spillover we had previously that looked at the earth is now looking at the sky. So the noise temperature pickup is considerably less than that of the standard focal-fed antenna. However, a little spillover does look at the 290 °K earth, but with both feed and subreflector available, the design, according to many engineers, can do a better job (Fig. 3-5).

Another comparison between the two is the transmission line's contribution. The shorter the transmission line, the lower the noise, and higher the G/T. So you want to keep the connection between the transmission line and the LNA as short as possible. The answer is to bolt the two together. With the feed at the rear of the antenna, you can get to the LNA to work efficiently. True, there are redundant LNAs (more than one) in most of the larger installations, but eventually someone has to climb the "rigging" and re-

Fig. 3-5. Cassegrain feed antennas perform for the WESTAR satellite control operation in Glenwood, New Jersey (courtesy of Western Union).

pair the defective one. So you don't want to take the station off the air whenever work is required on a low-noise amplifier. The cost difference, however, is said to be almost the same; so noise aspects and ease of repair would seem to favor the Cassegrain. Regardless, we know of *no* large earth terminals that are *not* Cassegrain-equipped anymore; only the lesser expensive, especially consumer type dishes are focal fed. When you've been in the business awhile, you can merely look at a satellite dish and tell who built it by its external characteristics.

Beam waveguide is an even better variation of the Cassegrain which eliminates the transmission line and puts all of the equipment on the ground where it is easiest to maintain. Where there is a feed in the usual Cassegrain-type reflector, there is only a hole, and energy passes through it and then to a flat mirror that is mounted at 45° parallel to the elevation axis. So this entire structure can rotate around the elevation axis without requiring additional recovery. The energy is intercepted by a second curved mirror which refocuses and strikes a third, flat mirror, then a fourth mirror once again refocuses the energy to the top of the building on which the antenna is mounted, and there is where the feed is placed. Energy travels parallel to both the elevation and azimuth axes. Efficiency of this system is greater than 98 percent. Most large installations are now so equipped and the mirrors are placed in a tunnel for protection. Mirrors are extremely reliable and don't have to be realigned for years. This is all done in the near field, rather than the traditional antenna far field.

The *offset-fed (low sidelobe)* is another type of antenna (Fig. 3-6) that has extremely good sidelobe performance but with some disadvantages. It can be either focal or Cassegrain, but still offset fed. The advantage is that energy coming in and out of the aperture has no blockage. The beginning is a large parabola, but then only the upper center section remains in use, and therefore no blockage—the lower part can actually be thrown away. Feeds are then focal point or with subreflector, just like any other ordinary antenna. But since this feed is in the lower area of the circle, there is no interference and sidelobes are almost nonexistent.

You can also put absorbing material around the subreflector and the main reflector, as well, to further restrict any sidelobe action. Here, peak sidelobes are at least 10 dB below the FCC specification of 29-25 log θ in the transmit direction. In receive, the same is true out to 12° in both elevation and azimuth.

This is a GTE development designed to compete with those now moving to K-band. Aperture efficiency is only a little lower than standard antennas so it has been necessary to build this antenna just a little larger. The cost is somewhat greater, however, and is not expected to be reduced until the low sidelobe dish is placed in mass production.

The *multiple-beam torus antenna* (Fig. 3-7) has been around for a long time. But it has not heretofore been used to try to combine

Fig. 3-6. GTE's low sidelobe antenna for the 4/6 GHz band (courtesy of GTE).

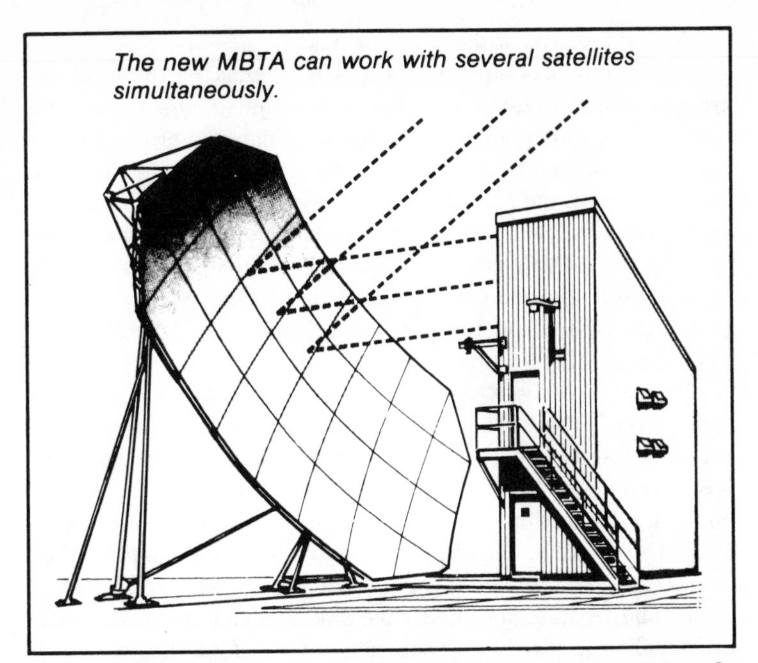

Fig. 3-7. COMSAT's torus can transmit/receive to and from 7 satellites at C-band (courtesy of COMSAT).

good sidelobe performance and a single aperture to look at more than one satellite simultaneously. We note, especially, that it can be used in both commercial *and* consumer TVRO installations and could become quite broadly based in many applications. It was originally built by COMSAT labs sometime ago in about a 30 × 50-foot model. The CATV people are aware of its potential and one 10-meter configuration recently has been installed in Alaska, and more are on the way, since it has a 20° field of view (or greater) depending on specific designs.

If you look in the vertical plane, this multiple-beam torus seems to be an offset-fed parabola. Instead of the usual axis rotation, this one rotates about an axis that is almost vertical and the focus also describes an arc. Any vertical cutting lines turn out to be a parabola, and horizontal lines become a circle. Each and every point on the focal arc uniquely maps into a single point for a geostationary orbit. If a feed is located at some special point on the arc, an antenna beam is directed toward a location on the synchronous orbit. If there are two or more beams, then several satellites may be "seen" simultaneously all from the same aperture. There are no frequency limitations, either; C-band or K-band may be mixed as desired or required without interference.

These antennas must be tilted to approximately the same degree as your latitude. The antenna, in order to look at multiple beams, must sweep about the satellites' anchored orbits. Consequently, the dish must be tilted at a specific angle to line up with the satellites in their 22,300-mile geosynchronous positions.

To estimate gain, look at the torus antenna and imagine a standard parabolic that's equal to the height of the torus. The gain of the torus will approximate the gain of a standard circular antenna having the same diameter, but just a little bit less (there's some small loss in efficiency).

In smaller versions, gain is as good as the parabolic, but in the larger models there is a difference, although it remains close to a standard Cassegrain. Cost, however, is almost double with the breakeven point slightly more than two beams, or two satellites. So if a CATV station, for instance, is working double or triple, it would pay to put in one of these multiple-beam torus units rather than buy several standard parabolas along with their expensive electronics and separate mounts. By the time this publication is in print, COMSAT expects to be in production on 4.5 meter dishes in TVRO. We're told, by the way, that the arc of coverage obtainable is limited by the long dimension of the reflector, with narrow dimensions governing gain. Thirty or 45 degrees are fairly easy.

That should hold you for the time being with enough information on the various types of antennas—the only one not being discussed for the moment is the *spherical*, which we'll take up briefly in the chapter on consumer products since that's where this type is sometimes used. Otherwise, the majors have been pretty thoroughly covered. What's needed now are a few equations to make you comfortable with antenna gain and efficiency followed by selected types of pedestal mounts and their applications.

GAINS AND APERTURES

To calculate the gain of any antenna aperture in dB: $G = 10 \log (4\pi A_e/\lambda^2)$ and a *parabolic dish*: $G = 10 \log \eta (\pi d/\lambda)^2$ with A_e as the effective aperture area and λ the wavelength. $A_e = \eta A$ becomes the total aperture area in identical measurement terms with lambda (λ). The aperture efficiency is then represented by the symbol η which is controlled by surface tolerance, taper and phase coherence illumination, and feed structure blocks.

If you had a 20 meter dish receiving at 4 GHz, with a typical efficiency of 55%, and lambda amounting to c/f which is calculated, $\lambda = 2 \times 10^8/4 \times 10^9 = 0.05$ m for a parabolic dish then, the full gain equation for a parabolic dish would appear:

$$G = 10 \log \eta (\pi D/\lambda)^2$$
$$G = 10 \log .55 (\pi \times 20/.05)^2 = 59.38 \text{ dB}$$

And to avoid an odd-ball negative number when taking the log, multiply the squared bracketed expression 2039184 by 0.55 *before* taking the $\log_{10}$ and then multiply by 10. That way, it's easy! For your parabolic aperture area (A): $A = \pi(D/2)^2$.

SIDELOBES

A contemporary rule of thumb says not to push efficiency too much if you want good sidelobe performance. Usually, 60-70% is good enough efficiency to expect decent sidelobes. Shrouds around the rims of some antennas do help, but they aren't too effective when there are multiple problems—only basic spillover interference. But shrouds do absorb water, so this practice is usually discouraged since you would have to change these shrouds every two or three years. There should be other ways to solve your interference problems than using shrouds. Aperture blockage, of course, is the major contributor to sidelobes. Offset-feed techniques, however, can eliminate completely and sidelobes due to spar and subreflector scattering. In smaller antennas, the peak gain of a sidelobe may be reduced by averaging its magnitude (or level) with adjacent sidelobes so long as the level does not transcend the gain envelope greater than 6 dB.

In measuring sidelobes, the amplitude of the envelope becomes a function of the angle (Fig. 3-8). The equation was empirically designed and resulted from international measurements of very large Cassegrain feed antennas. You have the main beam and the various sidelobes, and the equation is:

$$G_i \text{ (gain above isotropic point source)} = 29 - 25 \log \theta$$

with θ being the angle. The FCC, however, realized that this doesn't apply directly to smaller antennas, and so they permitted a process called "smoothing," which was given in the paragraph above. Both small and large antennas can usually meet this specification. This applies in transmit as well as receive.

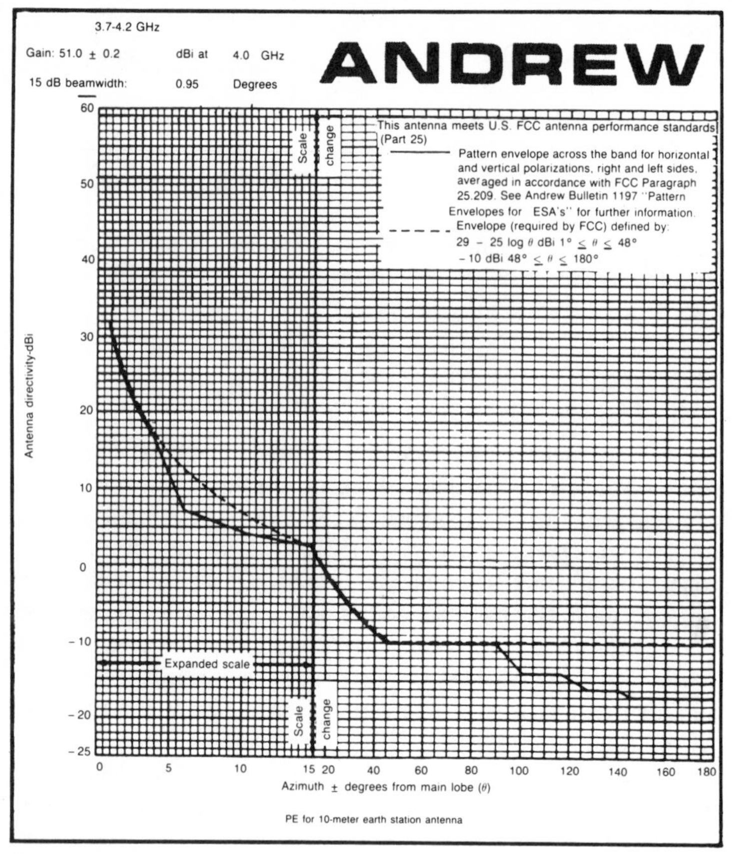

Fig. 3-8. Smoothed sidelobe diagram for 10-meter dish (courtesy of Andrew Corp.).

BLOCKAGE

In the middle of a Cassegrain antenna the back of the subreflector is visible, with perhaps 3 or 4 spars to support it. Those spars and the subreflector, of course, decrease the available area of the antenna, and this decreases the deficiency. Such energy is then lost to sidelobes in some sensitive areas. So blocking the aperture is a major factor in decreasing efficiencies of parabolic antennas.

Two other factors are amplitude and phase illumination and this returns to the theme that manufacturers can't build a perfect feed. There is both spillover and taper associated with amplitude illumination, which is a further loss of energy. Next, you need a good in-phase wavefront of outgoing rays parallel to the front of the antenna. Unfortunately, feeds are not perfect either, and this decreases the aperture.

To increase the efficiency there are three techniques:

1. Special feeds, in many varieties, independent of frequency.
2. Dielectric guides which attempt to recapture spillovers. It's a big fiberglass cone that's no problem to rf, filled with a special foam dielectric which is refractive to the subreflector. However, there is a maintenance problem because of moisture accumulation.
3. Surface shaping, the final technique, is used by almost all antennas today, both big and small.

PEDESTALS (MOUNTS)

Most large pedestals in common use today may be separated into two types: azimuth-elevation (AZ-EL) and X-Y configuration (Fig. 3-9). The AZ-EL has its azimuth axis closet to the mount and is vertical. Its second axis, above the azimuth axis, is the elevation axis and the two are perpendicular to one another. In the AZ-EL configuration you're trying to support a vertical shaft. One problem with this type is its keyhole tracking. An error signal is generated from a servo to drive motors to keep the antenna pointed. If you're in the center of the beam, the error signal can't be picked up. So if the satellite moves directly overhead, then the mount simply rotates on its own axis, the servo system won't generate an error signal, and the system locks up. Ordinarily, however, you aren't directly under the satellite since it's normally positioned over water and also at the equator.

The X-Y pedestal has a horizontal lower axis, with the second perpendicular, and they may not intersect. A horizontal mount can

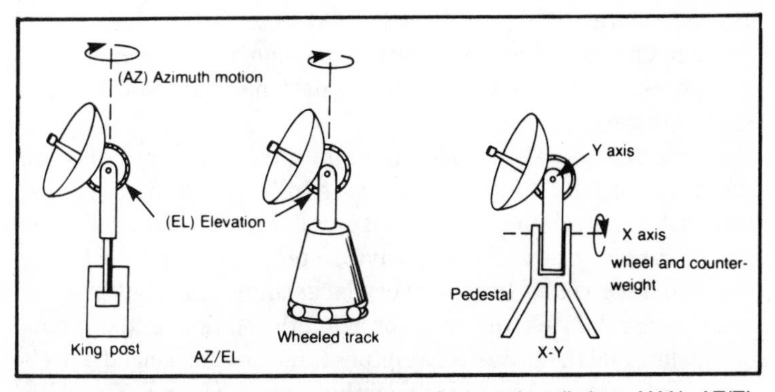

Fig. 3-9. Usual antenna earth station mounts for large installations. X-Y is AZ/EL turned on its side.

be more desirable than an AZ-EL, which has a vertical shaft. You can say that X-Y is really an AZ-EL turned on its side, so the X-Y keyholes (both of them) are on the horizon. Point the earth station in a direction where the satellite won't be, and there is no keyhole problem. The X-Y configuration is normally good for smaller antennas in commercial service.

Almost all large antennas can be adjusted by moving their panels one way or the other. Usually the factory tries to prealign them. The very large antennas, such as the standard A require shimming with a theodolite on top of the feed and there is a minimum of 120 targets around the structure.

The theodolite is a high precision transit that works between feed and into a reflecting mirror to establish precise panel angles. Such work is done only at night with the antenna pointing up. These targets are "shot" to make sure all surface tolerances are correct. Structural "sags" are not apparent on the small terminals as they are on the larger ones because of weight differential. In 10-meter stations, for instance, you should be able to install and walk away in less than 5 days. A standard A military will take months.

This method tries to improve the phase illumination for better efficiency. Antenna designers found they could build surfaces that are not true parabolas; it makes a considerable difference in the performance of the antenna. Shape reflectors, then, used on both the sub and main reflectors aid both phase and amplitude illumination, improving the aperture efficiency of the antenna with no additional cost. Corrugated feeds also help, but they're expensive.

Sidelobes, by the way, had no specs, until after 1971 since weak satellites required maximum gain and interference was avoided by finding a wide open space to locate the antenna. Today conditions are entirely reversed. You have to worry both about sidelobe and adjacent channel interference, not to mention microwaves near the ground. So, today's antennas are *not* designed, necessarily, to be super efficient.

Pedestals can also be identified in three other ways of which some are related to AZ-EL and others to X-Y. The first is a *king post* that has a relatively small base with a very rigid base structure to support bearings in the drive system. Another AZ-EL type, is called a *wheel and track* for very large antennas. That pedestal offers a circular railroad track for azimuth rotation on the ground or building and the pedestal has either three or four flanged wheels. The third pedestal which can be either an AZ-EL or X-Y is called a *limited motion pedestal.* Since the satellite has only an inclination

Fig. 3-10. Limited motion station designed for only 10° azimuth/elevation window.

angle (figure-8 motion), a 10° window in azimuth and elevation is more than adequate for a synchronous satellite. This pedestal is then driven by a linear actuator which is considerably less expensive and operates with a screw jack and small drive motor (Fig. 3-10).

All large antennas use dual drives because of their size and also to eliminate backlash. Such motors are counter-torqued so they can hold these antennas perfectly still as one motor bucks the other.

There is a requirement in antenna construction, however, that terminals must cover an entire range of the synchronous satellite arc. This is because the FCC reserves the right to relocate everyone's satellites and reassign orbit locations. So all antennas must be capable of movement from east to west when necessary. This can be done by making mount ball joints flexible so the antenna can easily be swung around.

There is also a *polar mount* that's been used for a long time by astronomers which has one axis mounted at an angle to the horizontal equal to your latitude, and this is parallel to the earth's axis of rotation. A clock motor makes one rotation in opposite direction so that the telescope remains pointed in one location corresponding to a fixed star. A geosynchronous satellite, however, is moving with the earth. If a polar mount is modified so that only one motor is required to swing the mount to any place on a synchronous orbit, then cost would be reduced by one motor and its drives. This antenna is mainly marketed to the broadcasters so they can quickly go from one satellite to another. The slew rate is very fast, in just a couple of seconds, and is very good for those in the uplink transmitting business. Microprocessor control, then, can quickly change the azimuth during various times of day so that programs are directed to each satellite on schedule.

TRACKING

Normally, antennas larger than 20 feet in diameter require tracking, and if you're looking at a satellite that is not exactly station-kept north and south you require tracking anyway. There are many types of antenna automatic tracking, but three are outstanding.

1. *Monopulse,* taken directly from radar installations and further adapted by antenna manufacturers to cut costs.
2. Pseudo monopulse.
3. Step track.

Closed loop automatic tracking systems are only used on the very large stations, they're expensive but very accurate. All are based on the technique of rather than having only one feed at the focal point of the antenna or subreflector, there are four feeds located around the focal point. These "squint" off in the opposite direction to the feed side. By using four such beams, signals received by all may be compared and if they are the same, then the antenna is pointed directly at the satellite. A misdirection point places more energy on some of the beams than others and comparison corrections are made.

During communications you want your main beams pointed directly at the satellite. These beams also form azimuth and elevation angles which travel through three separate low-noise receivers forming sum and difference patterns to keep the servo systems operating. In other words, one receiver is for communications and two more receivers are for tracking, which must have identical phase characteristics.

To eliminate two of these beams, a system called *pseudo monopulse* was invented to sequentially sample the four beams, store the results, and then make comparisons for accurate pointing. The two difference channels are sampled in and out of phase, and are summed and converted to AM and demodulated. *Step track* works only on the amplitude of the main beam and is an open loop (versus closed loop) tracking system. You have a meter and a pair of up/down and left/right switches. The switches, in turn, operate drive motors to move the antenna in its present position to another direction. So the AGC-operated system simply recognizes any change in beam direction and changes are shown on the meter. Operator responsibility is to keep the station on track as it peak detects any changes in direction away from the satellite. Step track is also used on 10 and 11-meter antennas, as well as the huge 105-foot standard A antennas and the system still works satisfactorily. Therefore, today, step track is almost universally used, but the other systems are still in operation among the older antennas. With the aid of microprocessors, these systems are now automatic and manual correction is only required if the tracking system fails. Such step track systems are all right on land, but ship motion makes satellite tracking considerably more complex and tri-axis mounts are required to keep the beams on target.

Program track is still another system that's now in vogue, consisting of a microprocessor or microcomputer, and you basically tell these little devices where the satellite is and the processor goes and finds it. However, the beams must pass through the ionosphere as well as the atmosphere where refraction takes place and this is not a straight line. But refraction corrections can be added with some trouble and the micro-device will respond and track. The process of working out pointing logistics for terminals in large systems leaves considerable room for error and the system, which evolved about the same time as step track, never really emerged as a going commercial enterprise.

Recently, another form of program track has been invented to back up the step track, and works quite simply. If the satellite's figure-eight motion during each 24 hours is continuous and not repositioned, the satellite's position is noted and stored in memory. But when satellite tracking wanders, and step track doesn't work, the program track microprocessor recalls the previously stored position and the antenna is once more directed toward the satellite's original position recorded during 24 hours past. Of course, if the satellite has been moved, then there is a problem, but usually this isn't the case since the satellite is nudged just a little bit at a time over some finite period.

LOW-NOISE AMPLIFIERS

The use of a good low-noise amplifier is always important to system earth station design. With a worthwhile unit, especially considering G/T, design of the remainder of the system can be fairly tolerant of a number of otherwise relatively tight parameters. There are basically three types:

Fig. 3-11. 100°K GaAsFET low-noise amplifier dc powered (courtesy of Channel Master).

1. Cryogenically cooled and uncooled parametric amplifiers.
2. Tunnel diode amplifiers
3. Field-effect transistors (Fig. 3-11).

Parametric amplifiers operate on the principle of a pump oscillator to vary the capacitance of a varactor diode, the input signal can be amplified and produce gain. Varactors, fortunately, generate little noise and are quite reliable. However, this particular system is very expensive and may require refrigeration down to almost absolute zero or at least temperature stabilized (the uncooled type).

Tunnel diodes permit electron tunneling through an energy barrier and amplify in the negative resistance region, thereby developing forward current just like normal rectifier diodes. Maximum voltage and maximum current occur both before and following tunneling, but the second set of peaks in the forward current/voltage range determines the diode's amplification. These tunnel diodes, however, are touchy, low voltage devices with quick responses but relatively small amplification swings. They are also used for very fast switching in older logic circuits.

Field-effect transistors in LNAs are usually found in cascaded, multiple stages since their gain in such applications is restricted. Normally called GaAsFETs, or gallium arsenide FETs, they must be carefully biased and kept close to the impedance of the waveguide. By using FETs, often coupled with bipolar transistors, the required 50 dB gain becomes a reality since amplification of each stage is *multiplied* (dBs are added) by the other to determine gain. But remember that noise also is being amplified along with the gain. These GaAsFETs are now so good that they have largely supplanted both parametric and tunnel-diode amplifiers in most, if not all modern LNA amplifiers. Final testing of these devices takes place only after they have been packaged and sealed. Then the price tag is added.

The noise figure is expressed as a simple numeric number, and not in dB. So, if this figure is stated in dB, you'll have to divide by 10 and take the antilog ($10^x/10$) to arrive at an ordinary digit (Fig. 3-12).

The actual numerical noise figure amounts to:

$$T = (F - 1) \times 290 \text{ °K}$$

and this is the noise temperature, where 290 °K is the earth's temperature and F − 1 as the noise figure. If your receiver is very noisy, digit 1 doesn't change anything. In modern earth stations,

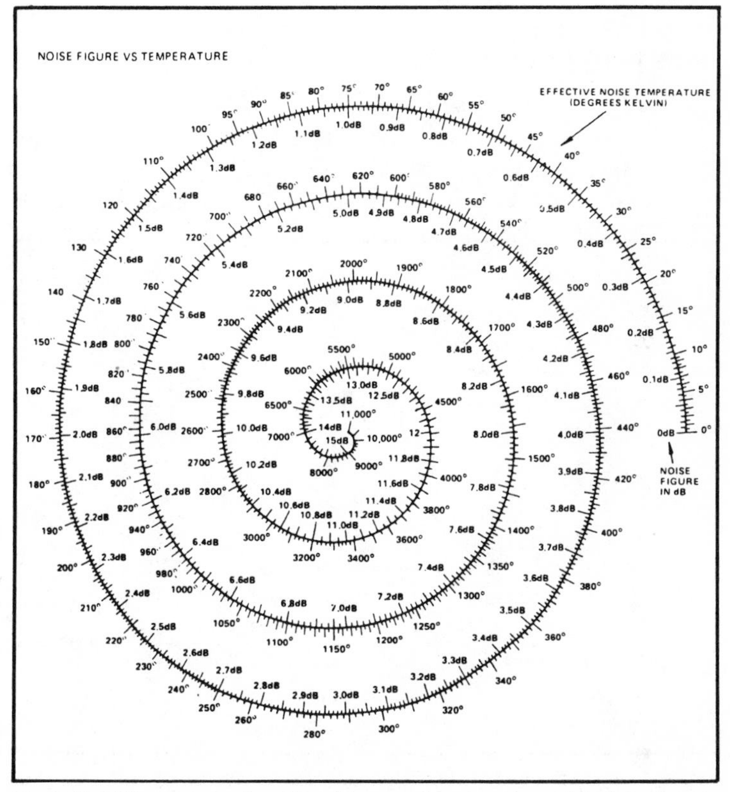

Fig. 3-12. LNA noise as a function of temperature (courtesy of Andrew Corp.).

the term "F" is rather small, and so "1" can become a considerable factor under certain circumstances. Three dB, for instance, becomes 2 when the antilog is taken, therefore 2 − 1 equals 1 which maintains the T equality at 290 °K. If that figure had been 6 dB, then 4 − 1 would have tripled the temperature to 870 °K, producing a considerably noisier result. With today's tighter tolerances, as the old song says, "little things mean a lot."

In cascaded stages, just as in the individual active devices, the noise temperature is always referenced to the input terminals. Therefore, the first low-noise amplifying stage should have a gain of some 30 dB and be located as close to the antenna terminals as possible. Under these conditions your LNA design is on safe ground because the resulting G/T noise factor is then negligible. As time goes on, GaAsFET gate widths will become even smaller than the 0.5 or 1 micron presently masked. As these narrow the device noise decreases proportionally. Cooling further decreases noise, but such techniques are not yet commercially available.

UP/DOWNCONVERTERS

Up/down converters are divided into units of single and double conversion. You could have more, but current terminology places them in these two categories (Fig. 3-13).

The single converter is simpler, since it has one local oscillator

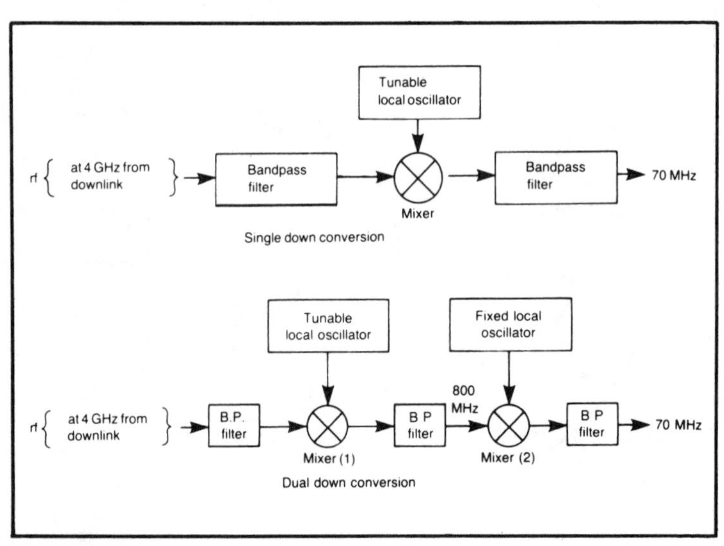

Fig. 3-13. Block diagram of both single and dual downconverters in the Fixed Satellite Service.

that's mixed with incoming rf to translate or heterodyne the usual 4 GHz downlink frequency to an i-f of some 70 MHz. Mixer outputs always contain the sums and differences of the two frequencies and the difference is selected by filtering and processed as the desirable intermediate frequency.

The double converter does the same thing in two steps, with the initial difference frequency close to 2.5 GHz, being and the second descending to the conventional 140 or 70 MHz. Local oscillators, of course, must be tunable (nominally the first stage) to select the precise downlink carrier required. A dual conversion arrangement is more desirable since it is more precise to beat somewhat similar frequencies together for their predictable outputs than to try wholly dissimilar sets of CW signals with inaccurate results.

The single unit, however, is subject to adjacent signals also, and these can be mixed in with the GHz to low MHz heterodyning and provide a problem in your normal satellite reception since there are sums and differences for the adjacents also. This process is often called image frequency which should be stopped by bandpass filter and that normally works. But, certain signals can get by and this causes the problems if for any reason the satellite frequency is altered. Then your bandpass filter must be changed since tunable units are not successful at microwave frequencies. Automatic switching between transponders or satellites is, then, negated since you can't retune and switch filters manually each time you're looking at a new frequency.

Double downconverters, of course, can be remotely switched because one bandpass filter will cover the normal 500 MHz range of the satellite since the first stage of the downconverter produces an output in the high MHz. The local oscillator may be easily tuned with the one bandpass filter doing its job as before. Crystals in the local oscillators may also be installed when the earth station requires just single frequency reception. Dual conversion also does not have the local oscillator "leak back" problem that single converters have, and so are considerably more desirable, although somewhat more expensive, depending on the associated electronics and first or second stage tuning.

NOISE TEMPERATURE RECEIVING SUBSYSTEMS

Let's now take a look at a receiving system with single downconversion, the usual low-noise filter, mixer, a pair of bandpass filters, and a 70 MHz i-f. The same equation as before applies to cascaded amplifiers, and the final i-f amplifier. Figure 3-14 shows

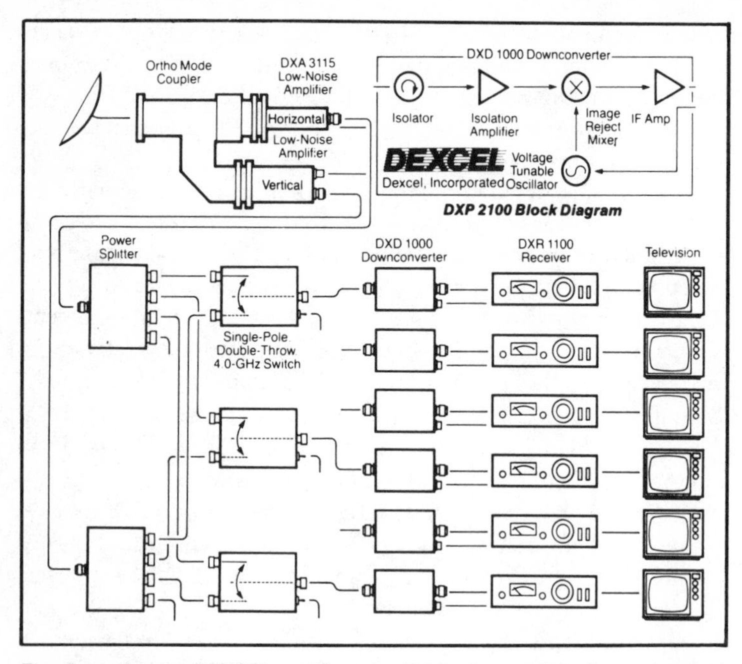

Fig. 3-14. A new DEXCEL multi-mode distribution system for up to eight receivers in horizontal and eight receivers in vertical polarization without extra line amplifiers (courtesy of Dexcel, Inc.).

an advanced single-conversion system that will handle as many as 16 receivers without additional amplification.

Say that LNA gain amounts to 25 dB, LNA noise temperature is 60 °K, downconverter mixer loss 9 dB, and i-f downconverter noise figure 2 dB. Therefore, the noise figure for the downconverter becomes 9 + 2 = 11 dB.

The effective temperature equation for cascaded amplifiers is given as:

$$T_e = T_{e1} + T_{e2}/G_1$$

where G_1 is the numerical gain of the first amplifier and T_e is referenced to initial amplifier input.

T_{e2} = (11 dB − 1) 290 °K or (12.58 − 1) 290 °K = 3358.2 °K (since dB must be converted to temperature by dividing by 10 and taking the antilog). We can find the effective temperature for the two cascaded amplifiers by using:

$$T_e = 60\ °K + 3358\ °K/316.2 = 60\ °K + 10.62\ °K = 70.62\ °K$$

This is just a little higher value than we'd like, and had the gain of the first amplifier been slightly more, say 30 dB, then the results would have been much closer to 63 °K, and that would have helped. So a reasonable gain in LNA electronics becomes quite important when you want to minimize temperature effects in the entire LNA. This is a simplified approach, as you might well imagine, but it does demonstrate what to look for and how to go about it.

AMPLIFIERS

These normally consist of several amplifying stages, including an intermediate amplifier, which can be thought of in terms of a driver for the final output. You will find such stages often have remote controls, variable attenuators, couplers, monitor points, power transmission monitors, and various filters to keep the outputs clean.

Microwave power amplifiers are available in both TWT (traveling-wave tubes) and solid-state. Traveling-wave tubes and klystrons are the tube types. Klystrons have resonant cavities, a cathode, focusing electrode, and a gap in the cavity resonator to change electron velocities and shock the cavity into oscillation. They are somewhat band limited, even with special switching. The TWT, on the other hand, is a wideband device and has an almost flat frequency response over the entire C-band, for instance. The klystron is slightly cheaper because of a less complex power supply, but both are used in uplink satellite transmitters, depending on mission, unless the extra klystron power is required above 5 kW. Air cooling is possible up to about 2 kW. Liquid cooling, however, is tricky and cooling systems have to be very carefully filtered to prevent particle stoppage.

Solid-state amplifiers will, eventually, probably take over much of these power amplifier-oscillator functions since they have both broad bandwidth and promise long life spans. At the moment, limited to some 10 watts output, IMPATT diodes for saturated operation and power FETs for saturated or linear operation are the prime devices available, but others are, undoubtedly, on the way.

In all amplifiers there are some intermodulation products, especially those of the third odd-harmonic order which filters will not stop. In transmission with just two carriers:

$$2f_1 - f_2 \text{ and } 2f_2 - f_1$$

are all in the band, and will be transmitted. Interference factors all depend on their amplitudes which have to be kept as low as possible to avoid outside pickup and retransmission. Usually after the 5th or 7th order, harmonic amplitudes in good transmitters tend to decline to insignificance. With multiple carriers, of course, all these factors increase and have to be dealt with by whatever signal cleanup methods are available. A low power transmitter with only a single carrier, therefore, is fortunate, indeed.

BASEBAND EQUIPMENT

At the input and output of up/downconverters you will find baseband (after demodulation or before modulation) intelligence processing equipment (Fig. 3-15). A large portion of this turns out to be multiplexers, especially in large stations. Baseband channels are multiplexed together so they may all be handled by the up/downconverters. The modems, however, are different since their demodulators are operated close to threshold so input signals must be maintained at full value, or the signal detectors will not operate as they should. Since satellite signals are weak, these demodulators have to work close to noise, and therefore can drop intelligence under abnormal conditions. Two principal types of demodulators (or modems) are in general use:

1. Frequency modulation for analog.
2. Phase-shift keying for digital.

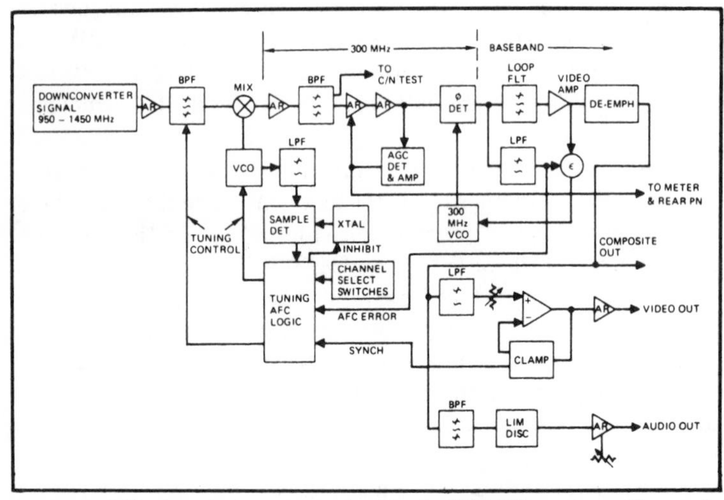

Fig. 3-15. A Hughes downconverter feeding its SVR-463 receiver with remote tuning capability and PLL demodulation. I-f (at second conversion) is 300 MHz (courtesy of Hughes Microwave Communications Products).

There may be exceptions, and the above are not entirely hard and fast categories.

Frequency modulation (FM), with the exception of its demodulators, is pretty much like any other FM system and is used in satellite communications for video, voice, etc. Here, carrier-to-noise ratios are very important, and signal quality is directly proportional to C/KT of the particular channel above threshold detection. We might add that carrier/noise cannot be arbitrarily read out on a spectrum analyzer but the analyzer's filter factors, in addition to detector degradation, must be included for an accurate measurement. Baseband, on the other hand, may be measured directly and its S/N recorded as a true measurement on the analyzer's CRT display.

In low C/KT conditions, however, the FM demodulator will degrade considerably faster than the carrier-to-noise does and this point is called the "threshold." It is defined as 1 dB less than a straight line signal, and in FDMA, the FM signal to be demodulated is placed at the same amplitude above threshold as the fade margin. So during rain conditions, your received signal should not dip below threshold, thereby maintaining the required level of intelligence. There are also threshold extension techniques which can offer operation at a lower carrier-noise-density than the usual detectors. Both types of normal and extended range FM detectors are in common use.

DIGITAL DEMODULATORS

Most commercial and military SATCOM systems use a form of digital modulation that's called phase-shift keying. The simplest variety being binary phase-shift keying. The carrier, here, begins at 70 MHz, and its modulation is entered as digital and controlled by phase into a mixer. This mixer also receives intelligence (baseband frequencies) converted to bipolar so that a "1" becomes a positive voltage and a "−1" a negative voltage. The mixer here is called a balanced modulator which multiplies the two inputs times each other. Positive and negative transitions out are specified as:

$$+ 1 \cos \omega t \text{ or } - 1 \cos \omega t$$

Between the two, each is phase-shift separated 180°. So, the equation can be written:

$$F = 1 \cos (\omega t + 180°)$$

This is the signal sent to the satellite in its phase-shifted equivalents of 1s and 0s. Remember the equations $y = \cos X$, and $y = \sin X$, $y = r \sin(\omega t \pm \theta)$, and the instantaneous value voltage $E = Em \sin \theta$, and $E = Em \sin \omega t$, or current $i = Im \sin \omega t$? They're all in the communications ballpark, and although most basically apply to sine waves, the applications are similar except that cosine is substituted for sine.

In the demodulator, we're detecting the data stream which passes to the user as baseband. With trigonometric identities this becomes:

$$\text{Mixer out} = \cos A \cos B = 1/2 [\cos (A + B) + \cos (A - B)]$$

and is called the product of two cosines.

If a 1 is sent: $\cos^2 \omega t = \cos (2\omega t) + 1$

If a 0 is sent: $\cos (2 \omega t + 180°) - 1$

So, by shifting the carrier 180° you're able to reconstruct the information in binary phase-shift keying. Intelligence recovery via suppressed carrier modulation takes place with $\cos \omega t$ and the mixer which multiplies a signal by itself and doubles the signal frequency. All this requires a very narrow band filter (with a phase-locked loop), down through a divide-by-two, and you have $\cos \omega t$, named "carrier recovery loop." This makes possible signal recovery in very noisy environments. The system is really called *coherent* phase-shift keying, and there's also a differentially coherent type which transmits a change only if the logic shifts from a 1 to a 0. In other words, if two bits are identical, there is no transmission change. Quadrature phase-shift keying is still another method where phases are handled by 90° changes. For the same data rate, only half the former bandwidth is occupied, but the system is more expensive to build, therefore it's used more in high data-rate systems.

Biphase and quadphase are the same for power, divided by noise density. There is a curve showing better C/N ratios with fewer errors. Qualifications are usually a certain data rate transmission versus a specified bit error rate.

Chapter 4

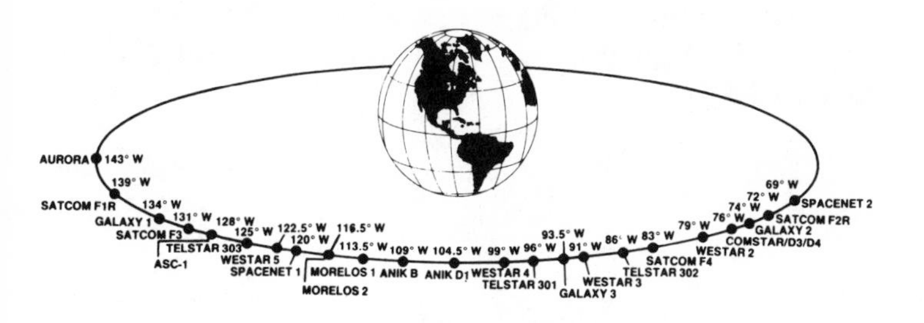

The Direct Broadcast Satellite Service

UNDAUNTED BY THE USCI "DISASTER" OF 1985, A NUMBER of moneyed and influential companies still believe high-powered, direct broadcast-to-home service (as now authorized) can not only survive but prosper, given proper incentives and program mixes. Despite dropouts of CBS, Western Union, The National Christian Network, and Satellite Development Trust, applications are still in the mill and more have received conditional approval as the various FCC "rounds" of selection continue. As many of you know, each of these DBS projected satellites will carry some 6 to 16 transponders delivering between 100 and 230 watts each. That's better than double the power of Ku-band's 45-W RCA K-series, and at least five to 10 times the power of SBS. With a downlink of 12.2 to 12.7 GHz, this additional power should easily cover large sections of CONUS, or even the entire 50 states, with a single satellite and enough signal to permit the use of 0.6- to 0.75-meter antennas almost anywhere. Even though DBS is specified as a prime video service, this does not mean that extra data carriers may not be added on some video channels to offer this service also. Further, the FCC has been responsive to the USSB petition for permission to deliver "other" than standard broadcast services. At any rate, for the record, we will list the various application "rounds" and their standings during mid 1986:

Round 1 (Conditional Contract Permit)

Dominion Satellite (in third round requesting two additional channels/satellite)—granted.

Columbia Broadcasting System (CBS)—withdrawn.

Graphic Scanning (Now Digital Paging Systems of Texas)—in third round.

RCA—Now in third round.

USSB—On course and also another third round applicant—granted.

Western Union—withdrawn.

Round II (Conditional Contract Permit)

National Christian Network—withdrawn.

Satellite Development Trust—withdrawn.

Satellite Syndicated Systems—delayed (now Tempo, Inc.)

Advanced Communications Corp.—now in fourth round.

Hughes Communications GALAXY—now progressing.

National Exchange, Inc.

Round III (Conditional Contract Permit)

Dominion Satellite (two additional channels)

RCA—Now 16-channel satellites at 100 W/transponder—much the same as Hughes.

USSB—Two additional channels (230 W/transponder)

Advanced Communications Corp.—two additional channels (total 8/per)

Antares Corp. wants 12 channels for two satellites—some half CONUS, remainder spot beams.

Digital Paging Systems (formerly Graphic Scanning).

Round IV (pending Applications)

DBSC (Pritchard)—much like Hughes design.

Advanced Communications—wants two satellites.

Tempo, Inc. (formerly Satellite Syndicated Systems).

Fully Approved

In contract for construction are COMSAT's STC, USSB,

Dominion Video Sat., and Hughes. However, if orderly sales negotiations proceed on schedule, Dominion Video Satellite, Inc. is now the conditional owner of the two STC satellites and their assets, pending completion of financial arrangements by December 15, 1986. Should all this take place on schedule, STC will probably fade out completely because even now it has no employees and simply maintains a corporate name. Selling price has not been formally disclosed with the initial press release, which was totally general in nature.

WARC-ORB No. 1 Approvals

In addition to other recent U.S. actions, the World Administrative Radio Conference on the *Use of the Geostationary Satellite Orbit and the Planning of Space Services Utilizing It*, has now approved the 1983 Regional Administrative Radio Conference allocations of 1983 for DBS. Consequently, the U.S. now has eight orbital positions at 61.5°, 101°, 110°, 119°, 148°, 157°, 166°, and 175° West Longitude (WL).

There are 32 channels assigned for main service areas with passbands of 24 MHz and channel separations of 14.58 MHz. Service area separations are, generally 10°, uplinks and downlinks are pegged at 17.3-17.8 GHz and 12.2-12.7 GHz, respectively, and as many as eight satellites may constitute a cluster if spaced no closer than 0.4° each. HTDT (high definition television) was much on the minds of the delegates, and they're considering both British time multiplexed MAC as well as the brute-force system proposed by Sony, Panasonic, and CBS. The latter would be made possible by strapping two 24 MHz channels together, supplying the necessarily increased bandwidth for 1100-line service, but requiring specially designed broadband receivers.

Formerly approved applicants for DBS include Satellite TV Corp., Direct Broadcast Satellite Corp., RCA Americom, U.S. Satellite Broadcast Corp., Video Satellite Systems, Western Union Telegraph Co., CBS, and Graphic Scan. FCC license terms are for five years, with one year from application approval to begin construction and six years to commence satellite service.

The Commission will permit approved applicants to operate as either or both broadcasters and common carriers, although each will have to adhere to the rules of the 1934 Communication Act (as amended) when rendering such services. And if one satellite, for example, does both, then each service must follow all appropriate regulations set forth by Congress in the Act.

DBS stands for Direct Broadcast Satellite Service and only reached the planning stage by 1982-1983, although some construction for STC was already under way by RCA Astro Electronics. The others were all standing by waiting the outcome of RARC.

In the interim decision, the FCC placed no restrictions on DBS ownership or control of DBS channels. Operators will be permitted to "determine the characteristics of their systems," and at a later date the Commission will use this experience in writing permanent rules. Commissioners feel that sufficient competition will exist between other DBS operators and outside sources to prevent abuses in executing the various public services.

Now, let's look at some of the material on DBS that has become public property since filings began with the FCC. It is by no means complete, but you can derive considerable information from its contents by reading what's there and making "guesstimates" of unwritten thoughts that might have appeared between the lines. At least you will have an idea of both ground and space stations planned for the service, complete with startup expectations, satellite translator projections, and indications of some of the ground terminals needed to receive and support it. Otherwise, the ensuing years and their developments will have to fill in the remainder.

We'll begin with COMSAT'S STC, followed by RCA Americom, and then some information on U.S. Communications, Inc., who used Canada's ANIK-C2 in 84-85 and beat the pack by several years. You are also advised to remember that this DBS service will probably be scrambled, is basically pay TV, and designed generally for consumer consumption, at least for the most part. Evidently, some common carriers may be operating also, but as to actual numbers we are uncertain because more applications are still pending.

STC ELECTRONICS

First of the direct broadcast (DBS) pay TV group to offer outside contractors a crack at receive only dish and electronics, COMSAT subsidiary Satellite Television Corporation, announced in 1982 that it would need 4 to 8 million units per year for a 5- to 7-year period beginning in 1985. Saying that it wanted complete units and no simply parts, STC required each manufacturer to provide his own development funds and combine with another source if he doesn't supply all designated components. Saying this, STC then offered a first cut at preliminary specifications.

These, of course, would have included a 2.5-foot antenna, its mount, feed, and a low-noise receiver/downconverter combination. The indoor assembly is to consist of a demodulator, channel selector, and AM *remodulator*. Downconversion from 12.2 - 12.7 GHz is expected to involve a factor of 12:1 so that the receive signal can be channeled into the house and its satellite receiver at about 1 GHz—at least that's the thinking of the moment, which may or may not change. Regardless, the received signal is specified as normal NTSC 525-scan line, 60-Hz TV with a 5.5 MHz digital subcarrier which is to be demodulated by an STC descrambler.

A block diagram in Fig. 4-1 illustrates the various indoor and outdoor electronics and the antenna. Note that the local oscillator in the home mixer must be tunable and a low-noise amplifier-mixer

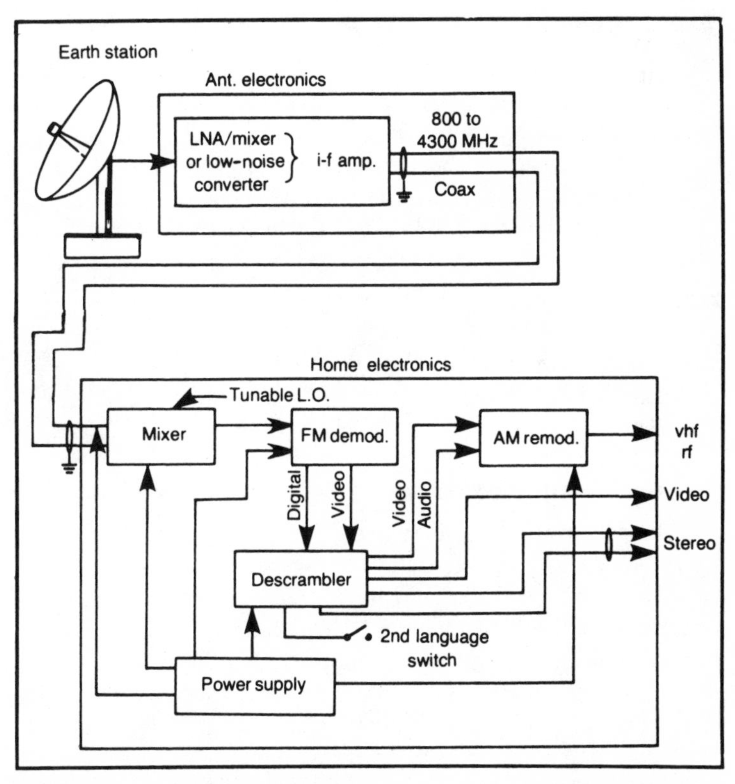

Fig. 4-1. Block diagram of STC's DBS antenna and accompanying electronics. Maximum coax lead-in is specified at 30 meters. (courtesy of STC now fully absorbed by COMSAT).

combination or low-noise downconverter, both with i-f amplifiers, are specified in the antenna electronics. Output of this array amounts to an i-f downconverter frequency of 800 to 1,300 MHz to be carried into the home by coaxial cable not to exceed 30 meters in length, or 98.43 feet.

Once there, the tunable local oscillator and mixer select a frequency for the FM demodulator, and this unit then supplies the special descrambler with both digital subcarrier and video/audio. Rf outputs may then be routed to an AM remodulator for video at vhf frequencies, or baseband outputs supply both video and stereo separately, with a second language switch available for multilanguage transmissions. A dc power supply furnishes operating potentials for both antenna and home electronics.

The Antenna

The antenna, its support, feed, and mount, must have a parabolic reflection with a maximum diameter of 0.75 meter, or 2.46 ft. Required gain amounts to a minimum of 37 dBi between 12.2 and 12.7 GHz, with cross and co-polar sidelobes as indicated on a special relative angle chart (not shown). Various mounting configurations are illustrated in Fig. 4-2, and installation and alignment of the antenna unit by an individual to the mast requires less than one-half hour. The mount must permit reflector positioning over 360° azimuth at elevation angles of between 10° and 70° minimum. Rf interface is specified as an EIA weatherproof standard WR75 waveguide flange with maximum VSWR (voltage standing wave ratio) of 1.25:1 over the receive band frequencies.

Outdoor Electronics (OEU)

The downconverter is designed to mount directly to the antenna with an equivalent VSWR at the interface, and handle signal levels between −100 and −130 dBW for all three carriers. Downconversion and amplification between 12.2 and 12.7 GHz must have a frequency translation error of not more than ±600 kHz over the ensuing 10 years with a noise figure of less than 4 dB at the input and a nominal gain of at least 50 dB. Gain changes are not to vary more than ±2 dB over "any operating day," nor more than ±4 dB for the entire life of the converter.

Third order intermodulation products have to be at least 45 dB down below the level of one of the output carriers when input carrier power is at least −80 dB and when the two carriers are spaced

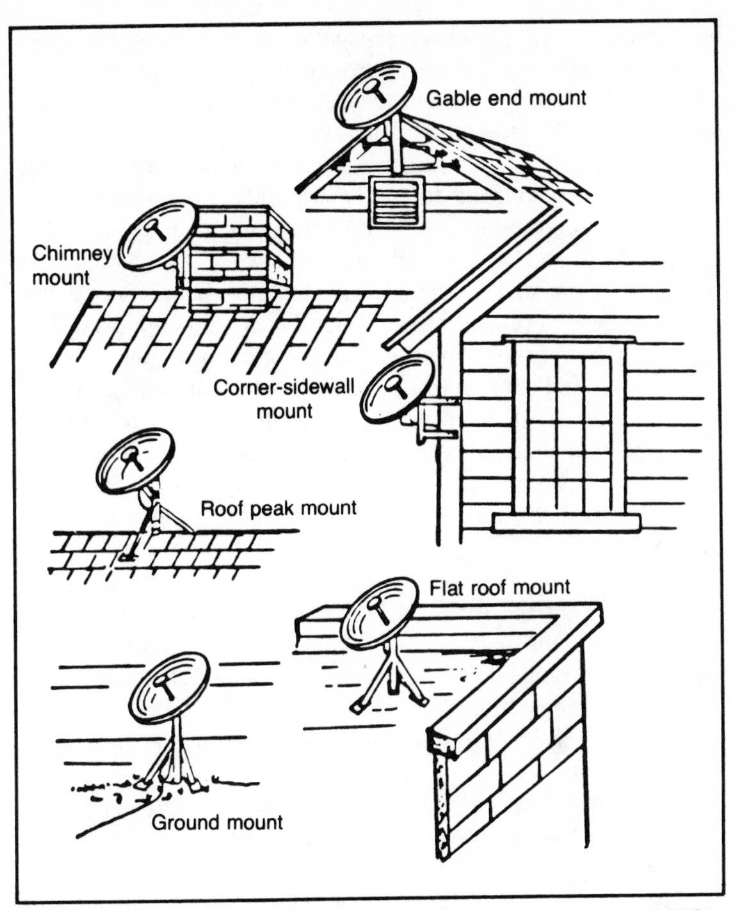

Fig. 4-2. Suggested DBS mast mounting configurations (courtesy of STC).

by 10 MHz or more in the receive band. Image rejection is specified by STC at 45 dB, and spurious radiation, including oscillator leakage, must now exceed −105 dBW at the OEU's input. Dc supply for the converter will come from the coaxial cable's center conductor at a nominal 15 V and not in excess of 5 watts. The downconverter must not weigh more than 1.1 lbs. and be no longer than 7.5 inches, nor more than 3.2 inches deep or high.

Coaxial Cable

The 30 meter cable (maximum length) is not part of the electronics package but does have its own set of specifications and will probably come from another supplier who can meet them. It is to be coaxial, with copper-clad steel conductor, with all-weather protection, and an overall rf loss of less than 0.35 dB/meter over the 800 to 1,300 MHz downconverted frequency range. For some reason, nominal dB external isolation was not included.

Environmental conditions for outdoor equipment are somewhat tougher than requirements for mechanical and electrical indoor components, as might be expected. Full performance is required between −30 °C and +45 °C and survival between −55 °C and +60 °C. Wind, snow, and ice factors have tentative specifications but most will be determined after consultations with suppliers. The wind, in any direction, however, must not disturb operations up to 50 km/hr. and survive to 120 km/hr.—which are not too tough specs. However, they must also withstand 75 and 160 km/hr. respectively in wind gusts which does translate to 100 mph in the final figure. Solar requirements are 949 kcal/m^2/hr., and shock/vibration are only those encountered during operation and transportation.

Indoor Electronics (IDU)

The indoor electronics you will obviously find more interesting. The owner-operator is to have a selection of channels 3 or 4 following FM demodulation, descrambling, and AM remodulation, with an input noise figure of not more than 15 dB, and nominal bandwidth of 18 MHz.

Amplitude linearity between IDU input and demodulator output must meet these intermodulation test specs:

1. Two unmodulated carriers at −30 dBW, separated by 25 MHz or more in the downconverted band, must have a third order intermod product within the desired channel at the input to the demodulator of at least 45 dB.
2. Image rejection of 80 dB or more.
3. Channel selection of any five in the 500 MHz receive band with spacings of 12 MHz or more.
4. An i-f bandpass filter of 18 MHz at 0.5 dB ripple P-P, shape factor rejection of at least 4 dB at 12.5 MHz; 18 dB at 18 MHz; and 31 dB at 25 MHz or above. Group delay requirements are 40 nsec P-P.

Signals are to be maintained within 1 dB over the dynamic range of 33 dB. The demodulator has a dynamic FM threshold of less than or equal to 10 dB and an AFC circuit included for accurate carrier tracking within ±200 kHz of the demodulator center frequency up to 4 MHz changes.

Baseband video/digital subcarrier filters are also required to separate these components into video baseband and a 5.5 MHz digital subcarrier, and fulfill the following analog requirements:

1. Less than ±0.5 dB insertion loss between 5 Hz and 4.2 MHz.
2. A return loss of 30 dB.
3. Maximum passband ripple of ±0.5 dB.
4. Stopband attenuation (min.) of 50 dB at 5.5 MHz.
5. Maximum group delay between 15 kHz and 3.58 MHz of 60 nsec.
6. Frequency cutoff 4.2 MHz.
7. Output 0-1 V P-P into 75 ohms at standard—40 to 100 IRE.

A *digital filter* also must separate digital from composite video within the following requirements:

1. Insertion loss less than ±0.5 dB at 5.5 MHz.
2. Center frequency 5.5 MHz and 3 dB bandwidth at 1 MHz.
3. Stopband (min.) 20 dB at 2 and 15 MHz.
4. Output level 0 dBm into 75 ohms.

The remodulator must be designed to accommodate video and mono sound and offer standard NTSC broadcast signal for either channels 3 or 4, with input of 0-1 V into 75 ohms, and 0 dBm into 600 ohms bridging audio. It will to conform to all FCC Subpart 4, Class 1 TV devices.

Power requirements originate with 115 Vac, 60 Hz, single phase, and offer a dc output of 15 V at 5 watts. Descrambler requirements are estimated at ±15 V at 1 amp and +5 V at 2 amps. Regulation at ±3% is required. The electrical and mechanical interface of the descrambler is yet to be determined.

Remodulated video and FM audio subcarrier are specified at an output of 7 dBm V ±3 dB into 75 ohms, with the channel transfer switch providing at least 60 dB isolation between IDU and the TV receiver antenna on any TV channel over which the IDU will operate. Working and storage temperatures for the IDU are from +5 °C to +50 °C and −25 °C to +60 °C, respectively.

Video Performance

Some of the toughest specifications in the entire package are devoted to video performance, especially at the output of the AM remodulator. Assuming a C/N (carrier-to-noise) ratio of 18 dB with 18 MHz bandwidth and 5 MHz peak video deviation at the IDU input, the signal-to-noise (S/N) ratio requirement has to be at least 46 dB or more. S/N (periodic noise) below 1 kHz to 4.2 MHz must be equal to or greater than 48 dB. Signal-to-impulse noise must measure more than 23 dB, with C/N of 14 dB at the IDU input. Insertion gain is not to exceed ±0.5 dB.

Luma/chroma distortion figures are tight, too. Luminance nonlinear distortion must be within 10 IRE units at 10, 50, or 90% of the average picture level, and chroma is not to displace the luminance component by more than 4 IRE units from input reference level. Chroma subcarrier amplitude cannot vary more than ±10% at the output with 20 to 80 IRE input units, and differential gain shall be within 5% with luminance variations between 10 to 90% of average picture level (APL).

Chroma subcarrier phase must remain within ±4° between luma variations of 10 to 90% APL, and system gains have a range of only ±4 IRE units over the same APL variations. In addition, peak overshoot is not to exceed 5 IRE units; tilt and rounding of window test signal must not exceed 4 IRE units; short time distortions are allowed 6% from the 2T pulse with half amplitude durations of 0.250 μsec; maximum chroma/luma difference gain is to be within 2.8 IRE units (or 0.25 dB); and chroma/luma delay inequality are to have maximum delay differences of only ±60 nanoseconds.

You may think these are somewhat long-winded specifications for home user satellite electronics, but even these *preliminary* starts at the home media space market will probably tighten up with both experience and time. After all, no one else has tried yet to put tiny receive dishes (TVRO) on all sorts of houses and under every type of condition over anything like the area they will cover in the U.S. True, Canada and its ANIK series have done some of this already, but conditions in the north vary somewhat from ours, as does transponder power, territory served, electronics, and general regulation or deregulation of the program. The vastness of this enterprise, not to mention its million and billion dollar expense hasn't yet dawned upon most of those casually considering the general effects. We're going to find huge amounts of capital tied up in this DBS venture, and the fiercest competition imaginable in securing rights to 12 and 17 GHz spectrums as the potentials develop. In addition, the prospect of broadcasting 1125-line high definition video (HDT) has more than one entrepreneur interested since this will certainly "be the start of something new" for both broadcasters and the receiver industry. When the public becomes acquainted with its splendid definition, resolution, and remarkable full bandwidth colors, the demand—even with attendant high prices—could be spectacular. As *some* industries have come to discover quality, especially in visible electronics, sells well every time.

HDTV is especially suited to cable television, particularly in systems with more than 50 channels both because of the number of potential and present subscribers and also because as many as five video channels may be required for this wideband service. In the satellite service, some are already proposing a considerable commitment to digital HDTV transmissions, including bandwidth compression at an 8:1 ratio, to gain maximum use of spectrum allocations. A digital "high compression" receiver has been demonstrated in Berlin as far back as 1981.

THE STC PROPOSAL

COMSAT's STC proposal before the Federal Communications Commission, Dated December 17,1980, initially offers DBS service to the Eastern time zone area of the U.S., followed by three other satellites for Central, Mountain, Pacific and Hawaii/Alaska zones. Each would be spaced 20° apart at 22,300 miles above the earth in a geostationary arc at west longitude positions 115°, 135°, 155°, and 175°. Uplink and downlink transmissions would be in the 17/12 GHz slots (Fig. 4-3).

The satellite would have three operating traveling-wave-tube amplifiers of some 200 watts each (rf) output, and dc power of between 1700 and 2000 watts. Two in-orbit spares are planned also for 115.5° and 175.5° WL and could be in position within minutes if either East or West coast service failed because of problems with original equipment. Transmissions, of course, are by frequency modulation to minimize complexity and reduce home equipment costs, especially that of the downlink receivers (not the ultimate TV sets showing pictures). Signal quality objectives for 99% of the worst month are: C/N of 14 dB, and S/N of at least 42 dB, with carrier-to-noise expected to be above threshold "all but about 4.4 hours per year" which means only slightly degraded video for those rare periods.

Video baseband is to meet CCIR standard M with NTSC color, a video deviation of 10 MHz, and an i-f bandwidth of 16 MHz. Au-

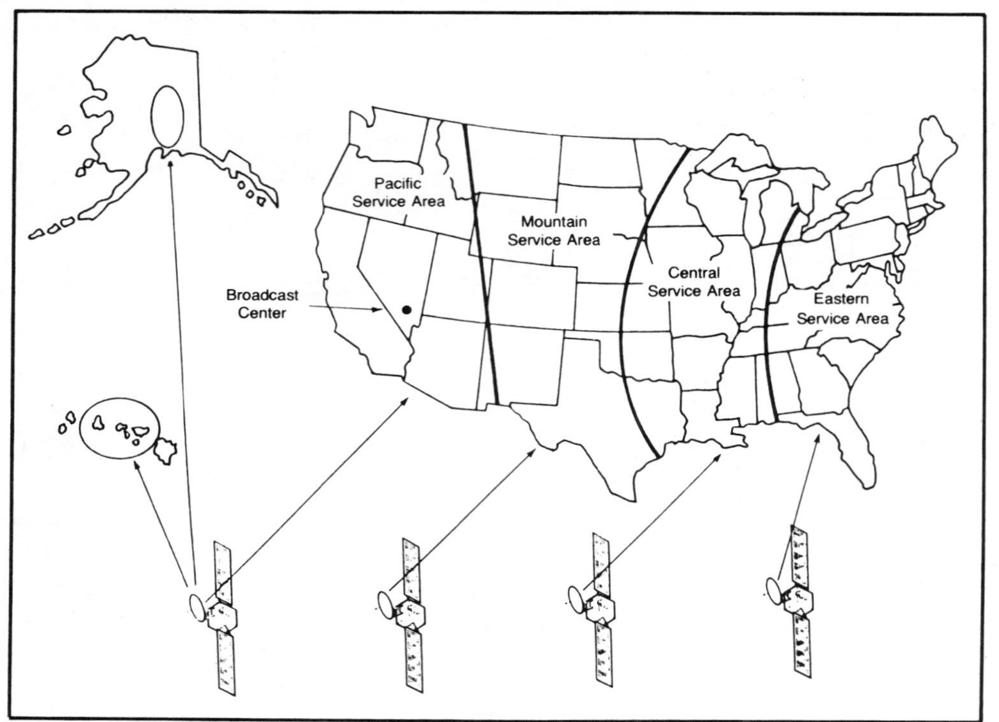

Fig. 4-3. Proposed coverage of all 50 states with four "birds" in original application (courtesy of STC).

dio encoding will occur at QPSK at a bit rate of 315 kbps/channel and a bandwidth of 13 kHz. The color video signal will then be combined with a digital subcarrier and FM modulated onto an rf carrier.

STC says the 16 MHz receiver bandwidth was selected to minimize satellite power requiring reasonable fade margins above threshold. Assuming antenna efficiencies of 50%, the company projects receive antenna home equipment peak gain of from 34.9 to 38.4 dB, at a noise temperature of 27.4 dB/°K and 0.5° pointing error loss in dB. Total noise temperature for the ground system is predicted at 546 °K. These figures are assuming antenna diameters between 0.6 and 0.9 meter. For rainy areas such as Florida and the Gulf Coast, a somewhat larger antenna may be required to remain within threshold specifications.

When normal TV transmissions have ceased on channels B or C, STC "intends to experiment" with a "a variety" of high definition television systems, using narrow and wideband filters for narrow to wideband switching. The company believed that an eventual market will develop for HDTV with "the equivalent of 1000-1500 lines" of horizontal resolution "coupled to a wide-screen display" having much larger dimensions than those currently used in available television receivers. A totally new receiver will be required since compatibility between HDTV and conventional sets appear unlikely. STC expects both single frequency digital and dual frequency analog transmissions to be evaluated. A "commandable 100 MHz bandwidth input filter has been added to the channel B transponder" for digital transmissions and this same transponder and filter can be used also for the luminance portion of analog broadcasts. Output filters will handle either normal or broadband outputs.

The two STC satellites have been conditionally sold to Dominion Video Satellite, Inc. (April, 1986).

Ground stations for DBS include a satellite control facility and broadcast center near Las Vegas, Nevada and backup transmission facilities at Santa Paula, California plus engineering support from Washington, D.C. Broadcast antennas are 11 meters in diameter. The U.S. Department of Labor and its Bureau of Labor Statistics forecasts that national STC (DBS) service will support some 23,000 jobs directly or indirectly by 1987. Other DBS carriers and broadcasters should add substantially to the total.

RCA FILED FOR DBS, ALSO

RCA American Communications, Inc., also applied to the FCC for authorization to "construct, launch, position, and operate" a group of DBS satellites and ground support facilities for television, audio, "and other services." Uplinks will transmit at 17.3 to 17.8 GHz, and downlinks between 12.2 and 12.7 GHz.

According to RCA, there are to be four satellites in orbital operation, plus one spare in orbit and another spare on the ground. Each satellite would have six transponders, four devoted to standard programming and two set aside for wideband experimental use "such as high-definition television transmissions."

The four operational satellites are to offer service throughout the U.S. individually covering a single time zone at minimum signal strength of 58 dBW from each standard transponder. Hawaii and Alaska are included as part of Pacific Time Zone operations. Channel bandwidths are specified at 24 MHz for standard operations with linear orthogonal polarization of each beam. By ground command, either of the two extra transponders may be switched from 24 to 72 MHz bandwidth. A four-year time lag is forecast from the date of final FCC approval to the time active Eastern time zone operations begin. Then another four year period will elapse from the beginning of eastern operations to full coverage of the entire 50 states. So, this enterprise will probably not really begin in earnest until sometime in 1986, even though the FCC is expected to approve responsible applicants during 1983, more later after RARC in Geneva. Terrestrial antennas for the system are scheduled for construction at RCA's Americom site at South Mountain, California.

RCA, in a market analysis of areas where CATV is not generally available, believes its system would be economically feasible at installation costs of less than $500 per home. The company estimates between 15 and 30 million DBS potential users exist throughout the U.S. Spacecraft in the system are expected to be deployed in geosynchronous positions so that receiving antennas will occupy advantageous elevations of at least 20 degrees above ground level when trained on the satellites. This has been projected to keep antennas from picking up the earth's 290 °K temperature in addition to offering minimum wind resistance. Such orbital stationing will also prevent satellites from entering an "eclipse" situation prior to 1:00 a.m. each morning.

RCA will offer only the *space* portion of its DBS system, which will include tracking, telemetry, and command operations from Vernon Valley, New Jersey, and South Mountain, California. Most uplink facilities are expected to be customer owned, as well as downlinks also. RCA, however, is prepared to furnish programming uplinks upon customer request.

The Spacecraft

As with other projected DBS systems, sharply increased power over existing transponders on C- and Ku-bands are protected by RCA. Americom's spacecraft are being designed to operate six 230-watt K-band channels, with on-board spares for 1:1 redundancy. Each satellite (Fig. 4-4) will generate 4,770 watts initially, tapering off to 3,600 watts at the end of its 7-year design life—all to be supplied by 400 square feet of solar arrays that are pointing at the sun. Traveling-wave-tube amplifiers are the transponders, and three of the four working satellites may receive transmissions from anywhere within CONUS (the 48 continental states). This will permit real time reporting of news and sporting events from any North American U.S. location.

Bandwidth has been fixed at 24 MHz because of channel numbers, threshold problems, and desirable signal-to-noise ratios. Transmissions are possible from mobile earth stations having antenna diameters as small as 5.5 meters, or 18 feet. Linear and

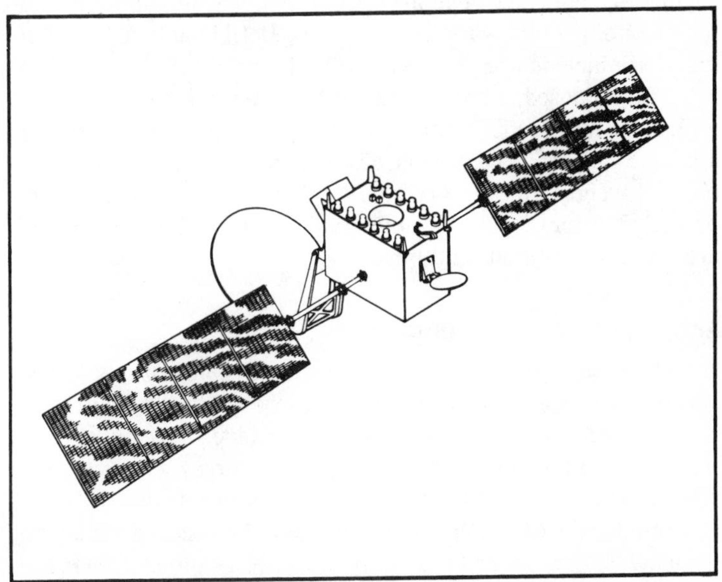

Fig. 4-4. RCA's direct broadcast satellite—an artists conception (courtesy of RCA Americom).

oriented polarization has been selected as the best defense against heavy rain areas because of its superior isolation. When high definition TV is shown, switchable filters, upon earth station command, will permit channel bandwidth increases from 24 to 72 MHz, although either channels 5 and 6 may also pass standard TV at 24 MHz, as required or desired.

The minimum beam edge EIRP (Effective Isotropic Radiated Power) has been designed for 58 dBW, with S/N video at 43 dB, S/N audio at 52 dB, and system availability of 99.9% for standard TV transmissions. Two high quality audio channels with top baseband response of 15 kHz will allow transmissions of high quality stereo, voice, and other sound programs. RCA says that "the ratio of the audio signal at average program level (+8 dBm) to the unweighted rms noise, (S/N) audio, will be a minimum of 53 dB."

Orbital locations for the four time zone regions are broadly fixed as follows:

Eastern:	95° - 125° W	Mountain:	125° - 155° W
Central:	110° - 140° W	Pacific:	140° - 170° W

Figure 4-5 details the general effect of these various areas, including an outline of the various time zones. Both longitudes and latitudes are included on the spherically drawn figure.

The proposed spacecraft are to be three-axis body stabilized with solid apogee kick motors, nickel-hydrogen batteries, deployable communications antennas, dc power of approximately 3.5 kW, on-oribt weights of over 2,000 lbs., and monopropellant hydrazine heated thrusters for north-south stationkeeping. Launch weights are now estimated at 4,620 lbs.

Antennas and Transponders

The *satellite* transmit antenna diameter is approximately 10 feet in diameter, while the receive antenna is expected to measure some three feet across. The transmit antenna includes an array of feedhorns and a switching network to optimize time zone coverages. A block diagram (Fig. 4-6) of each satellite setup is shown with both horizontal and vertical inputs, the various losses, and switching arrangements for the two high definition TV channels. Total frequency span accommodates 36 channels, each with active bandwidths of 24 MHz. Co-polarized channels have a 27 MHz sepa-

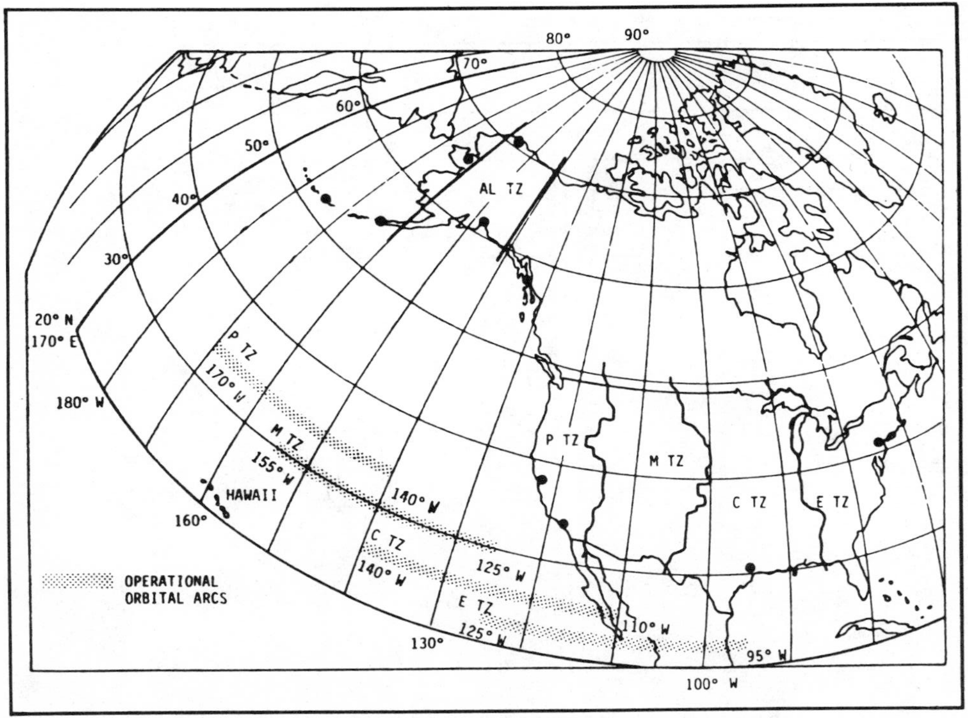

Fig. 4-5. RCA's DBS proposed coverage (courtesy of RCA Americom).

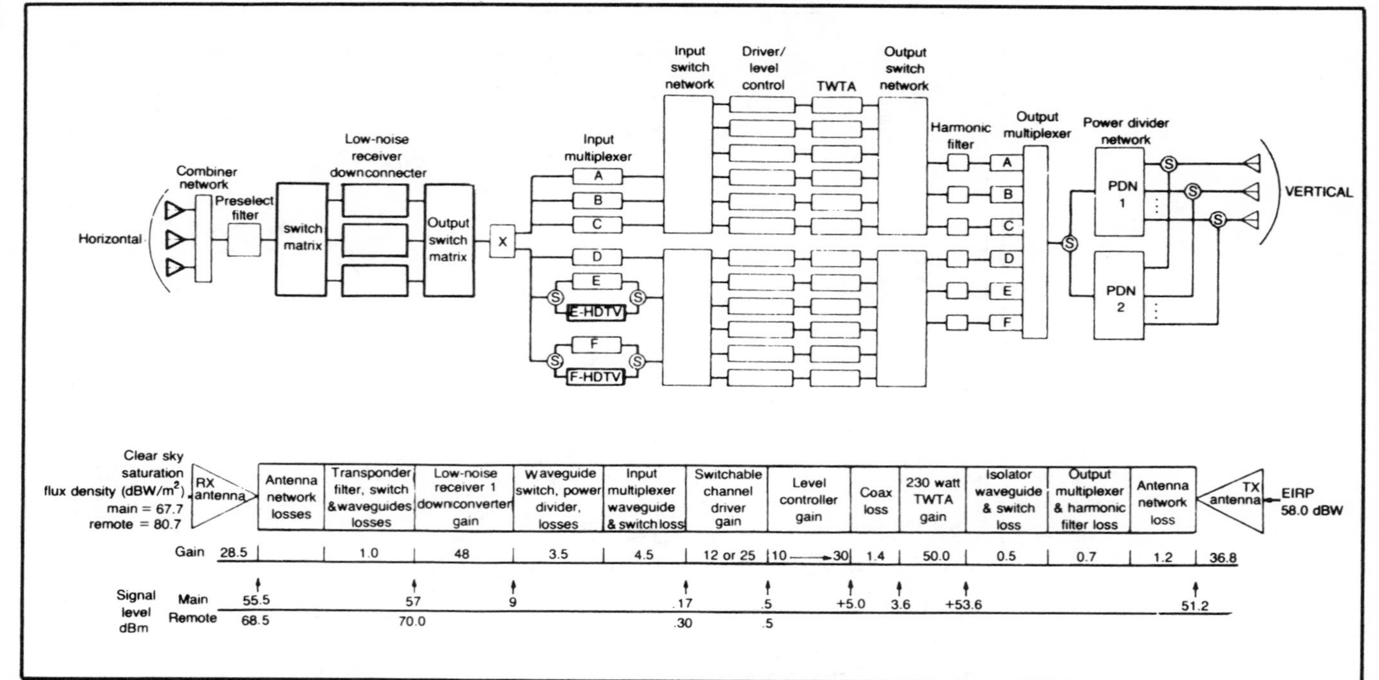

Fig. 4-6. A block diagram of the RCA satellite with both calculated losses and anticipated EIRP gain (courtesy of RCA Americom).

ration factor, with 6.8 MHz between orthogonally polarized channels. Shifting cross-polarized channels by 6.8 MHz permits assignment of HDTV channels in both bandwidth and frequency as long as they are multiples of the standard channels. And although precise bandwidths for high definition TV have not yet been assigned, RCA believes that two to four standard channels will be sufficient to carry this wideband service. Satellite *receive* antennas are to offer *full* CONUS coverage instead of just a single time zone provided by the transmit antenna. According to RCA, the satellite receive system noise temperature will amount to 1445 °K, at a satellite G/T of −3.6 dB.

Terrestrial Antennas

Two antenna diameters have been considered by RCA in analyzing ground requirements. One would measure 0.6 meter (1.97 ft.) and the other 1 meter (3.281 ft.). Receiver and LNA would be mounted on the antennas and antenna temperature forecast at 27 °K at lowest elevation angle anticipated and carrier frequencies of 12.5 GHz. In clear weather, therefore, antenna diameters and receive noise temperatures would amount to:

Antenna Measurement	Receiver Temperature	G/T Resultant
0.6 meter	300 °K	9.4 dB
	500 °K	7.2 dB
1 meter	300 °K	13.2 dB
	500 °K	11 dB

Generally, LNAs would have to be in the 300 °K temperature range for K-band use.

At the moment, there are all kinds of studies going on in attempts to nail down an HDTV set of parameters. However, even the K-band channel-spread frequency hasn't been decided upon, although a 27 MHz figure has been recommended to the FCC. But HDTV appears headed for either some sort of FM-modulated baseband or digital. All that can be said at this time is that a number of companies and their employees are working on it.

OAK AND U.S. COMMUNICATIONS, INC. FILED FOR DBS INITIALLY

The largest operator of STV—over the air subscription

television—and *the* major producer of STV decoders, Oak Communications, Inc. initially reached agreement with TeleSat of Canada to begin satellite pay TV broadcasting following the launch of ANIK-C2 in 1983. Filed, but not approved until 1983, the FCC eventually gave the go-ahead for this U.S.-Canadian venture (now with U.S. Satellite Communications) along with the other nine that have already been approved for consideration. ANIK-C2 will have GTE video channels already spoken for. With Oak not involved with the ANIKs, it has now applied for a 12-DBS channel system for 1994, preceded by a 14/12 Ku-band satellite to serve the Eastern U.S. Initial USSC-ANIK service would include New York, Boston, Detroit, Rochester, and Washington, with all areas in between. Reportedly, homeowners are the primary target, but SMATV systems and low power TV outlets, in addition to hotels, CATV, and hospitals could also be included. Initial investment is set at many millions, and homeowner cost is said to amount to $30 per month. The 14/12 Oak Ku-band system was to have been launched in 1986, with 12 channels of programming at K-band and DBS would get underway in 1988. Oak subsequently withdrew this application. U.S.C.I. service ceased March 31, 1985.

ANIK-C

There were three ANIK-C's placed in orbit between November 1982 and 1984. Figure 4-7 shows one beaming down information from one of its four contiguous spot beams which may be used individually or in combination for Canadian or (in this instance), U.S. coverage. The area shown is Canadian, for the most part. Joining ANIK-D-1, launched August 12, 1982 from Kennedy Space Center, Florida, the ANIK-C's are spin-stabilized and will operate in the 14 GHz uplink, 12 GHz downlink Ku-band, having 16 channels.

Each channel can handle two full color television programs and their associated audio, cue, and control circuits, offering 32 programs per satellite. TWT outputs are 15 watts and, because of their additional power, will require earth stations as small as 1.2 meters in diameter (4 feet). Launch dates aboard the NASA space shuttle scheduled for Nov. 11, 1982, June 1983, and April 1984.

The ANIK-C series has 20 TWTAs, a channel bandwidth of 54 MHz, and EIRP of 46.5, a design life of 10 years, and a power array capacity of 800 watts. Transfer orbit weight is said to be 1,140 kg, with height and body widths of 3.28 and 2.05 meters, respectively. Coverage will include Victoria, Edmonton, Calgary, Regina,

Fig. 4-7. Canada's ANIK-C looks down on the U.S. and Canada (courtesy of Telesat Canada).

Winnipeg, Montreal, Toronto, Halifax, and St. Johns. Boston, Detroit, Minneapolis, and Seattle, Washington are all located in the prime transmit areas. Earth stations for these satellites are found in Ottawa, Lake Cowichan, B.C., Allan Park, Ontario, Frobisher Bay, N.W.T., Harrietsfield, N.S., and Calgary, Alberta.

TWT Amplifiers

Despite rumors of impending doom to the contrary, the TWT traveling-wave tube still holds its own in generating microwave frequencies or amplifying those at or above uhf. Invented by Dr. Rudolf Kompfner, an Austrian refugee in 1944 for the British Admiralty, the device was not developed for practical use until the period between 1945 and 1950 by Bell Telephone Labs. At the time there was a requirement for considerably higher power, wide bandwidths, plus quick frequency shifts in radar applications, so TWT was the answer: and still is.

With the dawning of the space age, there was also a need for these same characteristics at considerable distances from earth. Applications, however, were far from radar jamming and countermeasures, mainly simple power amplification at gigahertz frequencies with bandwidths of 36 or more megahertz. Even with its rather demanding power supplies, requiring such disciplines as TWT cooling, heater current, grid bias, collector potentials, etc., regulation mut be only close enough to maintain rf performance and prevent electron beam defocusing. Advance work was done by Hughes Electron Dynamics Division of Hughes Aircraft beginning in the late 1950s and continuing to this day, especially in manned and commercial space applications. TWTs have now flown on military applications plus Syncom, ATS, Intelsat, DSCS, TDRSS, Space Shuttle, and many domestic communications satellites. Work is continuing on longer life, higher efficiencies, smaller size, and less weight.

The Way It Works

As you will see in Fig. 4-8, the TWT has a heater, grids, rf power input, a helix, barrel, attenuator, collector, isolator, and rf power output, in addition to a cathode, anode and focus electrode.

When the cathode is thermionically heated, it emits a continuous stream of electrons, just like any conventional tube, which are drawn through the anode and tightly focused into a narrow tubular beam. This is done by a magnetic field which makes these electrons pass inside the tightly wound helix and its support rods, eventually being dissipated in the collector.

Rf, at the same time, enters the power input and helix "slow wave" structure. At the speed of light, rf energy should progress through the helix, but because of the helical path, its progress is determined by both the diameter and pitch of the wire helix. A

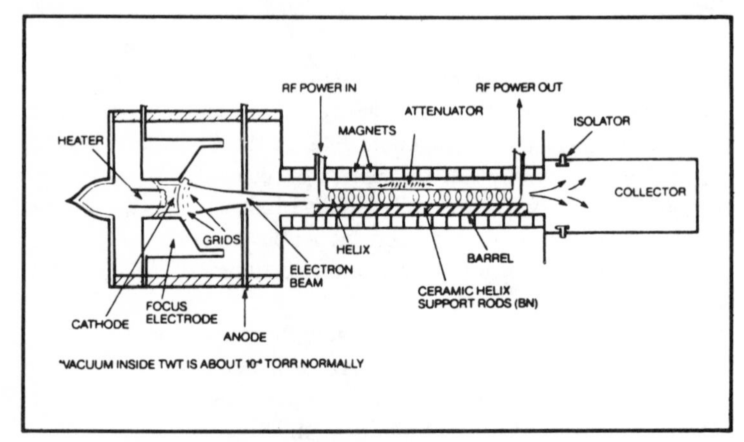

Fig. 4-8. What a traveling-wave-tube amplifier (TWTA) is all about (courtesy of Electron Dynamics Division of Hughes Aircraft).

somewhat synchronous effect results between rf and the electron beam which slows some of the electrons and accelerates others. Consequently, bunches are formed, overtaking and interacting with the slower helix rf, losing dc to it. This results in an exponential amplification of rf which has been as high in single TWTs as 70 dB, or 10 million.

Just as with standard tubes, life and reliability of TWTs depend mainly upon design and type of cathode material used. This primarily depends on current density beam power and projected life expectancy. TWT tubes usually have control grids to turn electron beams on and off with better results than with cathode modulated voltage.

The noise figure, as given by Hughes; amounts to:

$$NF = \frac{\text{Input S/N}}{\text{Output S/N}}$$

and is defined as a measure of degradation in dB as signals pass through a given tube.

Noise power output, then, becomes:

$$NP_0 = -114 + (BW) + (Gss) + (NF)$$

where −114 dBm/MHz thermal noise, input terminated, and system bandwidth in dB relates directly as frequency in MHz (say 1,000) to dB (30), Gss equals small signal gain, and NF the figure from our previous equation.

Hughes also has an excellent illustration (Fig. 4-9) showing the dynamic characteristics of a traveling-wave tube in terms of small signal, constant gain, and the small signal dynamic range linear region. Following saturated power output you will also see the area of nonlinear operation and saturated gain falloff. Signal threshold, of course, divides linear and nonlinear operations.

A REAL DBS ANTENNA

Prodelein, a division of M/A-COM, has now developed a real, live DBS antenna in the 1-meter range (give or take a little) which is illustrated in Fig. 4-10. The photo of this was just received and there's not much information available other than to tell you to pay attention to the offset feed. This has been done to remove any block-

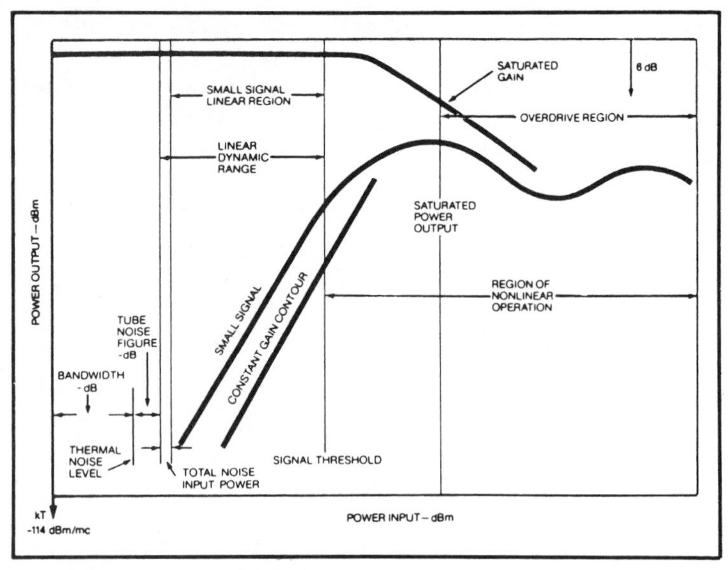

Fig. 4-9. Dynamic characteristics of a TWTA (courtesy of Electron Dynamics Division of Hughes Aircraft).

age problems associated with small antennas and the usual prime focus feeds, as well as permitting favorable ice and snow shedding in northern latitudes. The efficiency may also be somewhat better

Fig. 4-10. An operating 1.2 meter M/A-COM fiberglass Ku-band antenna with offset feeds (excellent results with SBS 3, K-1 and K-2).

than those with regular prime focus feeds. We do, however, know that all these DBS antennas from major producers are of the offset-feed, variety and are parabolas. Such offset feeds are also popular in 1.2 to 2.4 meters at Ku band.

DUAL-PURPOSE PRIME-FOCUS FEEDS

With patent rights still not totally established, several manufacturers are saying little but showing more as combined C- and Ku-band prime focus feeds are gradually making their appearance in several sections of the country. A really worthwhile concept comes from Boman Industries of Downey, California, which would accept orthogonal (H and V) dual LNB inputs for C-band and a traditional skewing arrangement for Ku, which would cover the two polarities by conventionally changing the position of its internal probe and, thereby, the horizontal/vertical phase. Boman's photo (Fig. 4-11) clearly illustrates the three inputs, which develop through a scalar feed with a single opening.

We would expect deliveries sometime during the fall of 1986 or shortly thereafter. By then, of course, there could be other competition, but most of it would offset either C- or Ku-feeds and this always poses a slight loss problem. Industry sources say that where you have a choice, aim the Ku-feed directly toward the reflector's center and offset C. With large, relatively high gain reflectors, how-

Fig. 4-11. Dual C- and Ku-band feed would permit a single receiver with polarity inversion to access all video-processing satellites within TVRO range of *your* location (courtesy Boman Industries).

Fig. 4-12. The future is now for Chaparral and its scalar or prime focus-mounted 12-GHz feeds available for f/D ratios of 0.30 to 0.45 and/or offset feeds. Buy them separately or buy both for dual-band reception from 8-foot or larger reflectors (courtesy Chaparral®).

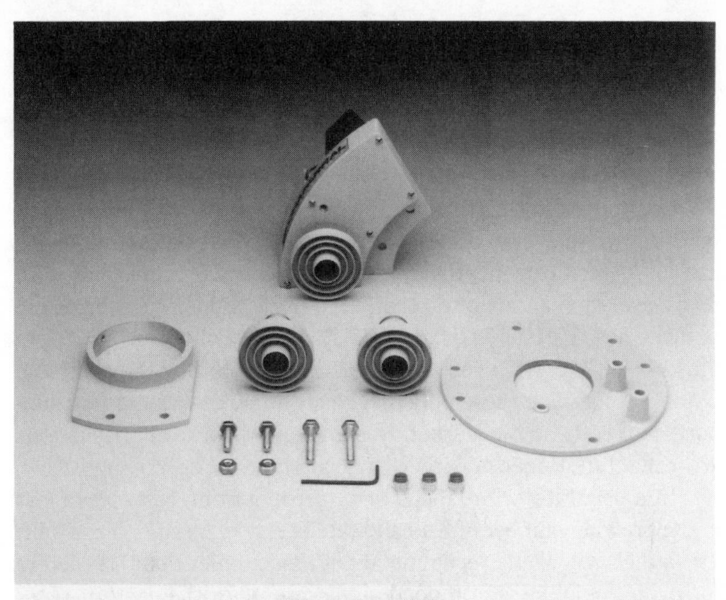

Pioneer Kit

Prime Focus Mounting

ever, we wonder if this makes any substantial difference except for narrower beamwidths, resulting in sharper directivity and more difficult aiming.

CHAPARRAL'S DUAL-BAND FEEDS

On the market and ready to mount on any f/D ratio between 0.30 to 0.45 (and also offset designs), Chaparral Communications can give you a retrofit, fixed or adjustable scalar mountings with 35 dB isolation and rf ports that are WR75 and CR62 compatible in these new 10.7-11.7 GHz or 11.7-12.7 GHz Polarotor® feeds. As illustrated (Fig. 4-12) you may use it as a kit, or buy the complete C and Ku combined assembly that easily attaches to most feed supports currently on the market. Recommended for dual installations are reflectors 8-feet or larger to accommodate both C and Ku.

In a separate prime focus or offset mounting to access most of the growing number of Ku-band satellites now on station or awaiting launch, we would recommend nothing smaller than 1.2 meters for prime results even in good signal areas. In the eastern U.S., for instance, you should have little problem with either SBS-3 or K-1, K-2 satellites, especially the latter with their 45-W output per transponder except SBS-3 during cloudbursts.

Chapter 5

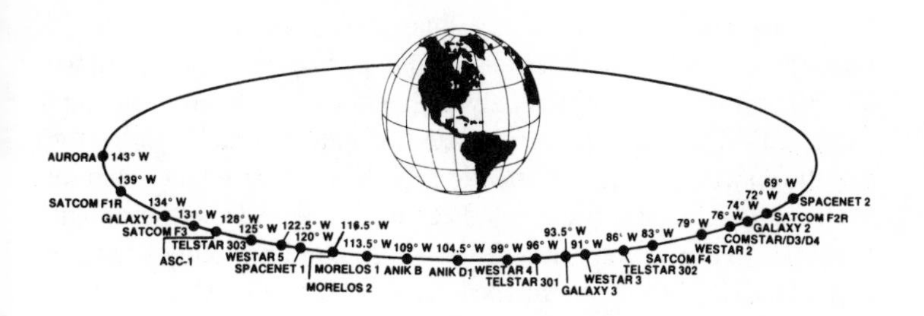

Transmitters and Trapping

THESE TWO SUBJECTS ARE NOT NECESSARILY ELECTRONIC bedfellows. The subject of transmission has been covered in varying degrees elsewhere in other chapters, and this one will be concerned with hardware only. Consequently, the subject matter is not too extensive, and we will have an opportunity to explain some of the interference problems encountered, especially in the C-band 6 GHz uplink and 4 GHz downlink regions, which are very important since terrestrial microwave frequencies also operating in this general band do cause problems, especially in congested areas of reception. Since many transmitter arrangements are also receivers, suggestions for reducing interference for these transceivers is placed in this chapter for well-rounded coverage. We begin, of course, with the topic of transmitters, and a good instruction manual supplied by Microwave Associates Communications, known as M/A-COM in abbreviated vernacular. This particular unit has been designed for satellite and cable television. But, the general principles are pretty much the same so much of the information applies to all transmissions. Further, it may also be used as a receiver, and so we'll actually work with a combination unit as our introduction, then switch to strictly satellite equipment to complete the discussion.

TRANSCEIVERS

The unit we'll be discussing is a "basic" MA-12X M/A-COM transmitter/receiver using PAC-6X modulators/demodulators and low-pass filters, designated as LPF in the simplified composite illustrations shown in Fig. 5-1. For cable use, receiver and transmitter are configured much as illustrated in A and B, while the hookups shown in C & D are for satellite purposes. In each instance, VR-3S or VR-4X satellite receivers have been added for downlink signals. PAC-6X modulators and demodulators remain. In the two satellite portions of the illustration, both composite video and 4.5 MHz audio enter transmitter and cable modulator from their points of origin. In the bottom satellite drawing, the same composite video and 4.5 MHz audio enter at the same ports, and PAC-6X modulators and demodulators pick up an additional channel of audio from the satellite. So as you see, the CATV and satellite setups are quite similar—at least in block diagram form—and the same principles apply with only modifications required for the unique satellite distribution.

Transmitter

The transmitter is of solid-state design requiring 20 watts of primary power between 105 and 130 Vac. It operates in the frequency band between 10.7 and 13.2 GHz, with an FM deviation of ± 4 MHz, an output power to the branching network of 13 dBm, and video input level of 1 V p-p. Audio subcarrier options are 4.5, 5.9, 6.2, or 6.8 MHz. Weight is 5 lbs. 13 ozs., and there is a S/N figure of 55 dB minimum. A photo of the assembly is shown in Fig. 5-2.

This particular MA-12X transmitter uses a solid-state Gunn local oscillator (lo), operating directly in the assigned frequency range between 10.7 and 13.2 GHz and is quite suitable for CARS band applications. The transmitter measures 4.7″ H × 4.5″ W × 8″ D and is designed to be mounted in a card frame system, occupying three vertical rack spaces in a 19″ rack. Power and fault indicator lamps are located on the front panel, in addition to a video gain control and optional audio subcarrier input connector (see Fig. 5-3).

Video inputs are coupled into a 525/625-line pre-emphasis network as well as a video presence detector, which also connects to a front panel lamp. Should video signals be absent, the detector grounds, shutting off dc to the Gunn oscillator, stopping its operation.

When normal signals are present the video gain control routes this voltage on to the Q9-Q13 video amplifier, which may also have an optional subcarrier input added. Along the way, the pre-emphasis network emphasizes higher video frequencies to improve transmission performance.

Power for the Gunn regulator and other portions of the transmitter originates from the A3 supply, which offers both high and low outputs ranging between +18 and +5 volts dc. Since the Gunn diode has a voltage-controlled negative resistance, it requires a constant voltage supply for normal operation. As video proceeds toward the Gunn oscillator, incoming audio is modulated and its subcarrier inserted above the video frequencies.

Both audio and video now reach the Gunn oscillator which is electronically tuned by a varactor diode to modulate the carrier. The exact control for this oscillator comes from a digital automatic frequency control (AFC) loop that is between the video amplifier and the rf assembly. The sequence begins with a crystal-controlled reference that is first doubled in the multiplier and then stepped up to times 60 for the AFC mixer. Here a sample of the Gunn oscillator and the reference are mixed and an output passes through a lowpass filter into the Q1 preamplifier. This 50 MHz i-f signal is then amplified, divided by 10, level shifted, then divided by 16, by 3, and once more by 16 until it reaches a 6.5 kHz phase comparator as sharp risetime pulses.

The second input into the phase comparator comes from a 6.666 MHz crystal-controlled oscillator and buffer which, in turn, is divided by 16 and then again by 64, reaching this same phase comparator with a second 6.5 kHz signal in the form of square waves. Any phase difference between the two is detected by the phase comparator, and an error voltage generated that is filtered and returned to the Gunn oscillator's varactor diode to correct any frequency drift. As shown on the diagram, the AFC input varies between ±0.5 V and +7 V, presumably augmented by corrective dc from the phase comparator and loop filter.

The Gunn oscillator, when biased above threshold and into the negative resistance region, operates in the dipole domain, moving at electron drift velocities of approximately 10^7 cm/second. The 10.7 to 13.2 GHz carrier, together with its video-audio modulation, then passes through an isolator and rf channel filter as a 20 mV output.

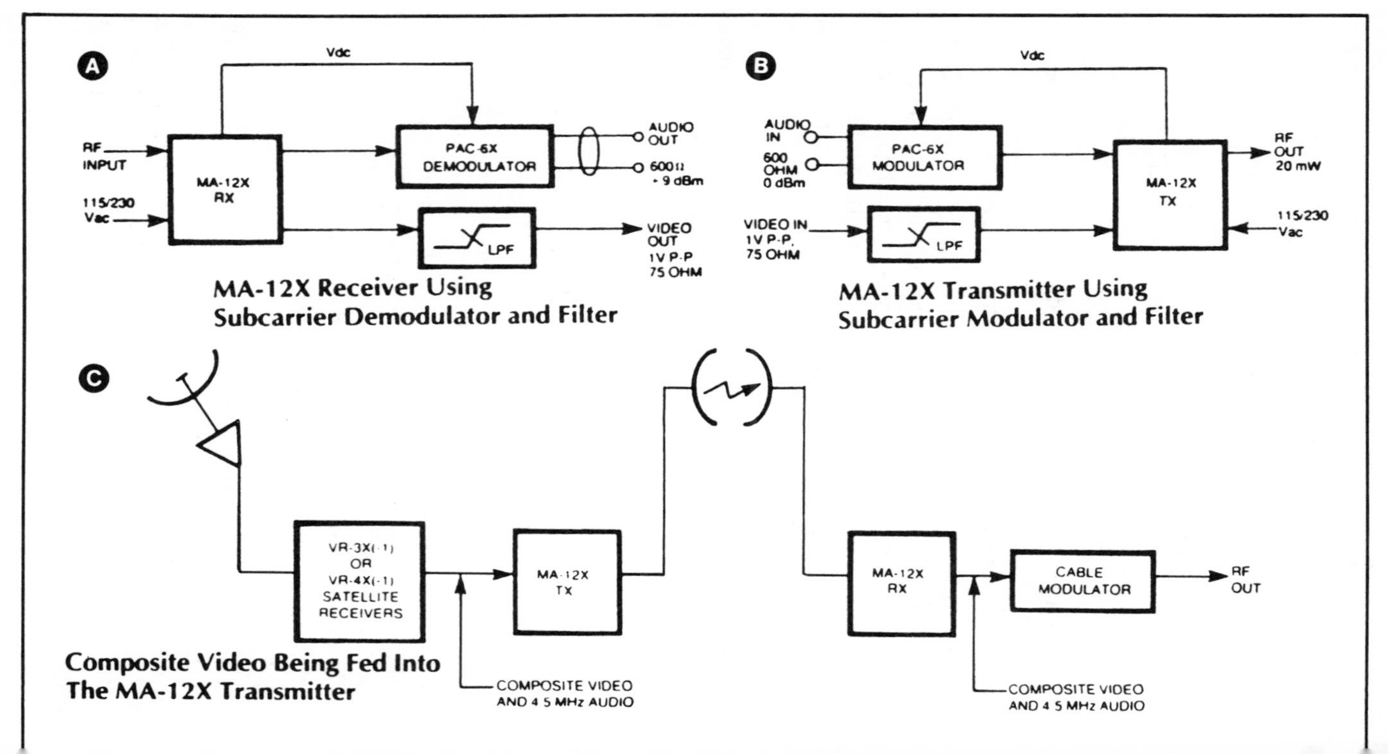

MA-12X Receiver Using Subcarrier Demodulator and Filter

MA-12X Transmitter Using Subcarrier Modulator and Filter

Composite Video Being Fed Into The MA-12X Transmitter

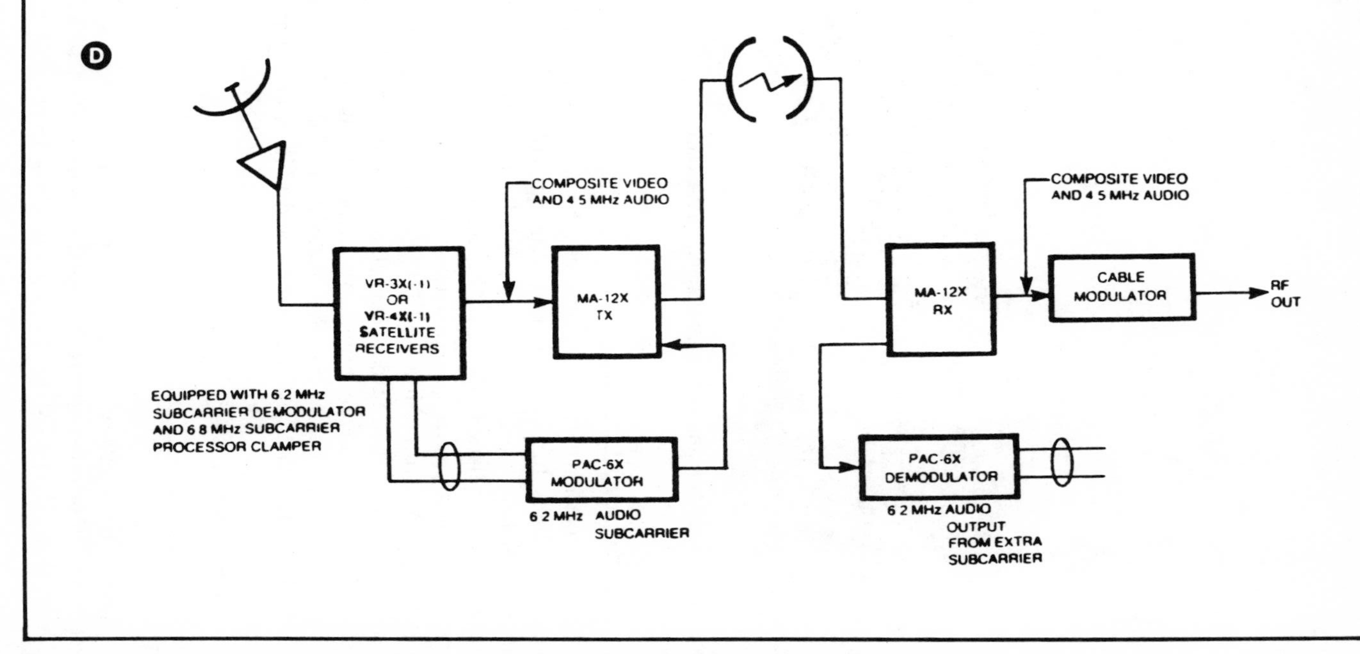

Fig. 5-1. Microwave and satellite configurations of the MA-12X transmitter (courtesy of M/A-COM).

Fig. 5-2. The MA-12X transceiver (courtesy of M/A-COM).

C- AND K-BAND UPLINKS

For this portion of the transmitter chapter we have chosen LNR Communications, Inc. as the equipment manufacturer with a broad range of electronic gear and many national and international customers. These include COMSAT, the Government of India, PTT France, Siemens, NIH, GTE, Turner TV Network, Christian Broadcasting Network, Morocco PTT, Page Communications, and many others. The Wall Street Journal, for instance, has its information distributed daily for printing throughout the nation via LNR equipment. Product lines are also available to INTELSAT and DOMSAT in the 11/12/14 GHz bands, and LNR Ku-band amplifiers are also designed for use with Advanced WESTAR, ANIK C, TDRSS, INTELSAT V, SBC, and others.

UC6

The UC6 upconverter with its redundant standby counterpart accepts video or other traffic signals (Fig. 5-4) at 70 ±18 MHz, with a bandwidth of 36 MHz and output frequencies between 5.925

and 6.425 GHz. Local oscillator leakage at output is – 65 dBm (max) and an i-f frequency which is above 1 MHz. There is dual conversion with an L-band upconverter and L-band i-f. Input impedances are 75 ohms and outputs 50 ohms, with optional frequency stability available between ±5 kHz to + 200 Hz per month. A photograph of the dual unit is shown in Fig. 5-5. Automatic switchover, as another option between on-line and standby units, plus manual override is also available if desired, in addition to remote monitor and control panels.

Baseband video enters the i-f input at a nominal 70 MHz, is amplified and then upconverted to 1112.5 MHz as it passes through the first upconverter, whose reference is the i-f xtal and phase-lock combination. An i-f local test point is shown that may be used to monitor this crystal output and adjust if necessary.

After initial upconversion, the i-f frequency is now amplified and sharply filtered at 1042.5 MHz after an amplifier gain of some 25 dB. A coaxial isolator provides suitable impedance matching. An L-band filter/equalizer reduces the i-f bandwidth to typically ±18 MHz with very sharp roll off (Fig. 5-6), followed by circuits to equalize group delay. Another coaxial isolator then matches this output and the final 6 GHz upconverter.

The rf oscillator is also crystal controlled with added phase lock and a pair of single-pole, double-throw switches and evident gain adjust for the mixer above. These switches permit introduction of an external local oscillator, if required or desired, in case of emergencies or the possibility of changing frequency ranges between 7 and 14 GHz, possibly higher. A local oscillator test point 20 dB is also furnished.

Once again these mixed signals, now at full-range rf, are impedance-matched through another coaxial isolator and routed to the rf filter and test coupler, which also has a –20 dB test port that can be used for a carrier alarm monitor. The filter/coupler offers low loss between 4.8 and 5.33 GHz and supplies 40 dB minimum rejection between 3.7 and 4.2 GHz. Carrier and modulation now go to the rf output at an overall system gain of 10 dB (min).

Both rf and i-f oscillator sources are powered by 24 Vdc at 600 mA. Should phase-lock voltage limits be exceeded or the phase-lock amplifier deviates into its search mode, alarm indicators will light to show a fault condition. At that point a 30 to 500 Hz square wave will become evident at the two test points. Power supply lamps also indicate any problems of drift or failure of the highly regulated +24 Vdc source.

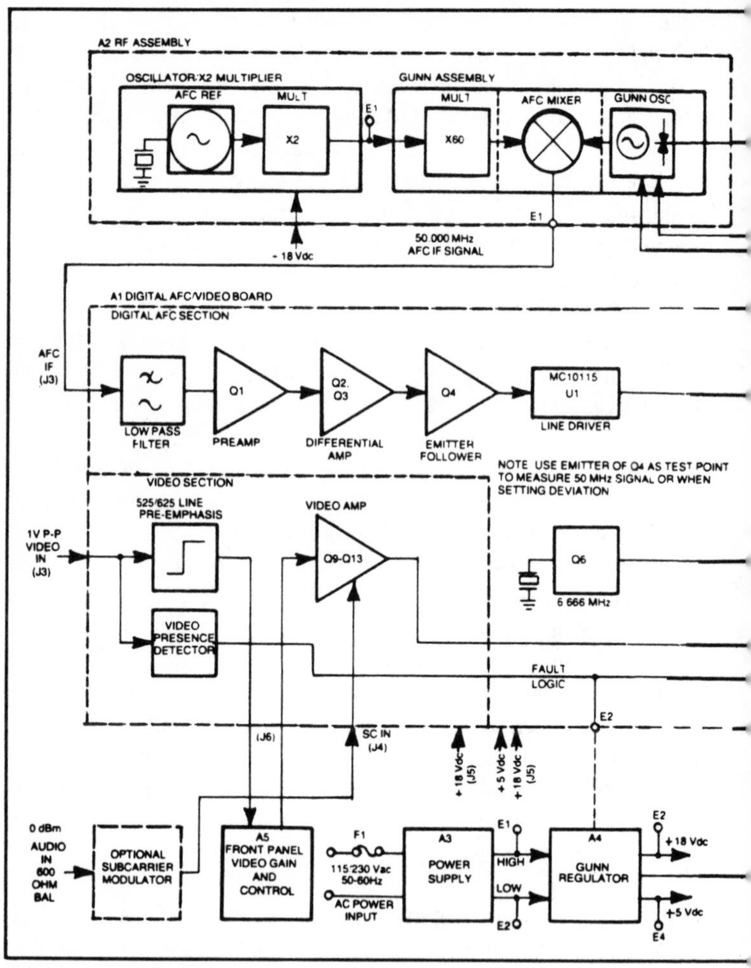

Fig. 5-3. M/A-COM's microwave/satellite transmitter block diagram.

THE FINAL AMPLIFIER

Now that two different transmitter and upconverters have been discussed, it's time to conclude the chapter with a short analysis of the final output amplifier. This is the stage preceding the transmitter dish where the very considerable power is generated to drive all the carriers and modulation we've been considering into deep space. As you may well imagine, high voltages are involved, along with forceful power. This particular unit has an output power level of 3 kilowatts, and a nominal gain of 40 dB by itself or, with optional TWT driver, a total gain of 80 dB.

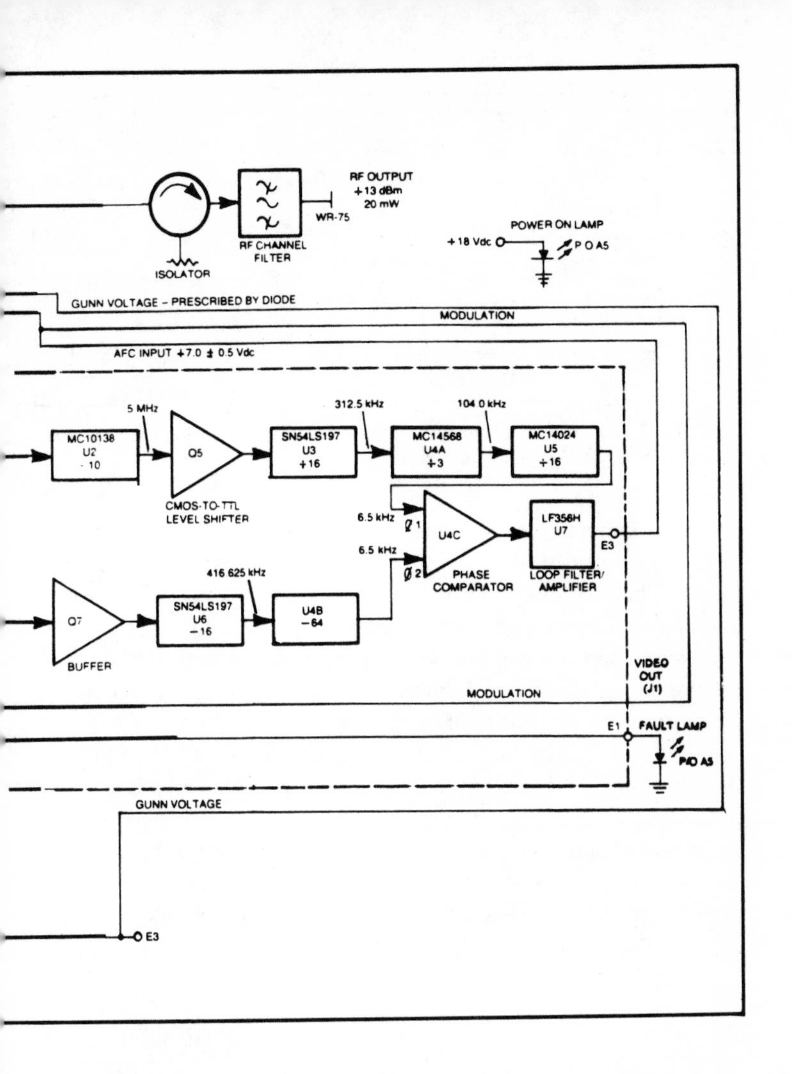

It will operate with single or multiple carriers using FDMA for video, teletype, or voice facsimile, and uses an air-cooled especially designed klystron amplifier which, with associated cooling and electronics, is contained within a single cabinet. This particular klystron may operate on either local or remote control with manual or automatic recycling. Frequency range is specified between 5.925 and 6.425 gigahertz.

The power supply generates both low and high voltage, is adjustable, and has circuit breakers as well as an auto/manual switch on the front panel. A transformer offers one phase to the bias sup-

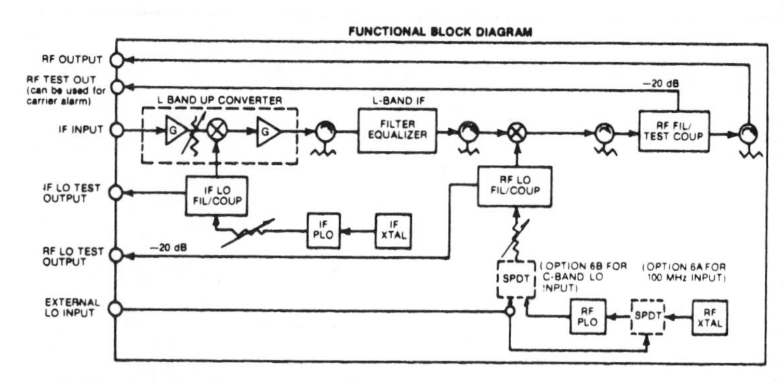

Fig. 5-4. LNR's UC6 dual-conversion upconverter using L-band as the processor (courtesy of LNR Communications, Inc.).

ply and filament regulator and transformer, a second goes to the low voltage transformer, and the third to the beam transformer and variac. The bias supply furnishes 500 volts dc and is used for surge and arc limiting. The variac in the beam transformer has a delta-fed connection, permitting an adjustable range between 0 and 6 kilovolts, and allows the high voltage to operate between 6 and 9 kV at a maximum current of some 24 amperes. High voltage is now rectified and filtered by diodes and capacitors and supplied to both the klystron's cathode and collector, allowing the tube operating potential. Arc limiting, whenever it occurs, can be handled by RLC circuits and diodes whose potentials are controlled by the bias supply. The trick here is to both shunt excess arc current as well as permit little or no overshoot when the circuit functions. The bias supply is also involved in beam shutdown whenever there's a problem. Beam voltage, by the way, is divided down at a ratio of 1000:1

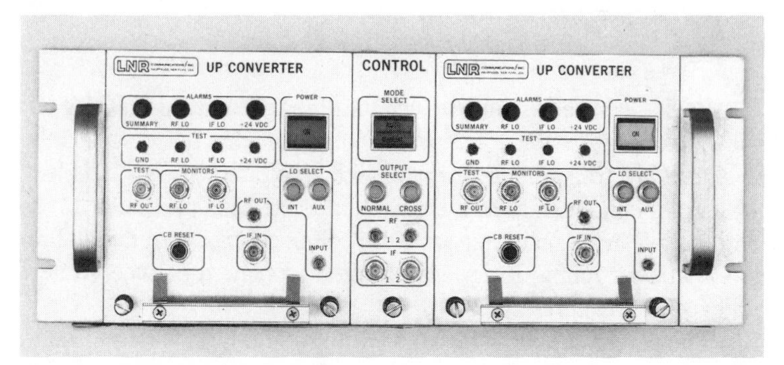

Fig. 5-5. The dual LNR upconverter for C-band frequencies (courtesy of LNR Communications, Inc.).

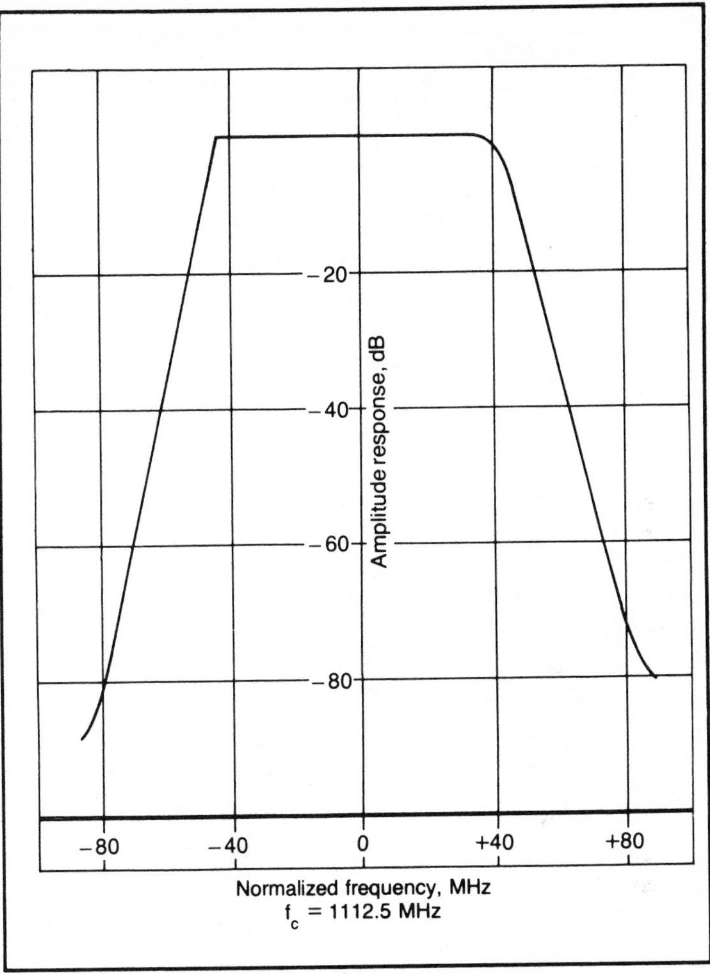

Fig. 5-6. Response of the 1042.5 MHz filter (courtesy of LNR Communications, Inc.).

and metered accordingly.

True, the foregoing has been a very abbreviated operational description of the inner workings of our final klystron high powered amplifier. But individual circuit descriptions are not that involved and require little in the way of interesting electronic discussion; therefore, they have not been included. Metered alarms include: air, filament, temperature, interlock, and arc. Helix and beam currents are constantly monitored by indicators on the TWT power supply, when used (see Fig. 5-7).

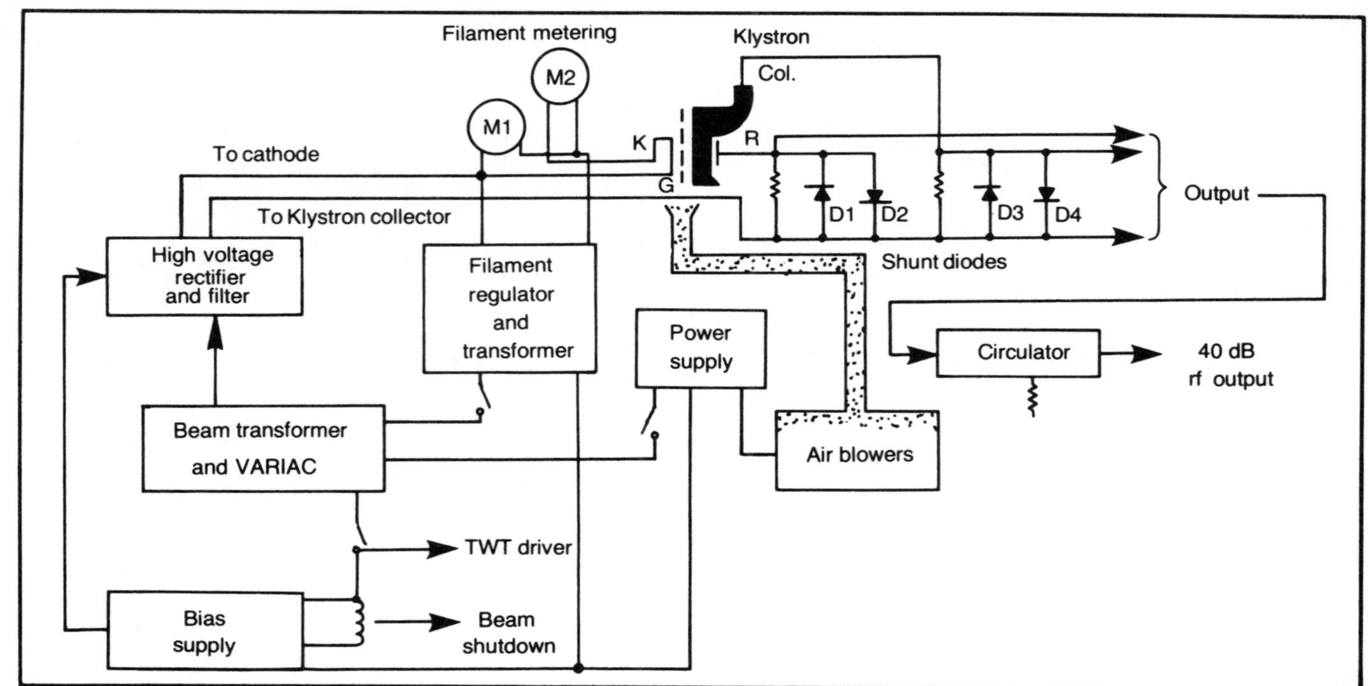

Fig. 5-7. Simplified block diagram of a klystron final amplifier with supply and control circuits (courtesy of LNR Communications, Inc.).

TRAPPING TERRESTRIAL MICROWAVES

Microwave signals between one and eight GHz can easily contribute total or partial interference to your receive earth station. According to the Microwave Filter Company, Inc. of East Syracuse, New York, they then become "part of your receiver noise level." MFC then prints Table 5-1 to prove just that. Beginning at 960 MHz and working up to 8.8 GHz, this table identifies offenders from the armed services, security, land mobile, maritime radio, radio astronomy, navigational, radar, TV pickup, and even airborne doppler radar. With a high frequency signal generator, of course, you can match these frequencies by introducing signals through a splitter and identifying them for entrapment and elimination. Of you can "see" these on-the-air-sources with a spectrum analyzer and identify each somewhat more easily, although not as precisely as though your calibrated signal source was handy. But, by either method, such interference can be detected and then Microwave Filter Co. (MFC) can go to work and help you out. As usual at these frequencies, 50 or 75 ohms are the impedances you're working with.

Glyn Bostic of MFC defines virtually all earth station frequency interference problems (in the C-band) as originating from the following terrestrial sources:

☐ In-band carriers are those within the usual 20+ MHz individual transponder slots.

☐ Out-of-band carriers plus noise operating below, above, and within the 3.7 to 4.2 GHz downlink of C-band.

☐ Band-edge carriers offset 10 MHz on either side from center transponder assigned frequency.

Such band-edge carriers may be eliminated by sharply set traps at 60 and 80 MHz since they will appear on either side of any 70 MHz i-f. They must, however, be placed *before* the i-f(s) in order to become effective with all transponders available for tuning. Should the i-f accept other frequencies, then trapping must be adjusted accordingly.

Out-of-band carriers are best handled by a bandpass filter that permits passage of the entire 3.7 to 4.2 GHz frequency band but sharply attenuates all other out-of-band sources. Best indications for location of this interference appears as "noise." Lower numbered transponders are affected most by frequencies below 3.7 GHz, and high numbered transponders show "sparklets" when in-

Table 5-1. Offending Frequencies which Could Cause Greater or Lesser Interference Problems at C-Band (courtesy of Microwave Filter Co.).

FREQUENCY (GHZ)	NATURE OF POTENTIAL OFFENDER
0.960-1.350	Land-based air navigation systems
1.350-1.400	Armed forces
1.400-1.427	Radio astronomy
1.427-1.435	Land-mobile: police, fire, forestry, railway
1.429-1.435	Armed forces
1.435-1.535	Telemetry
1.535-1.543	SAT—maritime mobile
1.605-1.800	Radio location
1.660-1.670	Radio astronomy
1.660-1.700	Meteorological—Radiosond
1.700-1.710	Space—research
1.710-1.850	Armed forces
1.990-2.110	TV Pick-up
2.110-2.180	Public common carrier
2.130-2.150	Fixed point-to-point (non-public)
2.150-2.180	Fixed—omnidirectional
2.180-2.200	Fixed, point-to-point (non-public)
2.200-2.290	Armed forces
2.290-2.300	Space—research
2.450-2.500	Radio location
2.500-2.535	Fixed, SAT
2.500-2.690	Fixed point-to-point (non-public) Instructional TV
2.655-2.690	Fixed, SAT
2.690-2.700	Radio astronomy
2.700-2.900	Armed forces
2.900-3.100	Maritime radio navigation
2.900-3.700	Maritime radio location
3.300-3.500	Amateur radio
3.700-4.200	Common carrier (telephone) Earth Stations
4.200-4.400	Altimeters
4.400-4.990	Armed forces
4.990-5.000	Meterological—radio astronomy
5.250-5.650	Radio location (coastal radar)
5.460-5.470	Radio navigation—General
5.470-5.650	Maritime radio navigation
5.600-5.650	Meteorological—Ground based radar
5.650-5.925	Amateur
5.800	Industrial and scientific equipment
5.925-6.425	Common carrier and fixed SAT
6.425-6.525	Common carrier
6.525-6.575	Operational land and mobile
6.575-6.875	Non-public point-to point carrier
6.625-6.875	Fixed SAT
6.875-7.125	TV pick-up
7.125-8.400	Armed forces
8.800	Airborne Doppler Radar

terference enters above 4.2 MHz. Usually when good bandpass filters are connected to the feedhorn, signal-to-noise ratios improve considerably and video cleans up quite convincingly. When interference is particularly heavy, it may be necessary to disable your automatic frequency control (AFC) and tune the receiver for best picture. At times, you may even have to move your antenna's location and then retune for maximum results, when all else fails.

MFC does have what are called tunable bandpass filters for individual transponders on in-band situations where interference or poor signal conditions interfere with part of the 500 MHz band but not all of it. With this type of filter, transponders between 1 and 24 may be selected,with an insertion loss of only 1.5 dB (max) and a 3 dB tuning bandwidth of 40 MHz. Should Ma Bell tend to "wipe you out" with strong signal override at one point or another, there are also custom microwave traps available for the various phone microwave company frequencies which will either aid or relieve the problem one way or the other.

All these filters, however, depend to a greater or lesser extent on the incoming signal, the effectiveness of your low-noise amplifier and antenna gain, and whether all components of the antenna system are operating at peak potential. All the traps of any description ever invented won't cure a dish, LNA, or acute receiver problem that has little or nothing to do with incoming interference. In this respect, earth stations are very little different from ordinary television receivers except that frequencies have moved from megahertz to gigahertz. You'll find that many of the techniques are somewhat similar, but because of frequency, the hardware is different. Doesn't a bandstop filter to exclude FM sound familiar? How about notch filters to squelch certain other interference at specific frequencies? And remember how half-wave and quarter-wave stubs helped remove standing waves in both coaxial cable and twin lead? Many of the same techniques can be of use in satellite transmission problems, except the tools to do the job are often more extensive and expensive, and insertion losses do have to be minimal because of the tiny incoming signals.

Implements to aid in such trapping and filtering are shown in Figs. 5-8, 5-9, and 5-10. These can be purchased either as standard items or especially made up for the occasion by writing or phoning in your requirements to the Microwave Filter Company whose toll-free number is 1-800-448-1666. Custom interference removal, they claim, is a house specialty for any and everyone.

Your best preventive cure for suspected interference is a good

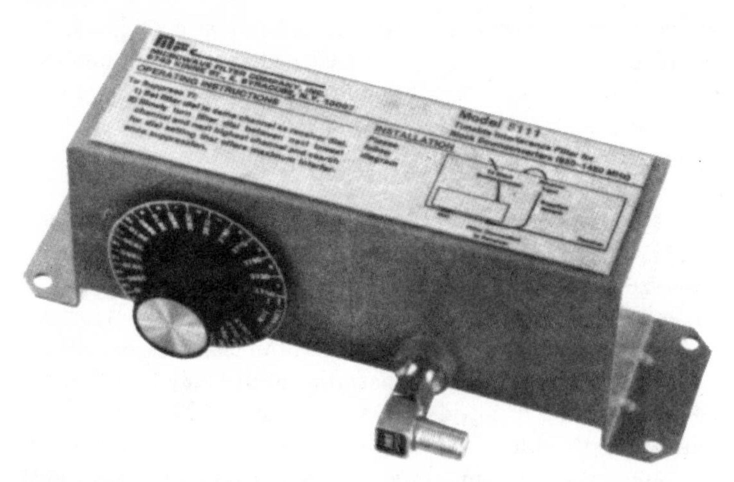

Fig. 5-8. The model 5111 terrestrial notch filter tunes between 910 and 1440 MHz, delivering a 15-dB notch and removing problems (courtesy Microwave Filter Co.).

Fig. 5-9. Out-of-band carrier suppressing T.I. Model 4352 is connected between LNA and downconverter. For LNB, LNC, use model 3716 (courtesy Microwave Filter Co.).

Fig. 5-10. The Sky Doc kit includes a terrestrial tracer (1); single channel bandpass filter (2); microwave bandpass filter (3); tunable/switchable i-f trap (4); 60.80 MHz i-f trap (5); notch filters (6); miniature i-f traps (7,8,9) and cabling (10 through 16). Very valuable in T.I. troubleshooting (courtesy Microwave Filter Co.).

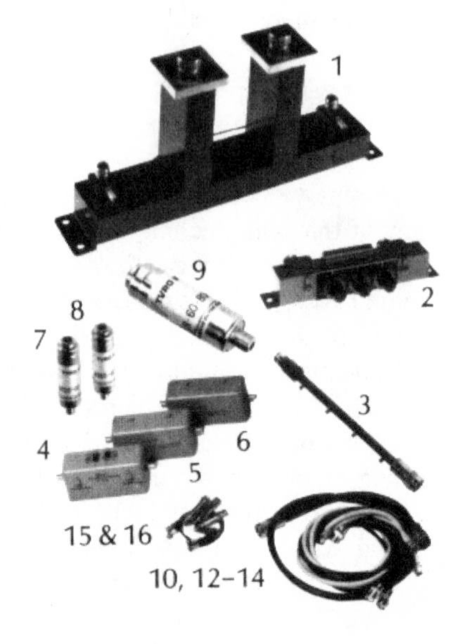

pre-installation survey to determine the presence of any and all possible stray signals inhabiting the dish-site area. Unfortunately, there's no way to tell when Ma Bell will erect one of its towers near your installation or the local airport's radar will spring a harmonic leak. In either event, you may either move your equipment or install filters, depending on circumstances. At least there are microwave traps to attenuate overloads for in-band problems, bandpass filters for out-of-band interruptions, and special 60/80 MHz i-f traps to reduce or eliminate telephone carriers. You also might remember that the greater the bandwidth of any incoming signal the greater the noise. Stray electromagnetic waves approximating frequencies upon which you're operating add to input noise power and decrease S/N ratios resulting in scintillating, weak, or fuzzy pictures.

Cable TV also requires various types of trapping to remove unwanted interference such as adjacent channel cutoff, narrow-band signal rejection, diplexers, and bandstops, but unique to CATV are the mass descrambling networks, off-the-air channel suppression and closed circuit TV insertion, co-channel eliminators, and co-channel erasure procedures. In these latter instances, a good directional coupler or variable phase shifter may well do the trick by helping with both phase and amplitude control to either null or avoid the impinging signals. If these don't work, then you may have to either move or change antennas, offering a sharply directional pickup with better than average front-to-back ratios and negligible SWRs. Usually, large antennas offer reduced beamwidths and little sidelobe penetration, in addition to considerably greater comparative gain. Directors, reflectors, and multi-parasitic elements all help.

Fig. 5-11. A highly versatile and accurate signal/sweep generator by Wiltron, Model 6647A, whose frequency extends from 10 MHz to 18.6 GHz (courtesy Wiltron Co.).

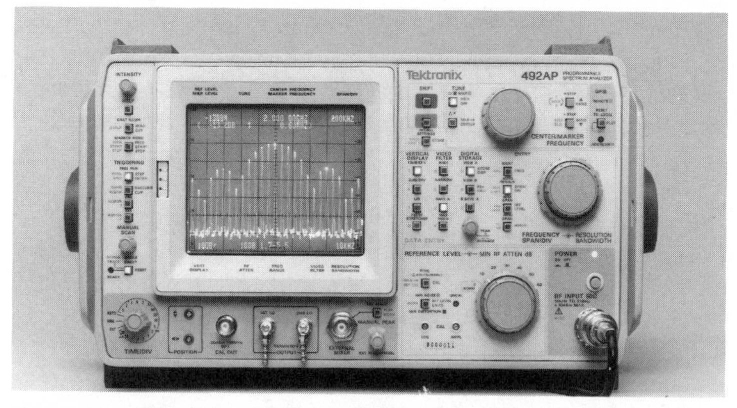

Fig. 5-12. A very new portable and programmable Tektronix spectrum analyzer with frequency markers and broad dynamic range (courtesy Tektronix, Inc.).

TVRO ANALYSIS

Using the foregoing or closely equivalent test gear, it is possible (with a few additions) to completely characterize antennas and analog electronics of virtually any TVRO earth station to within a very few dB. LNA/LNBs, of course, are included as well, in addition to traditional and very important carrier and sidelobes received (generated) by the reflector. As long as you watch dish elevation angles, far field ratios, and precise pointing, there's no reason that excellent field measurements can't be made. Naturally, there has to be a standard gain, calibrated feed horn, excellent cabling, clear sight lines from transmitter to reflector, and a valid license to transmit from the FCC. You'll also need a modulation source such as a color bar generator and a few other minor items, including an audio source, if you really want to dig for all parameters. Last, but not least, a handy compass with a large compass rose (compass card) calibrated in degrees will make horizontal measurement angles for the sidelobes fairly exact without having to resort to an X-Y recorder. However, just *any* spectrum analyzer and *any* sweep-signal will *not* do. That's why we suggest Tektronix and Wiltron—they will! See Figs. 5-11 and 5-12.

TROUBLESHOOTING VERSUS FILTERING

This may seem a peculiar heading when we've already devoted considerable attention fo filtering, but after a 1986 Syracuse session with operations analyst Ron Mohar of Microwave Filter Co. Inc., there are some additional suggestions worth including.

Channel Master®

SATELLITE RECEPTION EQUIPMENT

2-Way
Power Divider
Model 6215

Frequency	Isolation
950-1450 MHz	15 dB Between Output Ports
Insertion Loss	**Connectors**
4 dB	"F" Type

4-Way Model 6216
Power Divider

Frequency	Isolation
950-1450 MHz	15 dB Between Any Output Ports
Insertion Loss	**Connectors**
7 dB	"F" Type

WARNING: Both Splitters are power passive on all ports.
Do not terminate unused ports receiver will be damaged!

Fig. 5-13. Handy 75-ohm splitters for 2- and 4-way reception *following* LNB block down conversion. Note the insertion losses and *don't* terminate unused ports (courtesy Channel Master).

Try cranial and equipment analysis *before* filtering, since filtering often reduces bandwidths to about 15 MHz instead of 36 MHz per channel. In addition, if there are a number of interfering frequencies, the number of filters may be considerable and quite costly. To do this proceed accordingly:

1. **Equipment.** Contemplate your reflector, LNA/LNB, and receiver first. There may be subtle problems with one of these or the connecting cable you've overlooked that's causing the problem. For instance, changing from RG 59 to RG 6 with an 18 AWG conductor may well pick up as much as 2-3 dB signal strength and nullify outside interference if the new cable is well shielded.
2. **Location.** Is this location high or low, and does it have a clear arc from at least south to west? Check lines of interference and reposition mount and reflector in lowest suitable spot, for unless you have reflected (bounce) EMI, such interference must pass directly into the feed and reflector, not simply overhead.
3. **Shielding.** You may use a wall, building, or screen. At-

tenuations of up to 30 dB are possible with a screen, and as low as 10 dB with a wall.

4. **Spectrum Analysis.** With a good analyzer you can pinpoint disturbances and take even more effective measures. Microwave Filter has a Terrestrial Tracer (Mod. 4043A) that will permit 3.7 to 4.2 MHz tuning at an accuracy of probably better than 10 MHz in the 500 MHz band. This unit also produces a pair of 15 to 20 dB notches and works well on the usual analog signals but *not* digital. Obviously, the calibrated dials will not only locate the problem but also positively indicate if notch filtering will solve dilemma.

Remember, however, that these notch filters are approximately 3 MHz in bandwidth and may also give you picture problems, too, if not applied with discretion. I-fs most adaptable to filtering are at 70 MHz, 140 MHz, 142 MHz, and 510 MHz. As frequencies increase, the filters become more expensive.

For block down conversion units in the 950-1450 MHz category, it's best to use a separate LNA and LNC so that filtering may take place *between* the LNA and LNC. Here, a model 5111 tunable i-f filter (from Microwave Filter, of course) should find and/or eliminate your problem. Other block down filters are even now in development or may be made up on special order by calling the company.

Other Considerations

When determining whether to adopt one cure or another try and define your problem first. Is the interference *in-band* or *out-of-band*? Here, bandpass filters 4352 and 3716 will help this determination with the LNA group, while filter 3716DC works with the LNB/LNC between them and their feeds. In some instances, the AFC needs to be detuned when filtering is required *before* the downconverter. Phase-locked-loop receivers will overload if interference is extremely heavy.

Unfortunately, some of the cheaper receiver i-fs are not tuned precisely, and you may have to use a tunable notch filter to eliminate your problem. The Model 4616 is nominally for 70 MHz i-fs, but available in different i-fs, too. These are both tunable and switchable. The 6612 is for different i-fs.

We are also well aware of the crosspole problems, especially those augmented by obviously poor reflector/feed aiming. Tendencies are to add amplifiers, traps and all sorts of wrong substitutes when only the reflector and mount require precise north/south polarity adjustment, adequate elevation, and reasonable tracking. Some transponders, too, just aren't aimed in your direction and pickups that are crispy clean are often hard to come by. See Fig. 5-14.

We have a very good illustration on your own C-band equipment where winter upset some of the tracking and caused some pretty heavy sparklies about G1, but cleaned up just before D1. When such occurs, your only prayer can be fulfilled with a spectrum analyzer. We thought a couple of pictures showing poor carrier-to-noise and cross polarity problems among the H/V channels might help.

These photos are for real, and decidedly indicate the specific problem, nullifying all others (see Fig. 5-15). This is why using such an instrument with good 10 dB/division vertical and MHz/GHz horizontal frequency calibration (both are a must) is positively required to pinpoint problems with both reflectors as well as their LNB and receiver electronics. Guesswork and cheap signal meters won't do the job at all! But a handy, portable spectrum analyzer with ranges between kilohertz and 20+ GHz will! Just stick a power divider on the line between the LNB and receiver, tune to 1.2 GHz and let the signal flow. The two photos tell the rest of the story.

Fig. 5-14. The PP-1450 signal level meter can help aim, focus, peak, and polarity-align your TVRO from 70 MHz to 1500 MHz (courtesy Pico Products).

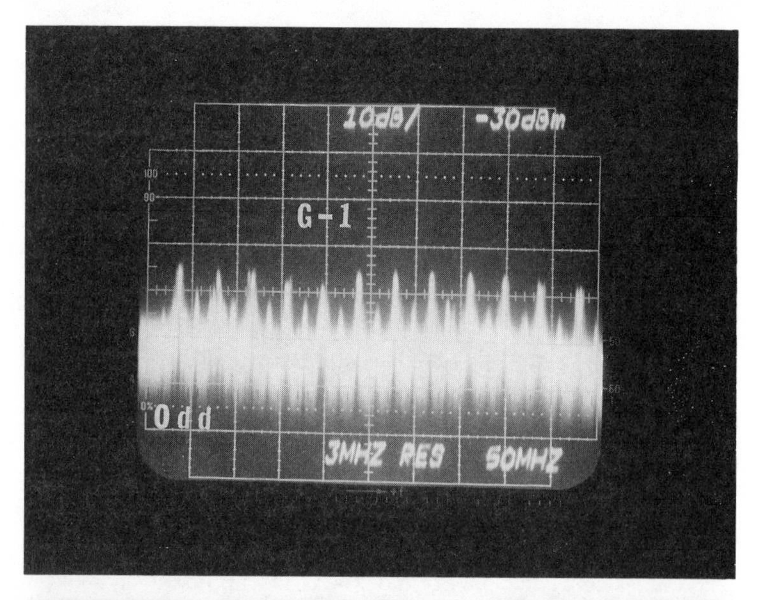

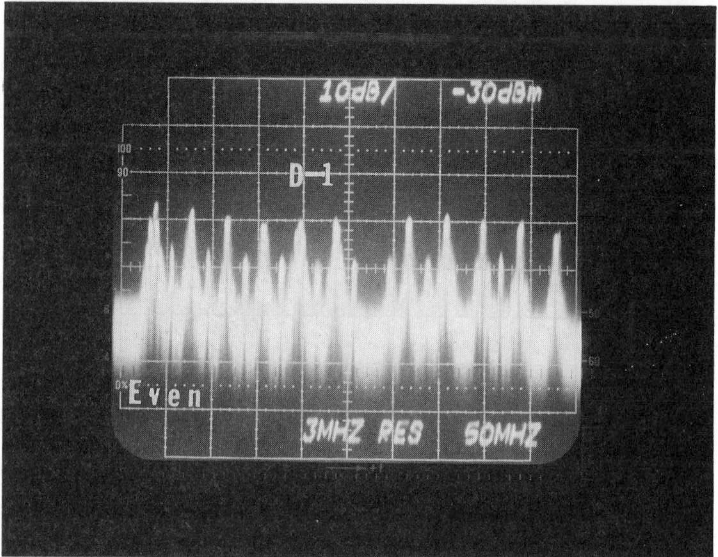

Fig. 5-15. Look hard at these two spectrum analyzer photos. The G-1 odd-numbered transponders are full of sparklies (top). The D-1 even-numbered transponders are pretty clean. What do you see? How about the amount of trash and crosspole in transponders 1 and 3, especially? Look at the very low carrier-noise ratios throughout. Then see how relatively clean D-1 (below) becomes on its even transponders with a good carrier-to-noise ratio of better than 20 dB and relatively clean crosspole at least 10 dB down from the others. Can you really survive without a spectrum analyzer in difficult situations?

Chapter 6

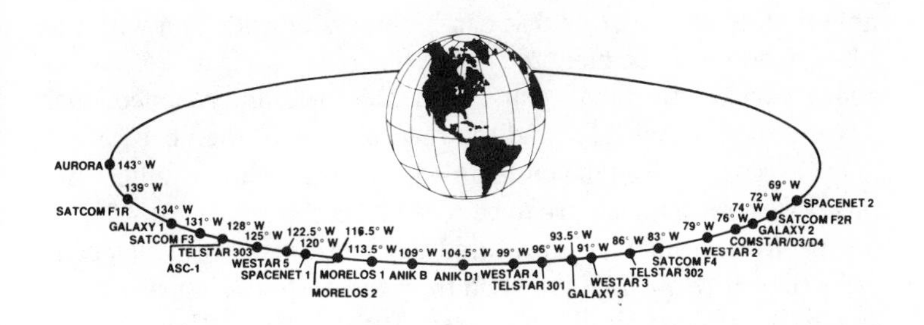

TVRO Receivers

As DAVID STOGNAR OF AVANTEK, INC. POINTS OUT, SINGLE downconversion from 3.7 and 4.2 GHz to 70 MHz is feasible only by using an image rejection mixer. Most receivers, however, have typical image rejection ranges of only 20 dB compared to 50 dB required for commercial installations. This, of course poses a problem, especially when the resulting i-f frequency is less than 250 MHz since it's probable that first order images do, indeed, exist. Further, should your earth station include more than one receiver for additional channels or another satellite, local oscillator leakage through the rf port can result in receiver crosstalk.

Dual downconverters, of course, normally mix to an i-f of about 1 GHz, and then proceed to second i-f frequencies of somewhere between 70 and 250 MHz, or even somewhat higher. Such higher frequencies, especially if bandspread, do permit multiple receiver tuning for the user, including selection of a number of single polarity channels. If dual LNAs with separate vertical and horizontal cross polarities are included, then all channels of any single satellite may be tuned when using good quality parabolic dishes. As satellite receive designs progress, torus antennas may permit channel tuning of several satellites without any antenna movement at all. Initially, this will be expensive, but it's not beyond state-of-the-art even now.

One suggested "fix" for a single downconverter is attachment of a tunable preselection filter ahead of 70 MHz, i-f mixer for required image rejection. However, if the receiver tunes 12 or 24 channels, the filter must be made frequency agile, and this can be expensive due to either varactor diode drive (with tracking) or mechanical motor drive with feedback servo.

Another alternative, according to Mr. Stogner, would be a tunable first downconverter that can be built alongside and with the LNA so both may be mounted at the feedpoint. This makes downconversion and the 50 dB gain attainable, fewer interconnections, fewer costly GaAsFETs, and the receiver, itself, then becomes a much less expensive uhf single-tuned, superheterodyne radio. Further, if several channels are to be received at the same time, a block downconverter may be substituted for the single LNC which converts the entire 500 MHz i-f band from the downlink into a tunable single conversion unit. Stogner predicts that such LNC design will become more popular toward the middle 1980s because of economy, performance, plus reliability, and the forthcoming 11.7 to 12.2 GHz K- and Ku-band transmissions where, with the LNC, only the downconverter has to be exchanged to convert the receiver to these higher frequencies. A single receiver, then, that can switch between two C- and K-band downconverters, could offer coverage of both bands as the 12 GHz downlink becomes attractive both at Ku and at K when more commercial and DBS begin active operations in the middle 1980s.

A lower cost approach to simultaneous reception of any six satellite-relayed programs with mixed polarities is already offered by Avantek with its AR1000 receiver and ACA-4220 dish-mounted LNA/downconverter (Fig. 6-1). The ACA-4220 heterodynes the complete 3.7 to 4.2 GHz downlink to 940 to 1440 GHz, which is outside the uhf TV broadcast band and in the spectrum of moderately powered on-the-air signals. The AR1000 receiver and its 6-way power dividers, may also include an ARC-1000 power-control unit, four ARA-1001 and one ARA-1002 i-f receiver modules that are all digitally tuned to the 940-1440 MHz LNA inputs. Four AR-1000 receivers may also be driven from the same feedlines for simultaneous reception of all 24 channels if these are active on a single satellite.

A handy chart, also furnished by Avantek, is shown in Fig. 6-2. Received frequencies between 3,600 and 4,200 MHz are shown versus 12- and 24-channel tuning, and their various center frequencies. The diagonal line, of course, should offer ready reference between the two sets of figures. Our thanks to editorial manager Northe Osbrink for this helpful material.

Using other Avantek equipment selections, you may also in-

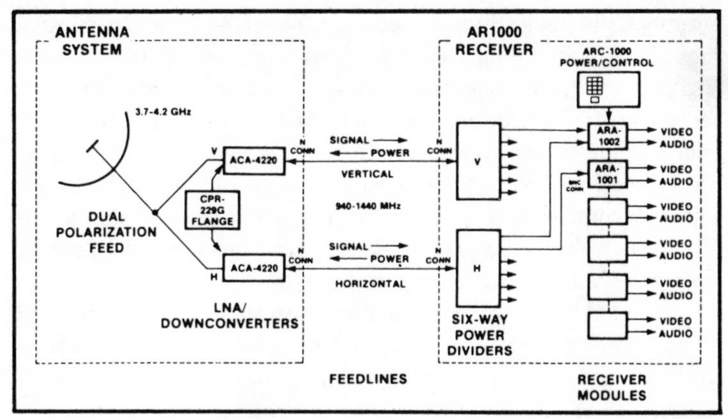

Fig. 6-1. Avantek's ACA-4220 LNA/downconverters and the 6 AR-1000 video-audio channels (courtesy of Avantek).

clude diplexers to reduce feedline costs, and even add a second earth station to existing facilities with additional diplexing and 6-way power dividers to produce an even dozen audio and video baseband

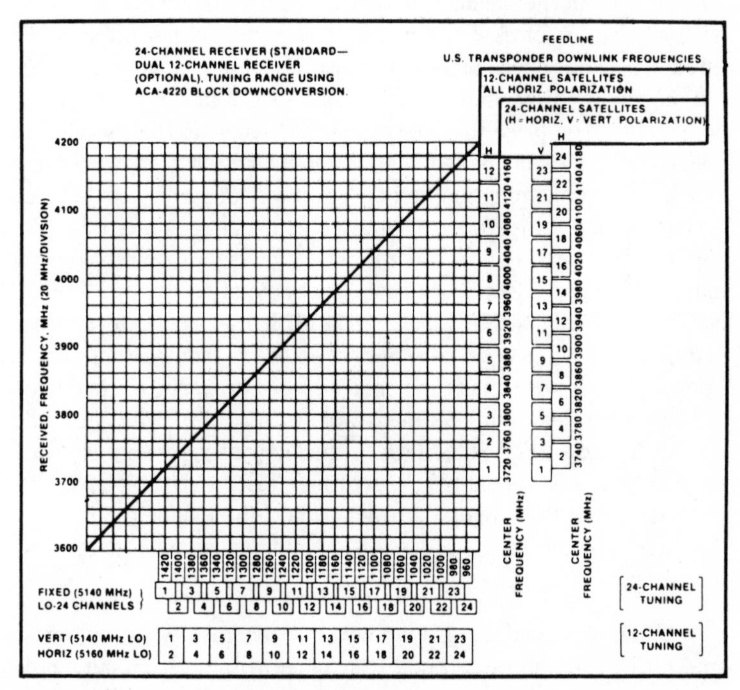

Fig. 6-2. This useful chart for C-band receiver frequencies shows 12-versus 24-channel tuning (courtesy of Avantek).

outputs that should fill many or even most CATV or commercial requirements in medium-sized or even larger installations. Needless to say (almost), many more possibilities are opening today to those who can appreciate and use the latest in satellite receiving equipment. Perhaps, as improvements continue, as many as a half-dozen satellites can all be accessed at the same time and all their signals accommodated. Or is this really too much to ask? How about simultaneously at K- and C-bands too? We think the day is coming soon when all this is highly possible.

Fortunately, downconversion remains the principal C/K-band mating problem since receivers will handle both when heterodyned to the same frequency—and so will a 12-foot or 10-foot dish.

GENERAL CONSIDERATIONS

Fancy or plain, quality equipment for either commercial or home is the primary consideration, even if the cost is a little more. Good signals now may become super signals later as improvements and state-of-the-art engineering continues. Microwave people who've been in the business for awhile really know best, and will have a selection of quality components to offer either directly or through package distributors who buy from them in quantity. There are, however, still a number of retail vendors in the woods who depend more on price than performance to market their wares in both communities and nationwide. Without being specific, we'd suggest a careful examination and check of *all* such products before taking the acquisitive leap. "That which is cheap is often dear," applies equally today as it did a hundred years ago, possible even more in the space electronics business. Some of us vividly remember the 1960 days at NASA when RCTL flatpacks of a few gates and D-type flip-flops cost fifty dollars apiece in thousand-lot quantities. But, they were hi-rel and once installed properly, worked like a charm after final tough temperature and humidity cycling in full system hookup. High reliability or not, however, investments of thousands of dollars should return their just rewards, relatively proportional to the money involved. If not, there should be a few questions. Here are some of the TVRO and receiver considerations which may be useful.

1. Do you need a polar mount for a general sky-sweep, or is an AZ-EL that is more difficult to manage with both azimuth and elevation requirements sufficient? In other words, are you going to look at several satellites or just one?

2. Should your system be straightforward and simple with a single LNA and mechanical polarity change, or do you have several TV receivers connected that might want to look at different channels simultaneously? In other words, a higher i-f frequency than just the usual 70 MHz and dual-polarized LNBs?

3. How about remote tuning? Do you want the dish swung by remote control, channels selected by pushbutton, LNB multichannel reception?

4. What about LNB and initial downconversion at the dish, with secondary downconversion at the receiver? Have you any need for both C- and K-band reception, or is one or the other satisfactory?

5. Is a multichannel torus of interest, or just the tunable parabolics with either AZ-EL or polar mounts?

6. Recall that most TVRO antennas have prime-focus feeds, and only the large radiators usually use Cassegrain. Be guided by both cost and antenna size—the big ones often require Cassegrain for serviceability and gain.

7. Don't forget that satellites will all soon go to 2° spacing, and to avoid interference because of broad beamwidths, you'll need a higher gain, more directive dish. A 12-footer at least will serve you the best over the long run, at least for the 4 GHz downlink service. You may well be able to swap or switch electronics for Ku and DBS, too.

8. Beware of local oscillator leakage levels, receiver channel tuning, transponder vertical/horizontal switching, receiver stability (drift), reliability, built-in of auxiliary test equipment, etc.

9. Your system will need a certain number of outputs. How about rf, baseband, dual audio (stereo) or voice, FM, 6.2 and 6.8 MHz subcarriers, multichannel sound, etc.?

10. What about expansion needs? Will you need more or different antennas, electronics, mounting space, remote controls?

11. Do you want your LNA and downconverter at the dish dc powered, and what kind of current do they require, and where will your supply originate?

12. Are there problems with local, county, or state permits, and how about microwave thru-ways? In short, has your site been checked for all these plus clear view to the satellites of interest?

These are just a dozen questions and suggestions to start the gray matter percolating. There are probably a dozen more that haven't been included, but with the above essentials you have a worthwhile start. Obviously there's considerably more to deciding

SATELLITE ORBIT BOOKMARK

KEY FOR BIRDWATCHER

1 — Vertical Polarization
1 — Horizontal Polarization
(R) — Repeat program
(CC) — Close-captioned/for the hearing impaired.
◇ — Restricted encrypted signal
♦ — Available encrypted signal

SIMULCAST SERVICES

CBN F3 **8** = G1 **11**
CBS T2 **20** = T1 **2**
PTL F3 **2** = G1 **17**

SERVICES BY SATELLITE

Sat.	Tr. #		Description
F5	**20**		Learn Alaska TV Network
F1	**8**		NBC—East
	13		NASA Contract Channel
G1	2		TNN (The Nashville Network)
	3		WGN—Chicago
	4		The Disney Channel—East
	5		Showtime—East
	6		SIN (Spanish Internat'l Network)
	7		CNN (Cable News Network)
	8		CNN Headline News
	9		ESPN
	10		TMC (The Movie Channel)—East
	11		CBN Cable Network
	12	◇	RequesTV
	13		C-SPAN
	14		TMC (The Movie Channel)—West
	15	◇	WOR—New York
	17		PTL Satellite
	18		WTBS—Atlanta
	19	♦	Cinemax—East
	20		GalaVision
	21		USA Network—East
	22		The Discovery Channel
	22		AVN (Alternate View Network)
	22		Deep Dish TV
	23	♦	HBO (Home Box Office)—East
	24		The Disney Channel—West
F3	1		Nickelodeon—East
	2		PTL Satellite
	3		TBN (Trinity Broadcasting Network)
	4		FNN (Financial News Network)/SCORE
	5	◇	Viewer's Choice
	6		Tempo TV
	8		CBN Cable Network
	9		USA Network—West
	10		Showtime—West
	11		MTV: Music Television
	12		EWTN (Eternal Word Television)
	13	♦	HBO (Home Box Office)—West
	15		VH-1/Video Hits One
	16		TLC (The Learning Channel)
	16		HTN (Home Theater Network)
	17		Lifetime
	20		BET (Black Entertainment Television)
	21		The Weather Channel
	22		HSN 1 (Home Shopping Network 1)
	23	♦	Cinemax—West
	24		A&E (Arts & Entertainment)
T3	1		CMTV (Country Music Television)
	2		SelecTV
	18	♦	American EXXXtasy
	22		KTVT—Dallas
W5	2		The University Network (The Unchannel)
	8		PASS (Pro Am Sports System)
	22		Meadows Racing Network
	24	♦	The FUN Channel
	24	♦	PPV—The Pay-Per-View Channel
S1	**9**		HIN (Healthcare Information Network)
	11	◇	HSN (Hospital Satellite Network)
	15		ACTS (American Christian Television)
	21	◇	BTN (Baptist Telecommunications)
	21	◇	VMT (Vanderbilt Medical Television)

Sat.	Tr. #		Description
M1	12		XEW—Mexico City
	14		XHITM—Mexico City
	22		XHDF—Mexico City
	24		XETV—Tijuana/San Diego
AD1	2		TSN (The Sports Network)
	4		Global Television
	6		MuchMusic
	8	◇	CHCH—Hamilton, ONT
	9	◇	WDIV—Detroit NBC Affiliate
	10	◇	WXYZ—Detroit ABC Affiliate
	11		CBC North—Pacific
	14	◇	TCTV—Sherbrooke, QUE
	15		CBC French—Eastern
	16		CBC Parliamentary Network (French)
	18	◇	CITV—Edmonton, ALTA
	19		CBC North—Eastern
	20		CBMT—Montreal CBC Affiliate
	21	◇	WTVS—Detroit PBS Affiliate
	22	◇	BCTV—Vancouver CTV Affiliate
	23	◇	WJBK—Detroit CBS Affiliate
	24		CBC Parliamentary Network (English)
	24		CBC French—Pacific
W4	**11**	◇	CTNA (Catholic Telecommunications)
	15		PBS A-Eastern
	17		PBS B-Central
	21		PBS C-Mountain
	23		PBS D-Pacific
T1	**2**		CBS—Central
	9		Wold Communications
	10		ABC
	12		ABC—Affiliate feeds
	15		CBS—West
	23		Wold Communications
T2	**10**		ABC—West
	16		CBS—Affiliate feeds
	20		CBS—Central
F4	1		HSN 2 (Home Shopping Network 2)
	2		Bravo
	2		The C.O.M.B. Value Network
	4		Nickelodeon—West
	5		Nightline Television Network
	6		BizNet (American Business Network)
	6		MSGN (Madison Square Garden Network)
	7		LBN (Liberty Broadcasting Network)
	8	◇	The People's Choice
	8		The Silent Network
	9		SportsVision
	10		AMC (American Movie Classics)
	11		HSE (Home Sports Entertainment)
	12		The Playboy Channel
	12		The Heartbeat Network
	12		NJT (National Jewish Television)
	13		NESN (New England Sports Network)
	14		Private Ticket
	15		SNL (Success-N-Life)
	18		Hit Video USA
	19		WPIX—New York
	20		Prime Ticket
	21		The Nostalgia Channel
	22		HTS (Home Team Sports)
	24		Sports Channel N.E.
F2	13		NASA Contract Channel
	22		AFRTS (American Forces Radio and TV)

Fig. 6-3(A). General Satellite Services. Reprinted by permission. Copyright 1986, ComTek Publishing.

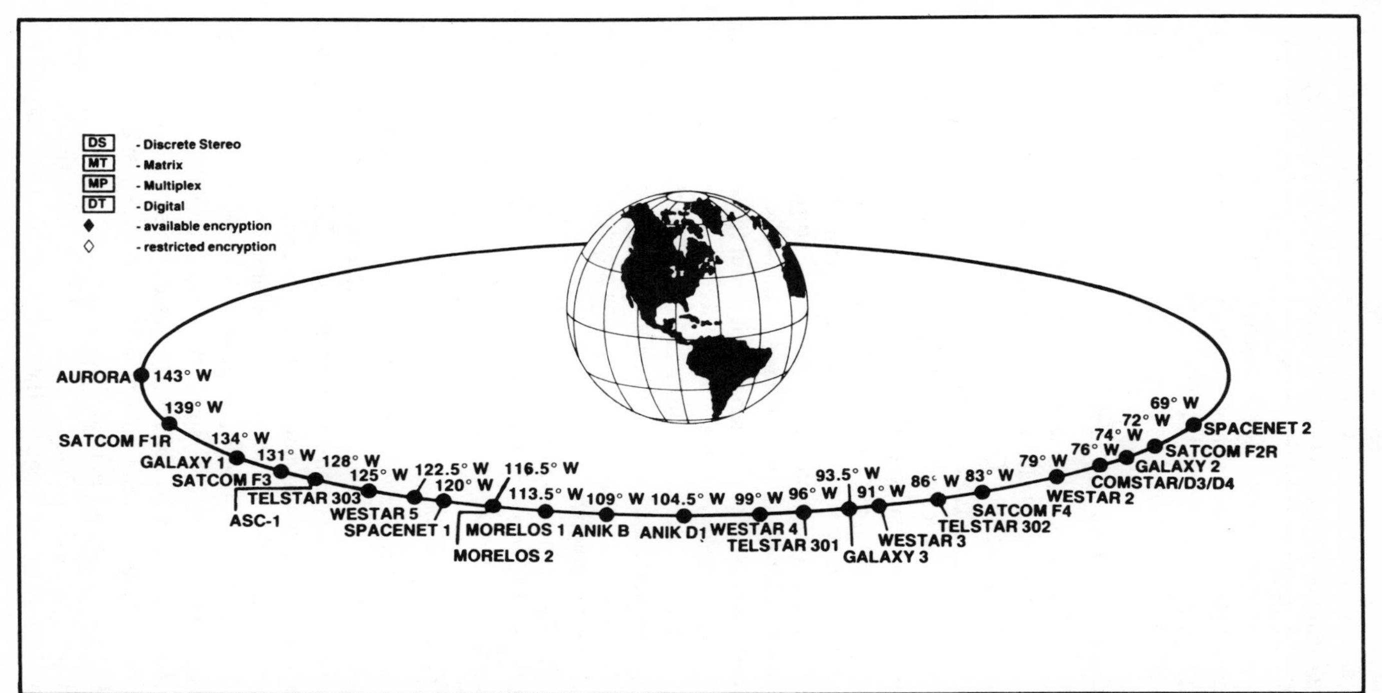

Fig. 6-3(B) U.S., Canadian and Mexican satellites in orbital positions. Reprinted by permission. Copyright 1986, CommTek Publishing.

if a satellite earth station would be enjoyable and what might be its ultimate home enjoyment or commercial worth. Many systems should be considered and their various merits compared before making that fateful decision. It's a choice you'll have to live with for some time before realizing any net return. (See Fig. 6-3 for a general satellite orbital chart and downlink frequency plan for 12 and 24-transponder satellites.

SOMETHING NEW IN TVRO

Let's look at a rather new and refreshingly packaged system built for utility that could be worth your consideration. Offered by Harris Corp. Satellite Communications Division of Melbourne, Florida, who says its deep dish 3-meter antenna, video receiver, and LNA costs at least 20-percent less than competitive systems. Called a "Delta Gain" system Harris also says it is 20-percent more efficient than competing prime-focus systems with an antenna efficiency of 78-percent which reduces terrestrial and sidelobe interference, has less antenna noise temperature, and with an appropriate LNA, its performance is almost equal to 4.5-meter prime-focus systems. Furthermore, it's said to be strong enough to withstand hurricane force winds. Offered in both AZ-EL and polar mounts, the antenna is shown in Fig. 6-4. Notice it is made in pie-shaped sections, has an unusually deep dish and what appears to be sealed electronics for both reflector and feed, which includes an electronic polarizer.

The G/T comparison (Fig. 6-5) with a low temperature LNA to a standard 4.5 meter prime-focus dish with standard 120 °K LNA is quite favorable, and is considerably above that with an 80 °K LNA and a "standard" 3-meter assembly. Graphs are obviously not given with 120 °K LNA for the three compared conditions—if that makes any real difference. Type 6303 LNAs are available from 70 °K to 120 °K in 10-degree steps or increments—said to be the largest range of any commercially available LNA. The feed is a modified Cassegrain monopod and the antenna, itself, has a considerative gain of 41 dB, with focus-to-diameter (f/d) ratio of 0.25. Because of the dish's depth, reduced terrestrial interference has been projected as well as spillover of radiation outside.

Model 6528 companion receiver features a sliderule dial showing all 24 channels of any geosynchronous satellite just like an AM or FM radio, and all functions are controllable with front panel switches (Fig. 6-6). If both vertical and horizontal channels need to be viewed simultaneously, the electronic polarizer may be re-

Fig. 6-4. A Harris' Cassegrain feed, high efficiency antenna (courtesy of Harris Satellite Communications Div.).

placed with an orthomode transducer in stations with two or more LNAs. As shown, there is front panel illumination, built-in power supply for LNA, a second audio subcarrier is optional, manual AFC, plus AGC, and on/off switching, signal strength terminals on back, filtered so that multiple receivers may be operated from a single antenna, and all units are burned in at 122 °F for 168 hours.

Although this is not necessarily an antenna chapter, a very new unit that would be appropriate for both consumer and commercial

SPECIFICATIONS

Diameter:	10 feet (3 meters)
Antenna gain (4 GHz)	41.0 dB
Antenna efficiency:	78%
VSWR:	1.3:1
Polarization options:	LP single polarization (manual rotation) Dual linear polarization (manual rotation) Electronic polarization rotation, ±75° remote-controlled
Antenna noise temperature at LNA flange:	Elevation: 10° 20° 40° 60° T_{ant}°k: 32 26 23 21
G/T (30 ° elevation):	LNA°K 80° 100° 120° G/T (dB) 20.8 20.0 19.4
Mount Adjustment range:	Elevation 10° to 70° Azimuth 360°
Operational winds:	60 mi/hr. (96 km/hr.)
Survival winds:	125 mi/hr. (200 km/hr.)
Mount Option:	Polar Mount

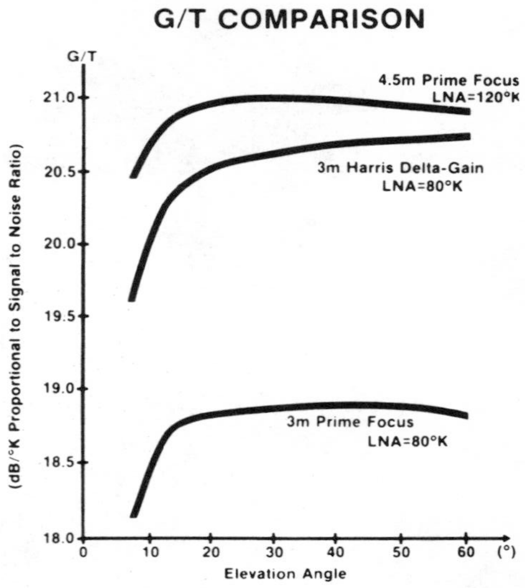

G/T COMPARISON

G/T
(dB/°K Proportional to Signal to Noise Ratio)
21.0
20.5
20.0
19.5
19.0
18.5
18.0
0 10 20 30 40 50 60 (°)
Elevation Angle
4.5m Prime Focus LNA=120°K
3m Harris Delta-Gain LNA=80°K
3m Prime Focus LNA=80°K

Fig. 6-5. Delta gain specs for Harris' newest and G/T comparisons with prime-focus feed counterparts (courtesy of Harris Satellite Communications Div.).

installations could be of some interest since it decidedly does indicate a trend among some manufacturers. And as more and more units are sold, we may all see prices fall and features rise to, perhaps, a rather considerable degree. Mass production, of course, should aid this trend even further. As a "matched" unit, this Harris offering has every appearance of being a winner if it performs as specified, and most manufacturers are conservative in their published specifications.

M/A-COM's NEWEST W/MONOPODS

Always a leader and never a follower in the TVRO industry, M/A-COM in Hickory, North Carolina, has now developed a new T-6 Spectra-Sat receiver/positioner in a single package with unitized remote control, and single strut monopod antenna feed support. In addition, there is a nine-favorite-channel memory that allows the satellite, its transponders and their polarities to be preset and recalled at the touch of a button—all automatically. There is also an additional programmed memory for 21 satellites as well as an automatic polarity and skew adjust for each satellite that may be committed to memory, too. Then, if there is some misadjustment for one reason or another in antenna positioning, a Correct-A-Trak circuit aligns actuator control when any one channel has been selected for best picture. See Fig. 6-6.

For programming assistance, you'll find a Spectra-Sat Command subsystem supported by code lights on the message center, plus parental supervision feature and full remote control for armchair audio, video, and satellite location programming. Both stereo channels are also included in addition to an audio select circuit for off-frequency sound.

Tuning is digitally synthesized with a block down i-f input fre-

Fig. 6-6. The latest M/A-COM receiver/positioner combination with dual-purpose remote control and satellite/transponder alphanumerical readouts and other indicators (courtesy M/A-COM).

quency from 950-1450 MHz and an input sensitivity from − 20 dBm to + 50 dBm. Bandwidth is specified at 27 MHz, with 24 channels for C band and 8 more channels for Ku. (Double, for instance, the 16 channels on RCA's K-1, K-2 newest 11.7 to 12.2 GHz downlink transmissions.) From this you may surmise that "splitting" these 54 MHz Ku-band channels is a possibility: we would say it's a probability for additional frequency re-use. Dual audio bandwidths are specified at either 150 kHz or 330 kHz, depending on narrow or wide BW settings. Audio format includes monophonic, stereo discrete, and stereo matrix, with baseband response between 30 Hz and 15 kHz.

Both baseband video/audio outputs are provided, either filtered and unclamped for VideoCipher II format, or filtered and clamped for monitor or VCR inputs. Pulse and drive rotor polarizer are included as is voltage for the video/horizontal polarity switch, plus positioner pulse. The Hall-effect transistor supply remains at + 5V. Motor drives are specified at ± 36 volts, and the circuit breaker is rated at 12-V, 2-amperes.

Antennas for the T-6 are supplied with linear actuators, and there is a mesh type with horizon-to-horizon worm gear drive. Low-noise block downconverters are specified at 85° with their omnirotors. There is also a low-noise switching block downconverter (LNSB), in addition to a dual-polarity low noise block down converter (DPLNB) that requires no polarity switching since the two LNBs are orthogonal.

Fiberglass antennas are a mainstay of the North Carolina facility and are made in a very new seven million dollar pressing plant located in Catawba County. Basic materials used in the process are a polyester resin, fiberglass, and calcium carbonate. Resin is pumped into a large mixer and added to the calcium carbonate plus a catalyst. Paste then floods out on the polyethylene film and fiberglass adder. This homogenous sheet of paste and "glass" consisting of the paste, chopped fiberglass, and film is pressed together in a process called *maturation* at temperatures of about 85 °C. Following this, two 2500 ton and/or one 1500 ton press(es) exert enormous pressure on the sheet molded compound. The plant delivers at least 100 units per day of the Ku-band reflectors or as many as 60 per day of the 10-footer C-band antennas. Complete reflectors or panels are then painted and passed through curing ovens on an endless belt preparatory to stacking, packaging and shipping. These are *excellent* fiberglass antennas, require little or no maintenance, and weather much better than expected, including super shape retention.

THE R.L. DRAKE COMPANY

This All-American TVRO receiver manufacturer is actually offering at least three new receivers, two or which are 950-1450 block-down i-f systems, and the other is a special 750 MHz frequency spectrum unit designed especially for the European market, including the 50-Hz line frequency and phase alternation line (PAL) German transmission standard. Identified as the ESR324E, Drake says several European satellite channels are now operating and there is a "niche" for this high-quality, reasonably priced receiver. It, too, has block-down conversion and may be included in a TV system or as a secondary receiver in deluxe arrangement. The 950-1700 MHz block-down LNB is compatible with all Drake LNBs, including those especially made for Ku band, one of which Drake has also recently introduced—a Model 2801 that mates well with Chaparral's 12-GHz feed. See Fig. 6-7.

There is also a stereo version of the popular ESR324 line, identified as the ESR324B—another 950-1450 LNB block system. Accommodating either matrix or discrete stereo, there is a dual-polarity input adapter for automatic horizontal/vertical channel switching and dual signal feeds. High gain AFC and wide-band discriminator circuits offer drift-free tuning and crisp video in addition to considerably improved sound over monophonic. A good TV/monitor attached will deliver very good audio/video results.

Although Drake, a leader in communications receiver manufac-

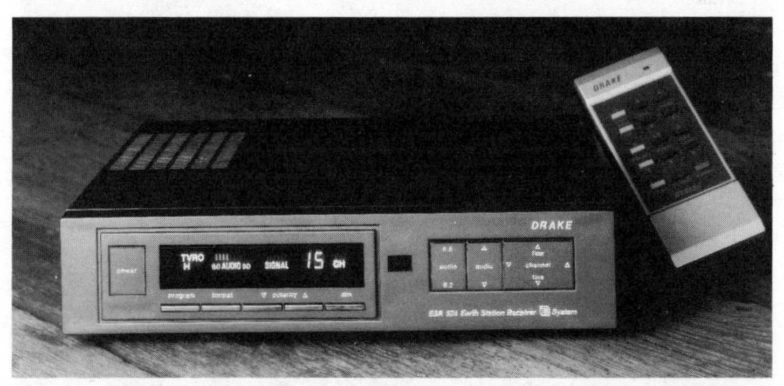

Fig. 6-7. A recent ESR earth station receiver system by Drake with program, format, polarity, and audio selection, plus channel fine tuning (courtesy R.L. Drake Co.).

turing since 1943, has a very good APS524 *separate* antenna positioner that has also been recently announced, it is also now marketing a Model ESR924i combined receiver and antenna positioner in a single package. Similar to some of its better competitors, this deluxe unit is both microprocessor controlled and Ku-band compatible with nine preprogrammed channels on *any* satellite for instant push-button tuning that includes both polarization and audio format.

There is also parental lockout for no-no "adult" channels that may only become unlocked by remote control, bandwidth stereo selection, 21 satellite-position memories locations, infared remote control, and a tasty charcoal grey case with easy-to-read LED channel and satellite name indicators. Being Ku-band compatible, video polarity output is reversible when switched from the C-band format.

HUGHES EXPANDABLES

Hughes Microwave™ Communications Products is now offering both 3.7- and 5-meter *expandable* antennas that will give you a bit more security against satellite separation angles as well as possible future signal loss (Fig. 6-8).

Based on a 360° rotation carousel mount, along with simple foundation, the 3.7-meter unit delivers 52% additional gain than a standard 3-meter antenna. Should antennas spacing become 2° as the FCC is projecting, expansion to 5 meters would probably avoid considerable adjacent satellite interference afflicting smaller

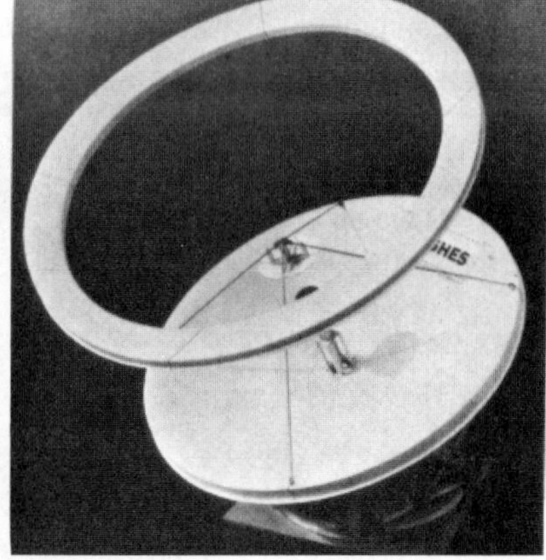

Fig. 6-8. Hughes "responsible" low-cost antenna system that may be expanded from 3.7 to 5 meters if required (courtesy of Hughes Microwave Communications Products).

and less well-designed TVROs. Since this antenna's AZ/EL carousel mount allows complete 360° rotation, coverage over the complete arc is permitted as well as elevation adjustments between 0 and 70°. It may also be either ground or roof mounted, and is focal fed.

The expandable portion is simply a ring that slips over the existing antenna frame, and aligns with the rest of the dish, increasing its diameter and circumference accordingly. Gain varies from 41.8 to 44 dBi at 4 GHz, and wind survival with no ice is 100 and 85 mph, respectively. The entire 3.7-meter antenna weighs 1,200 lbs., while the 5-meter expandable is 1,800 lbs. A crew of three should install this one in about four hours.

WINEGARD SATELLITE SYSTEMS

This Burlington, Iowa Company has now graduated from parabolic and rectangular fiberglass reflectors to perforated aluminum parabolics designed for both Ku- and C-band reception. Sizes range from 6-footers to 10 feet in diameter, with good bracing, adequate powder coating, and satisfactory mounting. Supplying others in the TVRO industry, these reflectors are exactly what you need in the way of see-through equipment that's charcoal black and blends readily with summer foliage. See Fig. 6-9.

Considerably superior to the mesh (extruded panel) variety but somewhat more expensive, such reflectors will hold their shapes under most conditions but are always subject to dents and bends from stray objects or excessive winds. Under such circumstances, either the affected panels must be renewed or the entire reflector replaced. Eyeball bending back into shape isn't practical because of the parabola's strict tolerance. The smaller the perforations, of course, the more any such dish appears as a solid piece of metal to the wind. However, soil earth temperature reflectance through the holes does *not* appear to be a factor since it must also reach the feed. In rocky formations, nonetheless, circumstances could differ to one degree or another, depending on existing surface angles.

Under ordinary circumstance, gains and characteristics of any well-made metal reflector are very similar to those for fiberglass and may be considered equivalent if parabolas, beamwidths, and f/D ratios are the same.

As a step in the do-it-yourself-direction, Winegard is already offering a Model RR-646 small 6-foot portable satellite system called a Mini-Ceptor™, that sells for a modest $995, complete with feed, block down converter, and manual receiver, Model

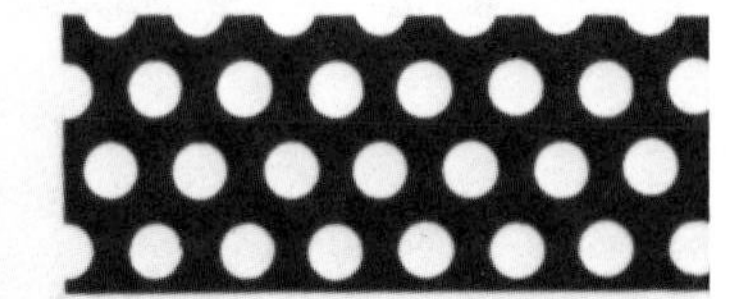

Fig. 6-9. Winegard's new C/Ku-band compatible aluminum perforated antenna with good specifications (courtesy Winegard Satellite Systems).

RF-90. This unit has a LED signal strength indicator, satellite format selection, polarity control, automatic fine tuning, and switchable modulator. System cabling is also included with terminations.

Such units and their larger brethren for C band have perforated hole diameters of 5/32 or 0.156 inch. The C- and Ku-band compatibles are half that, or 0.0078 in. and are called the Pinnacle™ line in smoked-chrome color.

Again, this is powder coating, which is actually a dry paint applied dry and more finely ground than common pepper. Each particle contains resin, pigments, modifiers, and possibly a curing agent, if reactive. It is applied by electrostatic charge in a closed chamber. In a reactive system (thermoset) curing occurs so that remelting won't take place. Typical thermoset powders consist of epoxies, acrylics and polyesters. Applications are usually electrostatically sprayed on since the powder, normally nonconductive, retains this electrostatic charge and clings to a grounded substrate

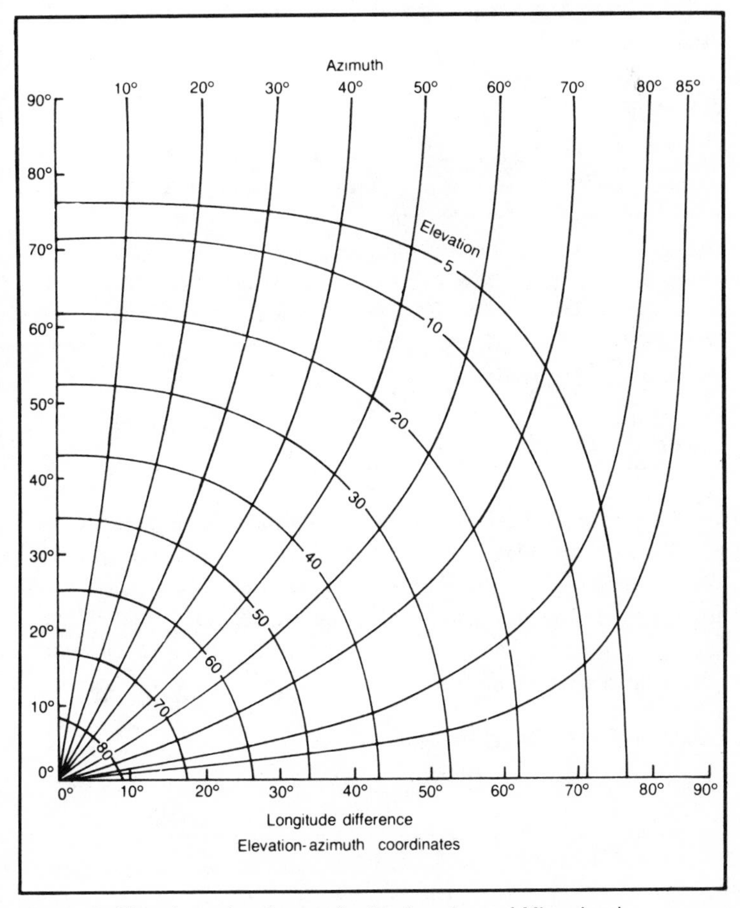

Fig. 6-10. Elevation-azimuth coordinates (courtesy of Microdyne).

until melted in the curing oven. Total material efficiency is 95 to 98 percent with "unlimited" color capabilities.

Both C and C/Ku compatible reflectors are manufactured in the Burlington plant at a tight surface tolerance of at least ±0.030 inch and on 1/4 in. centers. Patterns are staggered and 36 percent open and of 0.040 gauge, stretch-formed aluminum. According to Winegard, perforated hole sizes of more than 0.1 in. will degrade antenna performance by loss of gain because of reduced reflectivity and added noise temperature. Winegard recommends that the Ku-band feed (if you're using a dual) be centered and the C-band feed offset, unless the reflector is something like a 10 footer; gain, then, would be adequately compensated at Ku. F/D ratios for the 10, 8, and 6 foot reflector sizes are 0.278, and all have prime focus feeds with half power beamwidths of 1.6, 2.8, and 2.8 degrees, respectively.

A REVIEW

Before continuing, let's look back just briefly on the electronics of these TVRO earth stations and perhaps introduce a few new ideas you may not have previously considered. For instance, would you rather have a mechanical or electronic polarity changer, with its automatic action and accompanying 0.5 dB signal loss? And are you going to operate multiple receivers from this single installation or is just one sufficient? And what is the significance of the low-noise amplifier (LNA) anyway?

LNBs and Receivers

Here, the last should be first. The LNB receives its tiny microvolt signals from the receiving dish and is required to amplify carrier and intelligence at some level approaching 50 dB to overcome cable and splitter attenuations so that the downconverter and receiver can supply adequate baseband and rf to video/audio and home receiver inputs. All this must be done with minimum noise figures which are characterized thusly.

Temperature	Noise Figure	Gain
120 °K	1.5 dB	50 dB
100 °K	1.3 dB	50 dB
90 °K	1.2 dB	50 dB
80 °K	1.1 dB	50 dB

LNAs may be bolted directly to the receiving feed or to an orthogonal mode transducer so that two LNAs may be used for both horizontal and vertical satellite polarities. This is especially important in most instances in the event you're using multichannel transponder reception at the same time. LNAs may be either ac or dc powered; but if ac powered than a separate cable is required.

If the foregoing noise figures aren't enough temperature translations for you, let's pass on a convenient equation that can be used in all such computations, whether you're talking about low-noise amplifiers or Gurus. This one is supplied by Microdyne Corp. and is quite useful in working between the two parameters when you have one and not the other. Observe that the traditional 290 °K earth temperature is always part of the equation as well as the usual 10 log power conversion factor. Parametric amplifiers, by the way, exhibit temperatures from 45° to 120 °K, and are less expensive, but easier-to-work-with GaAsFETs nominally can be procured with ranges between 80° and 120 °K. Regardless, here's the simple conversion:

$$NF_{dB} = 10 \log (1 + T/290\ °K)$$

or

$$T\ °K = 290\ (10^{NFdB/10} - 1)$$

Examples are: if $T = 290\ (10^{0.7} - 1) = 1163\ °K$, where 7 dB NF – T °K or, if $NF_{dB} = 10 \log (1 + 135/290) = 1.66$ dB, where 135 $°K = NF_{dB}$.

When considering video receivers, remember they select your satellite transponder, isolate unwanted inputs, sometimes down-convert the information, place audio and video at baseband, and then usually modulate both on a channel 3 or 4 carrier for the ultimate display. Here, you'll look for high quality audio and video, low threshold performance, reliability, and ease of maintenance and positive tuning. Thresholds of good receivers with 30-36 MHz bandwidths are considered normal around 10 or 11 dB C/N. Some satellite transponders may require even more sensitivity and need threshold extenders for another two to five dB—but this technique certainly costs more, if you need or want it. Phase-locked loop modules as additions are not cheap.

Cable and Dividers

Coaxial cable connections between the satellite dish and its LNA or LNB can be of several types, depending on connections, specifications, and satellite downlink frequencies. But if working with standard TV U/V LNC *outputs*, a good quality, low loss, RG6/U type, preferably with both aluminum foil wrap and at least 60% braid shielding is preferred. Steel center conductors with copper coating are also desirable for strength as well as signal transmission characteristics. Belden 9116 or 9914 75/50-ohm, polyethylene jacketed types are good examples. Other companies offer air dielectric coax and foam core cable, both with 50-ohm impedance for the gigahertz connections. M/A-COMs Comm/scope, Inc., in Hickory, North Carolina, has a new line of Quantum Reach coaxial cables with considerably reduced attenuation and low mass dielectric that may be of interest. Regardless, losses should not exceed more than 3-8 dB per hundred feet, regardless of frequency. Air dielectrics, of course, must be pressurized. Cable N, F, or BNC connectors should present as little leakage as possible and be carefully installed.

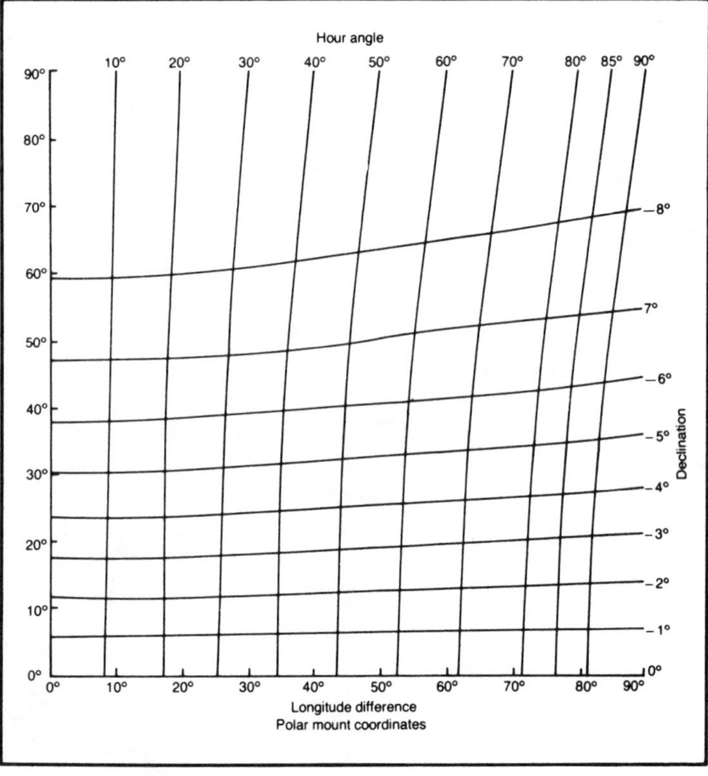

Fig. 6-11. Polar mount coordinates (courtesy of Microdyne).

You should also be very careful in evaluating signal attenuation of any power dividers used. A 4- to 16-port divider often doubles losses from about 7 to 14 dB, enough to easily move weak signals into noise, losing part of all of the incoming information. When using a 50 dB gain LNA, losses should not exceed 20 dB between LNA/LNC and receiver, and 15 dB is a more practical figure. Simply total up cable and other dB drops along the way so the more conservative figure isn't exceeded. Otherwise you'll have to install another expensive amplifier, with perhaps some degradation in system performance.

Lat/Long Coordinates

Obviously, for any installation to acquire satellite signals, the antenna must be pointed in the right direction. From the information in the text, you'll find the various satellite locations. Then, with the aid of three graphs generously supplied by Antennas for Communications, Inc., a subsidiary of Microdyne, Corporation of Ocala Florida, you should be able to obtain your general bearings and quickly find the satellites from there.

These figures are identified under the headings: Elevation-Azimuth Coordinates (Fig. 6-10), Polar Mount Coordinates (Fig. 6-11); and Polarization Rotation (Fig. 6-12). Latitudes are those of the earth station antenna in either hemisphere; longitudes describe differences between the longitude of the earth station antenna and the geostationary satellite's ground position longitude.

To find AZ-EL coordinates, use Fig. 6-10, determine the earth station's longitude and subtract that from the satellite's longitude. Note whether the satellite is East or West of the earth station. At the point of latitude and longitudinal differences read off the elevation and azimuth angles. For all existing satellites, do the same for extreme satellite longitudes and subtract the maximum antenna coordinates. For polar mount coordinates, repeat the same procedure but use Fig. 6-11. For given latitudes, of course, declinations change only slightly for large differences in longitude, consequently the Polar mounts may sweep the sky with only a 0.5° change for 40° of latitude. Don't forget the 180° addition if looking west of due south and subtract from 180° if east of south.

Obviously, feeds must be adjusted for best antenna gain and low crosstalk with orthogonally polarized transponders. Microdyne says this is best done by rotating the feed for a signal null, then back off precisely 90 degrees. But initially, you may use the information in Fig. 6-12 to establish a polarization setting. Most satel-

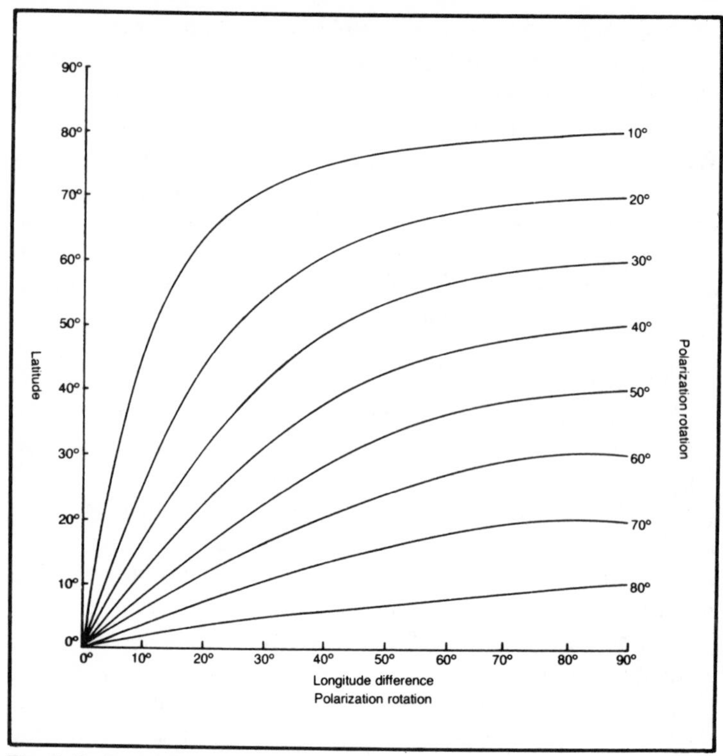

Fig. 6-12. Polarization rotation (courtesy of Microdyne).

lites are horizontally polarized to the equator, but some "may differ from this by as much as 25°." Where this occurs, the satellite's polarization angle is added to the polarization rotation as seen on the graph. Final settings for AZ-EL are done with accurately surveyed azimuth direction and elevation plumb line. Polar mounts require precise north/south directions and a clear east-west horizon.

In making these adjustments, do be careful that you're using the proper polarity, noise control is satisfactorily adjusted, AGC switch correctly set, all cable connections secure, there is adequate ac or dc power, and that any and all peak-level meters are operating correctly. Any one of these incorrectly applied or neglected can cause either excessive noise of total loss of picture. Cable problems may be confirmed or rejected by disconnection and monitoring of the LNA/LNC by a short jumper to a spectrum analyzer. At this point you may also determine C/N, power, interference possibilities, and other parameters that are forever useful in determining the operation of your earth station. In evaluating C/N, always remember that any bandwidth resolution must be taken into account as well as the addition of +2.5 dB for the analyzer's detector. Otherwise the measurement won't be correct.

MICRODYNE

One of the pioneers in terrestrial and space communications, Microdyne and its subsidiary, Antennas For Communications, offer excellent quality equipment for microwave communications, satellite television, and aerospace telemetry. As early as the 1970s, Microdyne was developing and manufacturing receivers for satellite TV, and in 1980 acquired Antennas For Communications (AFC), a precision fiberglass antenna manufacturer. The company now is a primary producer of all major components used in satellite earth stations, operates four manufacturing plants, and employs 480 people in Florida and Maryland.

Not only does Antennas For Communications offer the usual 3, 5, and 7-meter TVRO dishes with prime-focus feeds (Fig. 6-13), but it also has recently announced a multifeed system that will receive up to five satellites on one antenna (Fig. 6-14), provided they are adjacent or nearly adjacent. Microdyne states that existing antennas may be retrofitted in the field with this MSF-16 arrangement. Two or more adjacent satellites will show little or no loss in antenna gain, and even those within 4° spacing have only a boresight gain loss of about 0 dB. At 8° between satellites, gain loss measures less than 0.5 dB. Best of all, Microdyne projects that

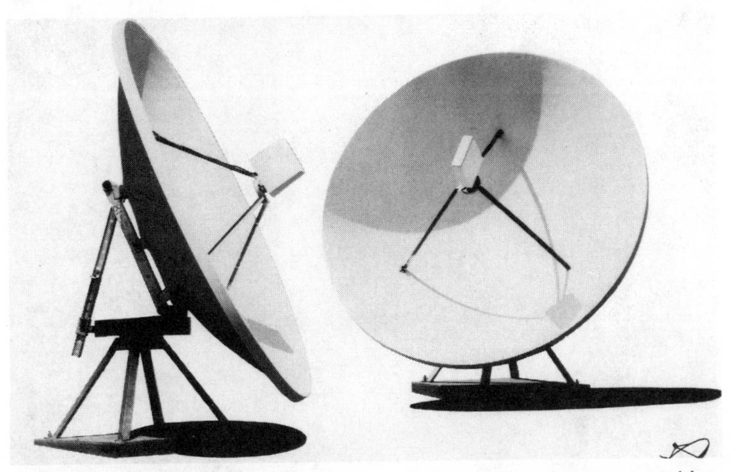

Fig. 6-13. Antennas for communications 3.6- and 5-meter transportables (courtesy of Microdyne).

Fig. 6-14. Antennas for communications dish of the future with multifeed inputs accessing as many as five satellites (courtesy of Microdyne).

this multifeed development can be applied to either 4 or 12 GHz downlink systems and, later, when a few additional facets have been solved, appropriate feeds may handle a mixed bag of the two frequencies with agreeable results. You will, however, need computer assistance to minimize any problem with fade margins, and this Microdyne has developed to assist you. You'll need to tell them your site locations, nearest city, satellite location and name, and measured EIRP for each transponder you wish accessed. In return, you will receive current footprint information, satellite link calculations, and computer readouts of predicted gains and patterns.

Standard Antennas

AFC's 3.66- and 5-meter antennas are both transportable on ordinary 4-wheel carriers and will handle 3.7-4.2 and 11.7-12.2 GHz downlink frequencies. Input flange gain measurements for the smaller of the two is 41 dB at 4 GHZ, and 50.7 dB at 12 GHz. The 5-meter unit exceeds these figures by about 3 dB in both categories, while its 1.1° beamwidth compares with 1.4° for the 3.66-meter dish. Wind loads are 40 and 60 mph (no ice), respectively, with very short installation times. Sidelobes are within the 32-35 log θ specified by the Federal Communications Commission. Trailers required for pickup should be 8-feet wide and 18 and 20 ft., respectively.

Multifeeds have already been adapted to the 5-meter dishes, and work is continuing on those measuring 12 meters. Other than torus types, these two Microdyne antennas are the initial parabolics we know of which can handle two or more simultaneous feeds successfully. Several other competitors, we understand, have already put their efforts on the back burner, or have given up entirely. We should be hearing considerably more about such techniques as U.S. Direct Broadcast Satellite (DBS) times approach in 1985 and 1986.

SATCOM F-3R

Meanwhile, you may want just a little additional information in locating one of the more popular satellites—RCA's SATCOM F-3R—courtesy Microdyne once more. Given in terms of true bearings for north latitude and west longitude, the graph shown in Fig. 6-15 should be relatively self-explanatory. Elevation is read via diagonal small numbers from 8 through 48, and azimuth from large numbers between 195 and 250. A straightedge, and your earth station position does the rest. SATCOM F-3R, of course, remains at 131° in equatorial orbit.

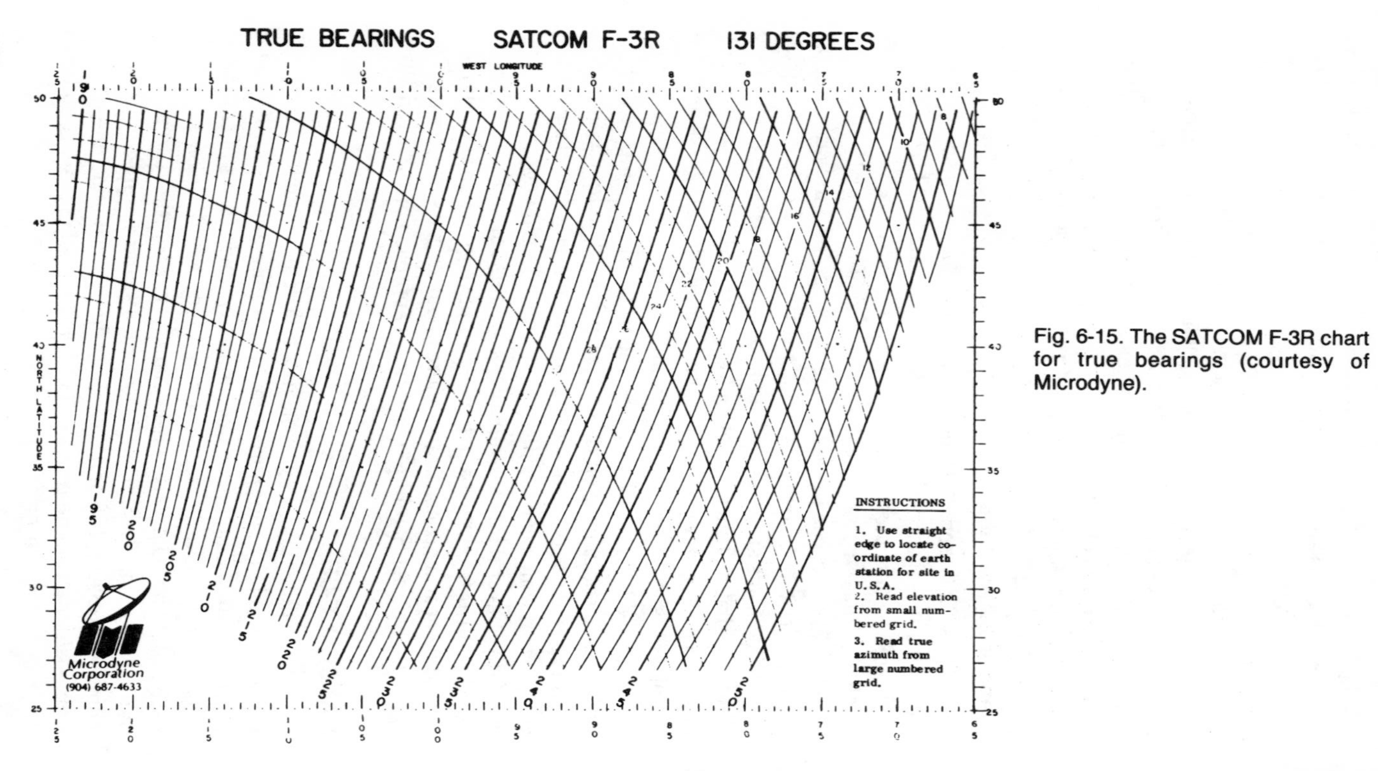

Fig. 6-15. The SATCOM F-3R chart for true bearings (courtesy of Microdyne).

WILKOM SUPPLIES MULTIFEEDS

One of the more successful suppliers of multifeed antennas is the Wilkom Company of Pocahontas, Arkansas (Fig. 6-16). Shown in a triple feed arrangement, the Wilkom equipment can handle F/D ratios between 0.3 and 0.6 in an off-axis set of ± 8 degrees before undesirable sidelobe skewing. In this arrangement, five satellites may be received on either an existing antenna or a new dish especially fitted for this purpose.

Not only is it possible to "see" C-band satellites, but those in the Ku- and K-bands can also be included as long as downconversions to available receivers are installed. At present, Wilkom's multifeeds permit an arc of 16 degrees which, it is said, will include SATCOM 3R, COMSTAR D4, WESTAR 5 immediately, and Galaxy 1 and Spacenet when they are launched. Canadian geosynchronous satellites between 104 and 117.5 degrees also may be captured.

Owner Bryan Wilkes says that "future feed developments will be important to the success of the multifeed systems . . . since with the advent of 2-degree spacing, present feeds will become inadequate." Feed designs up to a limit of 10 degrees to permit more efficient off-axis illumination are even now under development. Individuals, contractors, and manufacturers may all make use of this multisignal capture arrangement. Wilkes claims an off-axis attenuation of only 0.5 dB for 4 degrees; 1.5 dB for 6 degrees; and 3 dB for 8 degrees. Presumably wider angles would produce proportional losses.

MAGNETIC DECLINATION CHART

The illustration in Fig. 6-17 is another Microdyne aid in estab-

Fig. 6-16. A Wilkom multifeed antenna.

lishing East and West magnetic declination corrections for your compass readings in different parts of the U.S. If you lived in the West, for instance, and your north latitude and longitude is 37° and 105°, respectively, a straightedge would coordinate your earth station for an East correction factor of 12°. Then go back to Fig. 6-13 and use the straightedge again to locate the same true bearing coordinates. You will find by following the intersect point down that the bearing will appear as 219°. Now, since you've found an East correction of 12°, this must be subtracted from 219°, giving a true bearing (reading) of 270°. This will be the correct reading for F-3R if your coordinates are as given.

Let's try another one, this time on the other side of 0 for a West correction. How about 40° north latitude and 74° longitude. On the Magnetic Variation chart you'll find a correction factor of 11°, which must be *added* to the position location on Fig. 6-13. Reading down from the coordinates, this amounts to 247.2° and, with the addition of 11°, the total is approximately 258° bearing true. So without having to call the nearest airport or haunt some government office, you have your corrected bearings for true azimuth. The remainder of your pointing problems, including the angle of elevation to the satellite, may now be easily derived from Figs. 6-10 through 6-12.

From all of the information contained in the chapter, your satellite aiming difficulties should be over—at least we hope so. Thereafter, may the rest of your problems all be little ones. But, do remember that all compass readings and computations are done first then the correction factors applied as the final step in your AZ-EL or polar-mount installations.

MORE COORDINATES

But just to add a little more icing on the cake's top layer, let's also include a recent set of coordinates from M/A-COM showing not only declination angles, but elevations, azimuth and magnetic compass corrections for North American states and even many cities. With M/A-COM this chart is standard issue—but some of you may not have it at all. See Fig. 6-18.

VIDEO SERVICES

To give you a sample of the video services that are available via satellite TV, I have included several pages from *Tuning In Home Satellite TV*. See Fig. 6-19. This material is reprinted by permission of Commtek Publishing.

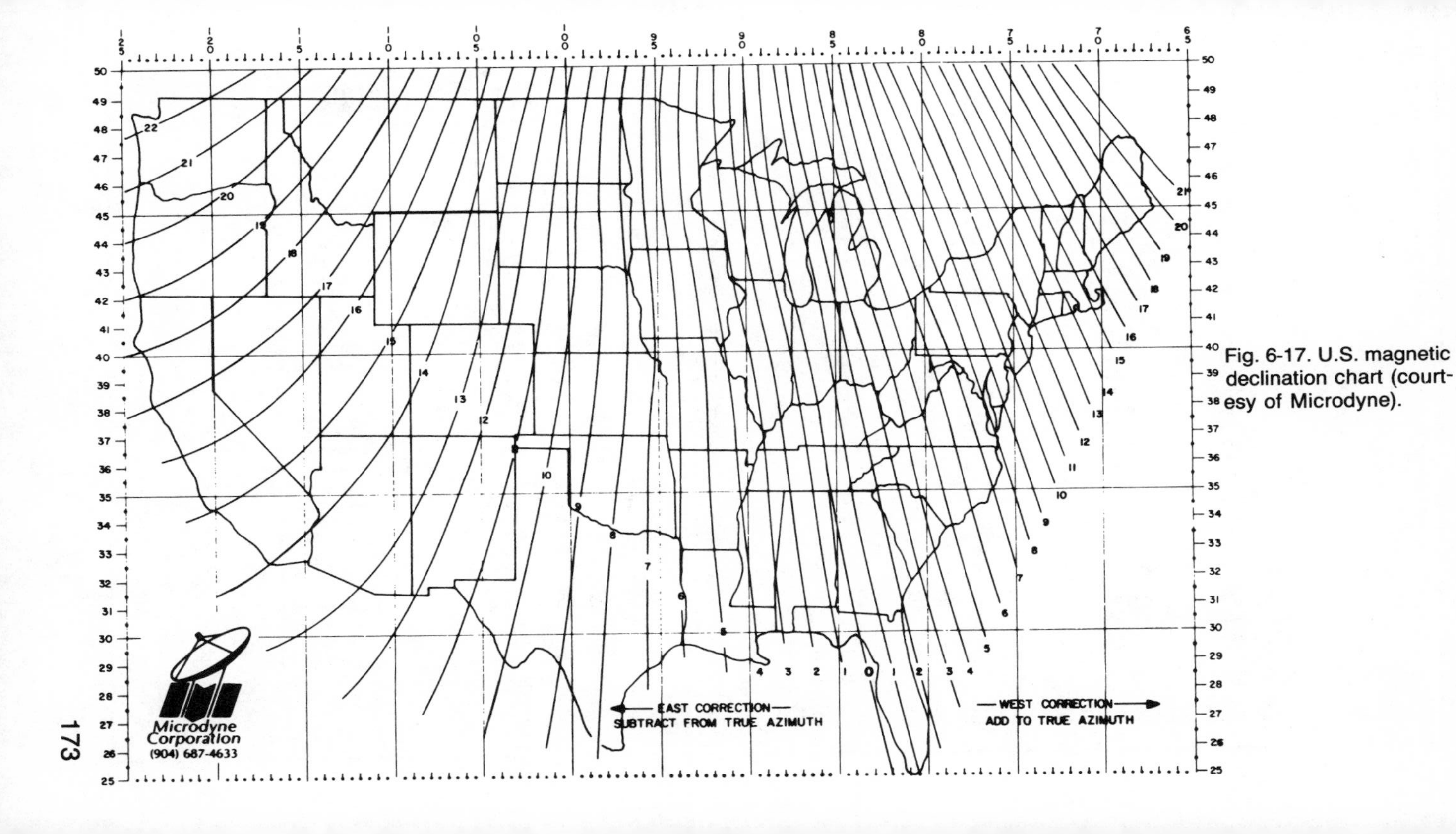

Fig. 6-17. U.S. magnetic declination chart (courtesy of Microdyne).

State	City	Declination Angle	GALAXY 1 (H1) Elevation	GALAXY 1 (H1) Azimuth	SATCOM 4(F4) Elevation	SATCOM 4(F4) Azimuth
1. Alabama	Birmingham	5.4	26.8	242.9	50.0	175.0
2. Arizona	Phoenix	5.4	44.5	216.1	41.0	135.0
3. Arkansas	Little Rock	5.7	30.4	237.4	48.0	165.0
4. California	San Francisco	6.1	44.6	198.5	30.0	127.0
5. California	Los Angeles	5.5	47.0	206.7	36.0	129.0
6. Colorado	Denver	6.3	35.2	220.9	38.0	148.0
7. Connecticut	Hartford	6.6	12.6	250.0	40.0	195.0
8. Rhode Island	Providence	6.6	11.6	250.9	40.0	198.0
9. Delaware	Wilmington	6.3	15.4	248.6	43.0	191.0
10. Maryland	Baltimore	6.2	16.4	247.9	44.0	190.0
11. Florida	Jacksonville	5.0	24.0	248.7	55.0	182.0
12. Florida	Miami	4.4	24.3	252.3	59.0	187.0
13. Georgia	Atlanta	5.5	24.8	244.7	51.0	178.0
14. Idaho	Boise	6.8	36.7	205.0	30.0	136.0
15. Illinois	Chicago	6.6	23.0	237.5	41.0	173.0
16. Indiana	Indianapolis	6.3	23.1	239.9	45.0	175.0
17. Iowa	Des Moines	6.6	27.1	232.0	40.0	164.0
18. Kansas	Witchita	6.1	32.1	230.6	44.0	158.0
19. Kentucky	Louisville	6.1	23.6	241.1	46.0	175.0
20. Louisiana	Baton Rouge	5.0	32.0	241.3	53.0	185.0
21. Maine	Bangor	6.9	8.8	252.0	36.0	199.0
22. Massachusetts	Boston	6.6	11.2	251.0	40.0	196.0
23. Michigan	Grand Rapids	6.7	21.1	238.8	40.0	176.0
24. Michigan	Marquette	7.1	20.0	235.3	36.0	175.0
25. Minnesota	St. Paul	6.9	24.5	230.8	37.0	186.0
26. Mississippi	Jackson	5.3	30.1	240.9	51.0	167.0
27. Missouri	Kansas City	6.2	29.3	232.5	43.0	163.0
28. Missouri	St. Louis	6.2	26.6	236.9	44.0	169.0
29. Montana	Great Falls	7.2	31.2	209.6	31.0	143.0
30. Nebraska	Omaha	6.5	28.8	229.9	40.6	160.8
31. Nevada	Las Vegas	5.8	43.6	210.1	36.0	134.0
32. New Hampshire	Manchester	6.6	11.3	250.5	39.1	196.7
33. New Jersey	Trenton	6.3	14.7	249.0	42.0	194.0
34. New Mexico	Albuquerque	5.7	40.0	222.0	42.0	142.0
35. New York	Syracuse	6.7	14.5	246.8	40.0	190.0
36. North Carolina	Hickory	5.7	21.3	246.4	48.7	183.4

37. North Dakota	Bismark	7.1	27.3	221.9	33.0	183.4
38. Ohio	Columbus	6.3	20.8	242.5	44.0	180.0
39. Oklahoma	Oklahoma City	5.7	33.7	231.9	47.0	157.0
40. Oregon	Eugene	6.8	38.1	195.5	26.0	130.0
41. Pennsylvania	Harrisburg	6.3	16.2	247.3	42.0	189.0
42. South Carolina	Columbia	5.5	22.0	247.1	50.0	184.0
43. South Dakota	Rapid City	6.8	30.7	220.6	35.0	152.0
44. Tennessee	Memphis	5.7	28.5	239.2	49.0	168.0
45. Tennessee	Knoxville	5.8	23.4	243.8	48.0	178.0
46. Texas	Abilene	5.3	37.4	231.8	48.0	152.0
47. Utah	Salt Lake City	6.5	37.9	211.9	34.0	140.0
48. Vermont	Montpelier	6.8	11.6	249.2	37.0	194.0
49. Virginia	Richmond	6.0	17.7	248.1	46.2	189.1
50. Washington	Spokane	7.2	33.0	201.9	27.0	137.0
51. West Virginia	Charleston	6.2	20.5	244.4	45.0	183.0
52. Wisconsin	Green Bay	6.8	21.7	235.9	38.5	187.1
53. Wyoming	Sheridan	6.8	31.9	215.9	33.2	147.8

Magnetic Compass Corrections for North America

Alabama	2E	Kentucky	1E	North Dakota	11E
Alaska	26E	Louisiana	6E	Ohio	3W
Arizona	14E	Maine	20W	Oklahoma	9E
Arkansas	6E	Maryland	8W	Oregon	20E
California	17E	Massachusetts	15W	Pennsylvania	8W
Colorado	14E	Michigan	3W	Rhode Island	15W
Connecticut	13W	Minnesota	6E	South Carolina	2W
Delaware	10W	Mississippi	5E	South Dakota	11E
Washington D.C.	8W	Missouri	6E	Tennessee	1E
Florida	2E	Montana	18E	Texas	10E
Georgia	0	Nebraska	11E	Utah	15E
Hawaii	11E	Nevada	17E	Vermont	15W
Idaho	19E	New Hampshire	16W	Virginia	6W
Illinois	2E	New Jersey	11W	Washington	22E
Indiana	0	New Mexico	13E	West Virginia	5W
Iowa	6E	New York	10W	Wisconsin	2E
Kansas	9E	North Carolina	6W	Wyoming	13E
Alberta	22E	Manitoba	10E	Saskatchewan	17E
British Columbia	23E	Ontario	8W	Quebec	17W

Fig. 6-18. Satellite coordinates.

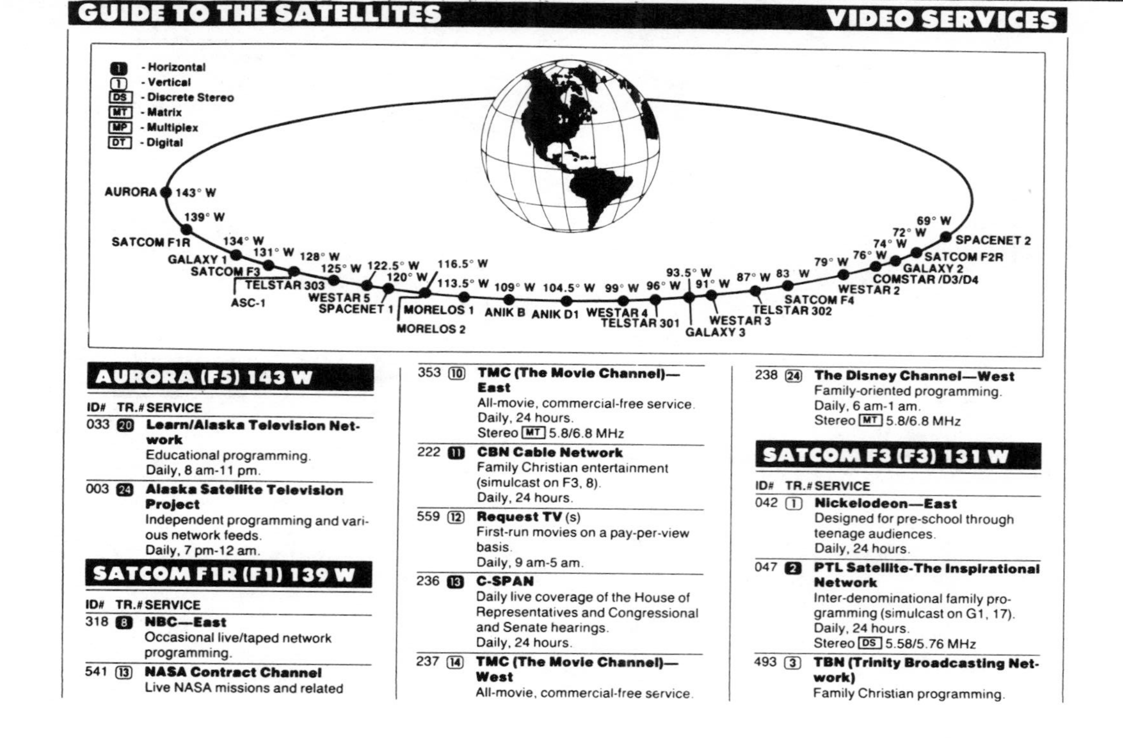

AURORA (F5) 143 W

ID#	TR.#	SERVICE
033	20	**Learn/Alaska Television Network** Educational programming. Daily, 8 am-11 pm.
003	24	**Alaska Satellite Television Project** Independent programming and various network feeds. Daily, 7 pm-12 am.

SATCOM F1R (F1) 139 W

ID#	TR.#	SERVICE
318	8	**NBC—East** Occasional live/taped network programming.
541	13	**NASA Contract Channel** Live NASA missions and related
353	10	**TMC (The Movie Channel)—East** All-movie, commercial-free service. Daily, 24 hours. Stereo MT 5.8/6.8 MHz
222	11	**CBN Cable Network** Family Christian entertainment (simulcast on F3, 8). Daily, 24 hours.
559	12	**Request TV** (s) First-run movies on a pay-per-view basis. Daily, 9 am-5 am.
236	13	**C-SPAN** Daily live coverage of the House of Representatives and Congressional and Senate hearings. Daily, 24 hours.
237	14	**TMC (The Movie Channel)—West** All-movie, commercial-free service.
238	24	**The Disney Channel—West** Family-oriented programming. Daily, 6 am-1 am. Stereo MT 5.8/6.8 MHz

SATCOM F3 (F3) 131 W

ID#	TR.#	SERVICE
042	1	**Nickelodeon—East** Designed for pre-school through teenage audiences. Daily, 24 hours.
047	2	**PTL Satellite-The Inspirational Network** Inter-denominational family programming (simulcast on G1, 17). Daily, 24 hours. Stereo DS 5.58/5.76 MHz
493	3	**TBN (Trinity Broadcasting Network)** Family Christian programming.

events.
Occasional during NASA missions.

356 15 **Yonkers Horse Racing** (s)
Occasional live horse racing.

GALAXY 1 (G1) 134 W

ID# TR.# SERVICE

219 2 **TNN (The Nashville Network)**
Country-oriented videos and programming.
Daily, 9 am-3 am.
Stereo DS 5.58/5.76 MHz

443 3 **WGN—Chicago**
Featuring movies, sports, specials, and syndicated programs.
Daily, 24 hours.

235 4 **The Disney Channel—East**
Family entertainment.
Daily, 6 am-1 am.
Stereo MT 5.8/6.8 MHz

354 5 **Showtime—East**
First-run movies, sports, and specials.
Daily, 24 hours.

051 6 **SIN (Spanish International Network)**
Spanish-language programming.
Daily, 24 hours.

220 7 **CNN (Cable News Network)**
Live round-the-clock news coverage and features.
Daily, 24 hours.

410 8 **CNN Headline News**
Continuously updated 30-minute wheel of hard news.
Daily, 24 hours.

424 9 **ESPN**
Professional, college, and amateur sporting events.
Daily, 24 hours.
Daily, 24 hours.
Stereo MT 5.8/6.8 MHz

224 15 **WOR—New York**
Sports, movies and syndicated shows.
Daily, 24 hours.

470 17 **PTL Satellite—The Inspirational Network**
Interdenominational family programming (simulcast on F3, 2).
Daily, 24 hours.
Stereo DS 5.58/5.76 MHz

346 18 **WTBS—Atlanta**
Family-directed programming including sports, movies, syndication, national and international news.
Daily, 24 hours.
Stereo DS 5.58/5.76 MHz

218 19 **Cinemax—East (s)**
First-run movies, sports, and specials.
Daily, 24 hours.

029 20 **GalaVision**
Spanish-language programming.
Weekends, 11 am-4 am. Weekdays, 4 pm-4 am.

592 21 **USA Network—East**
Sports-focused network also featuring special-interest programs.
Daily, 24 hours.

542 22 **AVN (Alternate View Network)**
Educational and religious programming focusing on religious and ethical issues.
Sundays 10:45 am-2:45 pm.

223 23 **HBO (Home Box Office)—East (s)**
First-run movies, sports, and specials.
Daily, 24 hours.
Daily, 24 hours.

259 4 **FNN (Financial News Network)/SCORE**
Live business and financial news.
Sports news and occasional sporting events.
FNN-Weekdays, 6 am-8 pm.
SCORE-Weekdays, 8 pm-6 am.
Weekends, 24 hours.

592 5 **Viewer's Choice** (s)
First-run movies on a pay-per-view basis.
Daily, occasional.

388 6 **SPN (Satellite Program Network)**
Movies, entertainment, how-to's, financial self-help programs, and video music.
Daily, 24 hours.

442 7 **ESPN Blackout Network**
Occasional programming that can be used in place of sports events that are blacked out.

015 8 **CBN Cable Network**
Family Christian entertainment (simulcast on G1, 11).
Daily, 24 hours.

058 9 **USA Network—West**
Sports-focused network also featuring special-interest programs.
Daily, 24 hours.

104 10 **Showtime—West**
Commercial-free movies, sports, and specials.
Daily, 24 hours.

037 11 **MTV: Music Television**
Advertiser-supported video music channel.
Daily, 24 hours.

Fig. 6-19. Video services. Reprinted by permission. Copyright 1986, CommTek Publishing.

VIDEO SERVICES — GUIDE TO THE SATELLITES

Reprinted from Satellite ORBIT magazine

Stereo DS 5.58/6.62 MHz
Stereo DT 7.4 MHz

027 12 **EWTN (Eternal Word Television Network)**
Catholic spiritual growth programming with family entertainment.
Daily, 8 pm-12 am.

105 13 **HBO (Home Box Office)—West (s)**
Movies, sports, and entertainment specials.
Daily, 24 hours.

586 14 **Open-net**
Public affairs programming from North Carolina.
Fridays, 8 pm-10 pm.

382 15 **VH-1/Video Hits One**
Advertiser-supported video music service.
Daily, 24 hours.
Stereo DT 5.8 MHz

032 16 **HTN (Home Theater Network)**
Family programming, including movies, specials, and travel.
Daily, 4 pm-4 am.
Stereo MP 6.8 MHz

002 16 **TLC (The Learning Channel)**
Learning programs and college credit courses for adults.
Daily, 6 am-4 pm.

216 17 **Lifetime**
Programming about health, relationships, self-development, and other self-help topics.
Daily, 24 hours.

048 18 **Reuter Monitor Service** (s)
News and price retrieval in commodity, money, and investment markets.
Weekdays, 4 am-8 pm.

programming.
Daily, 24 hours.
Stereo DS 5.58/5.76 MHz

499 2 **SelecTV**
Feature films and adult movies.
Daily, 24 hours.

496 18 **American Extasy**
Advertiser-supported X-rated movies.
Daily, 11 pm-4 am.

501 22 **KTVT—Dallas**
Featuring outdoor and specialty sports, movies, and syndicated programming.
Daily, 24 hours.

WESTAR 5 (W5) 122.5 W

ID# TR.# SERVICE

199 2 **The University Network (The Unchannel)**
Nonsecular programming with Dr. Gene Scott.
Daily, 24 hours.

292 7 **CBS—Contract Channel**
Occasional live/taped network programming.

494 8 **PASS (Pro Am Sports System)**
Occasional sports service from Michigan, Ohio, and Indiana.

593 20 **Telebet Racings** (s)
Occasional live horse racing.

291 22 **The Meadows Racing Network**
Harness racing
Fri., Sat. 7 pm-11:30 pm. Sun 6:30 pm-11 pm.

497 24 **The FUN Channel** (s)
XXX-rated adult entertainment.
Daily 11 p.m.-6 a.m.

539 14 **XHITM—Me**
Spanish languag
programming origi
Mexico.
Daily, 24 hours.
Mono audio at 6.2 MHz

538 22 **XHDF—Mexico City**
Spanish language programming originating from Mexico.
Daily, 24 hours.
Mono audio at 6.2 MHz

537 24 **XETV—San Diego/Tijuana**
CBS network affiliate.
Daily, 9 am-5 am.
Mono audio at 6.2 MHz

ANIK B (AB) 109 W

No Active Video

ANIK D1 (AD1) 104.5 W

ID# TR.# SERVICE

320 2 **TSN (The Sports Network)**
Covering a variety of sports programming.
Daily, 24 hours.

334 6 **MuchMusic**
Videos, concerts, and special events.
Daily, 24 hours.
Stereo DS 5.41/6.17 MHz

017 8 **CHCH—Hamilton** (s)
Independent service with varied programming.
Daily, 6 am-2 am.

248 9 **WDIV—Detroit** (s)
NBC affiliate feeds and independent programming.
Daily, 24 hours.

313 18 **NJT (National Jewish Television)**
Family Jewish programming.
Sundays, 1 pm-4 pm.

441 20 **BET (Black Entertainment Television)**
Black-oriented movies, specials, and sports.
Daily, 24 hours.
Stereo DS 5.58/5.76 MHz

059 21 **The Weather Channel**
Live, constantly updated, national weather service.
Daily, 24 hours.

491 22 **HSN (Home Shopping Network)**
Designed for shopping at home via television (Simulcast on F4,1).
Daily, 24 hours.

057 22 **USA Blackout Network**
Occasional programming that can be used in place of sports events that are blacked-out.

106 23 **Cinemax—West (s)**
Commercial-free movies, sports and specials.
Daily, 24 hours.

387 24 **A&E (Arts & Entertainment)**
Featuring Broadway plays, dramatic and comedy series.
Daily, 8 am-4 am.
Stereo DS 5.58/5.76 MHz

ASC-1 (AS1) 128W

No Active Video

TELSTAR 303 (T3) 125 W

ID#	TR.#	SERVICE
500	1	**CMTV (Country Music Television)** Country-oriented videos and

498 24 **PPV—The Pay-Per-View Channel** (s)
XXX-rated adult movies and other special events on a pay-per-view basis.
Fri. and Sat. 9 pm-11 pm.

SPACENET 1 (S1) 120 W

ID#	TR.#	SERVICE
326	11	**HSN (Hospital Satellite Network)** (s) Medical programming and uplifting movies. Daily, 24 hours.
336	15	**ACTS (American Christian Television System)** Family Christian entertainment. Daily, 24 hours.
451	21	**VMT (Vanderbilt Medical Television)** (s) Medical programming for doctors and hospitals. Tues., Wed., Fri., 9 am-10 am. Thurs. and Sat. 9 am-10:30 am.
460	21	**BTN (Baptist Telecommunications Network)** (s) Religious programming for churches. Daily, 11 am-4 pm.

MORELOS 2 (M2) 116.5 W

No Active Video

MORELOS 1 (M1) 113.5 W

ID#	TR.#	SERVICE
540	12	**XEW—Mexico City** Spanish language programming originating from Mexico. Daily, 24 hours. Mono audio at 6.2 MHz

408 10 **WXYZ—Detroit** (s)
ABC affiliate feeds and independent programming.
Daily, 24 hours.

249 11 **CBC North—Pacific**
English-language network feeds.
Daily, 11 am-3 pm.

056 14 **TCTV (Telemedia Communications Television)** (s)
French-language TVA network programming.
Daily, 8 am-1 am.

250 15 **CBC French—Eastern**
French-language network feeds.
Daily, 11 am-3 am.

264 16 **Canadian Parliamentary Network**
Occasional French-language live coverage from the Canadian House of Commons.

213 18 **CITV—Edmonton** (s)
Independent service with varied programming.
Sundays, 8 am-5 am.
Mon.-Sat., 8 am-6 am.

344 19 **CBC North—Eastern**
English-language network feeds.
Daily, 8 am-12 pm.

251 20 **CBC—Montreal**
CBC English-language affiliate.
Daily, 8 am-2 am.

252 21 **WTVS—Detroit** (s)
PBS affiliate feeds and independent programming.
Daily, 24 hours.

007 22 **BCTV—British Columbia** (s)
CTV affiliate feeds and independent programming.
Daily, 9 am-5 am.

253 23 **WJBK—Detroit** (s)
CBS affiliate feeds and independent

Reprinted by Permission. Copyright 1986, CommTek Publishing.

programming.
Daily, 24 hours.

014 (24) **Canadian Parliamentary Network**
Occasional English-language live coverage from the Canadian House of Commons.

254 (24) **CBC North—Eastern**
French-language network feeds.
Daily, 8 am-12 pm.

WESTAR 4 (W4) 99 W

ID#	TR.#	SERVICE
416	6	**Atlantic City Horse Racing** (s) Occasional live horse racing.
415	6	**Word of Faith Satellite Network** Occasional interdenominational family programming.
282	11	**CTNA (Catholic Telecommunications Network of America)** (s) Family Christian programming. Daily, 1 pm-4 pm.
045	15	**PBS (Public Broadcasting Service)** Eastern programming. Sun. 9 am-2 am, Mon.-Fri., 10 am-2 am, Sat., 8 am-12 am.
332	16	**CNN (Cable News Network)** Occasional incoming regional bureau news feeds.
242	17	**PBS (Public Broadcasting Service)** Central programming. Sat., 12 pm-4 am, Mon.-Fri., 8 am-1 am, Sun., 9 am-11 pm.
301	20	**ABC—Contract Channel** Occasional live/taped network related events from the Senate Daily, 3:45 pm-4:45 pm

WESTAR 3 (W3) 91 W

ID#	TR.#	SERVICE
339	5	**CNN (Cable News Network)** Occasional incoming regional bureau news feeds.
303	21	**TIN (The Independent Network)** Occasional national newscasts for independent stations.

TELSTAR 302 (T2) 86 W

ID#	TR.#	SERVICE
492	10	**ABC—West** Live/taped network feeds.
457	16	**CBS—Affiliate feeds** Occasional live/taped network programming.
458	20	**CBS—Central** Live/taped network feeds (simulcast on T1, 2). Sun. 10 am-12 am, Weekdays, 8 am-5 am, Sat., 8 am-12 am.

SATCOM F4 (F4) 83 W

ID#	TR.#	SERVICE
550	1	**HSN (Home Shopping Network)** Designed for shopping at home via television. (Simulcast on F3, 22). Daily, 24 hours.
551	2	**The C.O.M.B. Value Network** Shopping at home via television.
425	11	**HSE (Home Sports Entertainment)** Regional sports service serving Texas, Louisiana, Arkansas, Oklahoma, and New Mexico. Daily, 6 pm-1 am.
046	12	**The Playboy Channel** Adult-oriented game shows, music specials, and films. Daily, 8 pm-6 am.
548	12	**The Heartbeat Network** Religious and family-oriented programming. Daily, 6 am-8 pm.
328	13	**NESN (New England Sports Network)** Occasional regional service covering the New England states.
425	14	**HSE (Home Sports Entertainment)—Affiliate feeds** Occasional sports feeds.
560	14	**Playboy Private Ticket** Occasional adult programming.
557	15	**SNL (Success-N-Life Network)** Educational, exercise, and variety Christian programming. Spanish audio on special telecasts. 6.2 MHz.
389	16	**The Silent Network** Programming in sign language and voice. Sat. 9:30 am-11:30 am.
588	18	**Hit Video USA** Contemporary hit music videos. Daily, 24 hours.
289	19	**WPIX—New York** Family-oriented programming, movies, and sports. Daily, 24 hours.
547	20	**Prime Ticket**

programming.

243 21 **PBS (Public Broadcasting Service)**
Mountain programming.
Sun., 8 am-2 am, Mon.-Fri., 7 am-4 am., Sat., 1 pm-4 pm.

244 23 **PBS (Public Broadcasting Service)**
Pacific programming.
Sun., 10 pm-11 pm, Mon.-Fri., 11 am-12 pm.

TELSTAR 301 (T1) 96 W

ID# TR.# SERVICE

319 2 **CBS—Central**
Live/taped network feeds (simulcast on T2, 20).
Sun., 10 am-12 am, Weekdays, 8 am-5 am, Sat., 8 am-12 am.

263 9 **Wold Communications**
Occasional news, sports, network, and syndicated programming.

001 10 **ABC—Central**
Live/taped network feeds.
Weekends, 8 am-12 am. Weekdays, 8 am-5 am.

302 12 **ABC—Contract Channel**
Occasional live/taped network feeds.

377 23 **Wold Communications**
Occasional news, sports, network, and syndicated programming.

GALAXY 3 (G3) 93.5 W

ID# TR.# SERVICE

593 8 **Senate Republican Conference News**
Republican conference news and
Weekdays, 2:30 pm-7:30 pm.
Weekends, 12:00 pm-5:00 pm.

011 2 **Bravo**
International films, award-winning film festivals, and the performing arts.
Weekends, 5 pm-6 am.
Weekdays, 8 pm-6 am.
Stereo MP 5.8/6.8 MHz

428 4 **Nickelodeon—West**
Designed for pre-school through teenage audiences.
Daily, 24 hours.

379 5 **ABC—Affiliate feeds**
Occasional live/taped network feeds.

478 6 **BizNet, the American Business Network**
Business and public affairs programming.
Weekdays, 6 am-2 pm.

546 6 **MSG (Madison Square Garden)**
Sports and variety programming.
Weekdays, 7 pm-4 am. Weekends, approx. 2 pm-4 am.

041 7 **NCN (National Christian Network)**
Multi-denominational religious programs.
Daily, 24 hours.

564 8 **The People's Choice (s)**
Occasional first-run movies on a pay-per-view basis.

284 9 **SportsVision**
Occasional regional sports network from the Chicago area.

329 10 **AMC (American Movie Classics)**
Covering movies made before 1970.
Daily, 8 pm-6 am.

Occasional nightly sports programming from the Los Angeles Forum.

471 21 **The Nostalgia Channel**
Movies, comedy, and variety.
Daily, 24 hours.

585 22 **HTS (Home Team Sports)**
Sports service from the Capitol region.
Daily, 24 hours.

558 24 **Nightline Television Network**
Religious and variety programming.
Daily, 8 pm-11 pm.

WESTAR 2 (W2) 79 W

No Active Video

COMSTAR D3/D4 76W

No Active Video

GALAXY 2 (G2) 74 W

Occasional video only on transponders 15, 16, 17, and 18.

SATCOM F2R (F2) 72 W

536 13 **NASA Contract Channel**
Live NASA missions and related events.
Daily, 24 hours during NASA missions.

453 20 **AFRTS (American Forces Radio and Television)**
Independent programming and various network feeds.
Daily, 24 hours

SPACENET 2 (S2) 69 W

No Active Video

Reprinted by Permission. Copyright 1986, CommTek Publishing.

Reprinted from Satellite ORBIT magazine

SATCOM F1R (F1) 139 W

ID#	TR.#	SERVICE
440	3	**CNN Radio Network** Complete news updates in 30 minute cycles, 24 hours daily in monaural 6.3 MHz.

GALAXY 1 (G1) 134 W

ID#	TR.#	SERVICE
454	3	**Satellite Radio Network** Gospel music, news, and ministries available 24 hours daily. 6.2 MHz. (simulcast on F3, 2)
445	3	**Moody Broadcasting Network** Variety religious programming available 24 hours daily. Stereo DS 5.4/7.92 MHz
446	3	**Seeburg/Lifestyle Music** Monaural commercial-free music broadcast 24 hours daily. 7.695 MHz.
450	3	**Seeburg/Lifestyle Adult Contemporary** Monaural commercial-free vocal hits broadcast 24 hours daily. 8.145 MHz.
447	3	**Cable SportsTracker** Alphanumeric sports information transmitted 24 hours daily.
448	3	**UPI DataCable** 24 hour text news service in English and Spanish. 7.257 MHz.
449	3	**WFMT—Chicago** Fine arts/classical music broadcast 24 hours daily. Stereo DS 6.3/6.48 MHz
426	15	**The Greek Network** Music, news, sports, and specials in Greek. Broadcast 24 hours daily. 7.335 MHz.
427	15	**The Italian Network** Music, news, sports, and specials in Italian. Broadcast 24 hours daily. 7.515 MHz.
479	15	**WQXR—New York** Classical music broadcast 24 hours daily. Stereo DS 6.3/6.48 MHz
584	18	**Sastellite Data Network** News, business, information, stock and commodity quotes in text.

SATCOM F3 (F3) 131 W

ID#	TR.#	SERVICE
094	2	**Satellite Radio Network** Gospel music news and ministries available 24 hours daily. 6.2 MHz. (simulcast on G1, 3)
535	3	**Genesis Cable Storytime** Children's storybooks in color computer graphics with silent text. Available 24 hours daily on 7.237 MHz.
495	4	**Satellite Jazz Network** Jazz programming from KKGO, Los Angeles, broadcast 24 hours a day. Stereo DS 5.58/5.76 MHz
366	6	**StarShip Adult Contemporary** Traditional MOR 24 hours daily. Stereo DS 5.58/5.76 MHz
475	6	**StarShip Country and Western** Country music 24 hours daily. Stereo DS 5.4/5.94 MHz.
067	14	**CNN Radio Ne** Complete news up ute cycles, 24 hours monaural. 6.3 MHz.
068	16	**Commodity Commun Corporation** (s) Scrambled quote service avai am-3:30 pm in monaural. 6.2 M
487	19	**Studioline/Rhythmix** Urban contemporary/dance music broadcast 24 hours daily. Stereo DS 5.76/5.94 MHz
488	19	**Studioline/Album Tracks** Selections from hit albums broadcast 24 hours daily. Stereo DS 5.40/5.58 MHz
489	19	**Studioline/All That Jazz** Big band and jazz music broadcast 24 hours daily. Stereo DS 6.12/6.3 MHz
490	19	**Studioline/Specials Channel** Specials and variety music broadcast 24 hours daily. Stereo DS 7.38/7.56 MHz

TELSTAR 303 (T3) 125 W

ID#	TR.#	SERVICE
534	18	**FM America** Easy listening format. Daily, 12 am-9 pm. Dish owner's talk show. Daily, 9 pm-12 am. 6.2 MHz.
587	18	**North America 1** Contemporary hits, easy listening, albums, and talk shows available 24

466 3 **Rock 'n' Hits**
Rock music broadcast 24 hours daily.
Stereo DT

463 3 **Country Coast-to-Coast**
Country music broadcast 24 hours daily.
Stereo DS 5.94/6.12 MHz

464 3 **Stardust**
MOR from the 1940s to the 80s available 24 hours daily.
Stereo DS 7.38/7.56 MHz

465 3 **StarStation**
Adult contemporary broadcast 24 hours daily.
Stereo DS 5.58/5.76 MHz

397 7 **CNN Radio Network**
Complete news updates in 30 minute cycles, 24 hours daily. 6.3 MHz.

482 13 **Studioline/All Time Favorites**
Adult contemporary music broadcast 24 hours daily.
Stereo DS 5.22/5.40 MHz

483 13 **Studioline/Easy Listening**
Easy listening broadcast 24 hours daily.
Stereo DS 5.58/5.76 MHz

484 13 **Studioline/The Classics**
Classical music broadcast 24 hours daily.
Stereo DS 5.94/6.12 MHz

485 13 **Studioline/Hit Country**
Country and western music broadcast 24 hours daily.
Stereo DS 6.30/6.48 MHz

486 13 **Studioline/Super Hits**
Top 40 music broadcast 24 hours daily.
Stereo DS 7.38/7.56 MHz

084 6 **StarShip Comedy**
Comedy broadcast 24 hours daily.
Stereo DS

474 6 **StarShip '50s, '60s, and '70s**
Hits from the '50s, '60s, and '70s broadcast 24 hours daily.
Stereo DS

086 6 **StarShip Big Bands**
The big band era broadcast 24 hours daily.
Stereo DS

476 6 **StarShip Hits**
Progressive rock broadcast 24 hours daily.
Stereo DS

477 6 **In Touch**
Reading from daily newspapers and leading publications 24 hours daily. 7.875 MHz.

074 7 **ESPN Informational Network**
Daily program schedules and advance event information. Available 9 am to 12:30 pm.on 6.2 MHz.

203 8 **StarShip Jazz**
Vocal and instrumental jazz, interviews, and discussions available 24 hours daily.
Stereo DS

205 8 **StarShip Contemporary Christian**
Vocal and instrumental programs with spiritual vignettes.
Available 24 hours daily.
Stereo DS 6.30/6.48 MHz

204 8 **StarShip Easy Listening**
Contemporary easy listening available 24 hours daily.
Stereo DS

hours daily.
7.56 MHz.

WESTAR 5 (W5) 122.5 W

ID# TR.# SERVICE

461 18 **Sheridan Broadcasting Network**
News, sports, and MOR 24 hours daily.
5.4, 5.58 and 5.76 MHz.

ANIK D1 (AD1) 104.5 W

ID# TR.# SERVICE

533 4 **CFNY-FM, Toronto, Ontario**
Alternative contemporary and variety rock music 24 hours daily.
Stereo DS 5.4/5.58 MHz

267 8 **CKY-AM, Winnepeg, Manitoba**
Occasional contemporary and MOR music. 7.605 MHz.

269 14 **CITE-FM, Montreal, Quebec**
French-language M.O.R. 24 hours daily. 6.17 MHz.

270 14 **CKAC-AM, Montreal, Quebec**
French-language M.O.R. 24 hours daily. 5.41 MHz.

431 16 **CBC-FM—Eastern (French)**
CBC French programming 24 hours daily.
Stereo DS 5.4/5.58 MHz

435 16 **CBC-FM—Eastern (English)**
CBC programming 24 hours daily.
Stereo DS 5.76/5.94 MHz

Reprinted by Permission. Copyright 1986, CommTek Publishing.

273 18 **CIRK-FM (K-97), Edmonton, Alta**
Progressive rock 24 hours daily.
Stereo MP 6.17 MHz

562 18 **CKO-FM, Toronto, Ontario**
News and information. 24 hours daily. 6.17 MHz.

274 20 **CBM-AM, Montreal, Quebec**
CBC programming available 24 hours daily. 6.12 MHz.

275 22 **CBC-FM**
CBC programming available 24 hours daily.
Stereo DS 5.76/5.94 MHz

276 22 **CFMI-FM, Vancouver, B.C.**
Soft album rock available 24 hours daily. 6.8 MHz.

537 23 **CKRW-AM, Whitehorse, Yukon**
Country and western music at 5.41 MHz. Daily, 24 hours.

561 18 **VOCM, St. Johns, Newfoundland**
Adult contemporary music at 6.17 MHz. Daily, 24 hours.

438 24 **CBC-FM—Eastern (French)**
French language programming available 24 hours daily.
Stereo DS 5.41/5.58 MHz

437 24 **CBC-FM—Eastern (English)**
CBC programming available 24 hours daily.
Stereo DS 5.76/5.94 MHz

SATCOM F4 (F4) 83 W

ID# TR.# SERVICE

396 3 **Georgia Radio News Service**
News, sports, and information available 24 hours daily.

075 7 **Family Radio Network (East)**
24 hour Bible-centered religious format.
Stereo DS 5.58/5.76 MHz

109 7 **Family Radio Network (West)**
24 hour Bible-centered religious format.
Stereo DS 5.94/6.12 MHz

481 21 **National Broadcast Museum Superstation**
Vintage music and talk shows.
Broadcast 24 hours daily at 6.2 MHz.

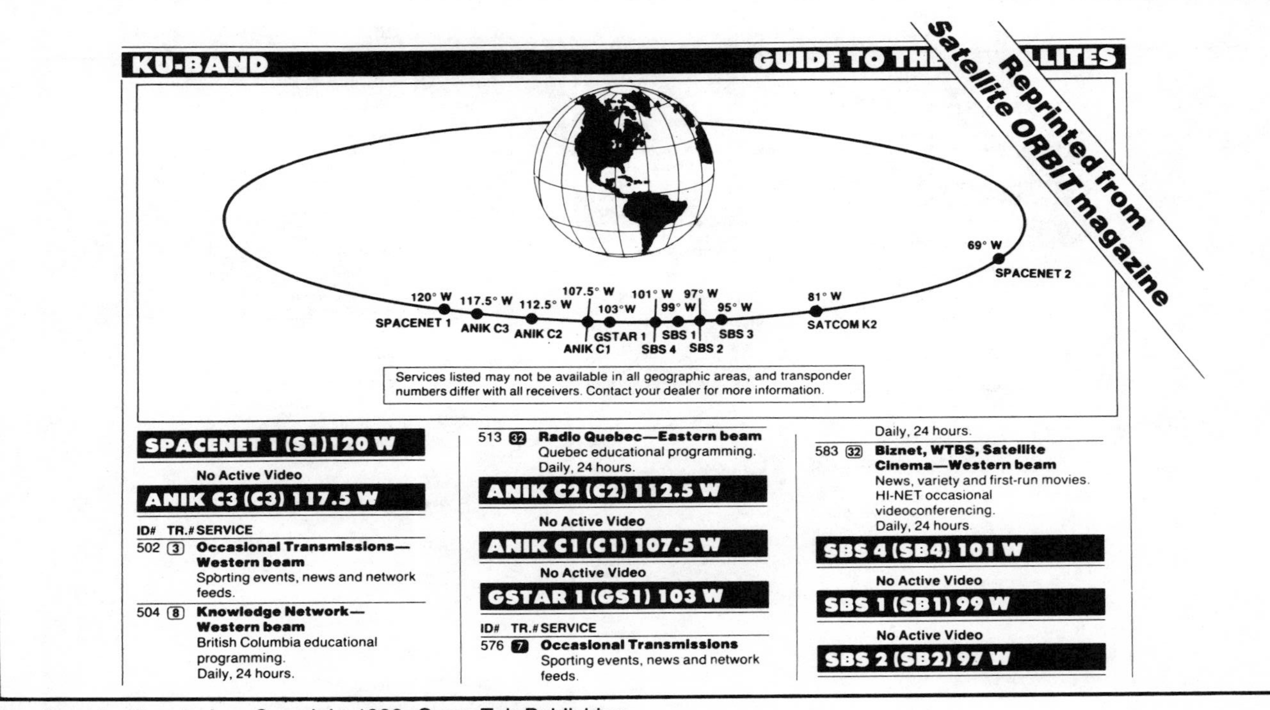

KU-BAND — GUIDE TO THE SATELLITES

Services listed may not be available in all geographic areas, and transponder numbers differ with all receivers. Contact your dealer for more information.

SPACENET 1 (S1) 120 W

No Active Video

ANIK C3 (C3) 117.5 W

ID#	TR.#	SERVICE
502	3	**Occasional Transmissions—Western beam** Sporting events, news and network feeds.
504	8	**Knowledge Network—Western beam** British Columbia educational programming. Daily, 24 hours.
513	32	**Radio Quebec—Eastern beam** Quebec educational programming. Daily, 24 hours.

ANIK C2 (C2) 112.5 W

No Active Video

ANIK C1 (C1) 107.5 W

No Active Video

GSTAR 1 (GS1) 103 W

ID#	TR.#	SERVICE
576	7	**Occasional Transmissions** Sporting events, news and network feeds. Daily, 24 hours.
583	32	**Biznet, WTBS, Satellite Cinema—Western beam** News, variety and first-run movies. HI-NET occasional videoconferencing. Daily, 24 hours.

SBS 4 (SB4) 101 W

No Active Video

SBS 1 (SB1) 99 W

No Active Video

SBS 2 (SB2) 97 W

Reprinted by Permission. Copyright 1986, CommTek Publishing.

ID#	TR.#	SERVICE
505	9	**Alberta Access—Western beam** Alberta educational programming. Daily, 24 hours.
507	10	**Premier Choix/TVEC—West—Western beam** First-run movies, sports, and specials in French. Daily, 24 hours.
506	12	**Superchannel/First Choice—Western beam** First-run movies, sports and specials. Daily, 24 hours.
565	17	**Atlantic Satellite Network—Eastern beam** CTV network programming. Daily, 24 hours.
566	18	**Occasional Transmission—Eastern beam** Sporting events, news and network feeds.
568	19	**Premier Choix/TVEC—East—Eastern beam** First-run movies, sports and specials in French. Daily, 24 hours.
509	20	**Occasional Transmissions—Eastern beam** Sporting events, news and network feeds.
510	24	**La Sette (TVFQ-99)—Eastern beam** Taped French network programming. Daily 24 hours.
511	26	**TVOntario—Eastern beam** Ontario educational programming. Daily, 8 am-12 am.
512	31	**Superchannel/First Choice—Eastern beam** First-run movies, sports, and specials. Daily, 24 hours.
575	9	**NTU (National Technological University)** Occasional college courses focusing on technological developments.
574	10	**AMCEE** Occasional college courses focusing on technological developments.
572	13	**Campus Satellite Network** Occasional entertainment for college campus distribution.
573	14	**Occasional Transmissions** Sporting events, news, and network feeds.
584	15	**Occasional Transmissions** Sporting events, news, and network feeds.
577	19	**ESPN** Professional, college and amateur sporting events. Daily, 24 hours.
578	20	**CNN Headline News** Continuously updated 30-minute wheel of hard news. Daily, 24 hours.
579	23	**Showtime—East—Eastern beam** First-run movies, sports and specials. Daily, 24 hours.
580	24	**Biznet, WTBS, Satellite Cinema—Eastern beam** News, variety and first-run movies. HI-NET occasional videoconferencing. Daily, 24 hours.
581	29	**Occasional Transmissions** Sporting events, news, and network feeds.
582	31	**Showtime—West—Western beam** First-run movies, sports and specials.

No Active Video

SBS 3 (SB3) 95 W

ID#	TR.#	SERVICE
517	1	**Occasional Transmissions** Sporting events, news, and network feeds.
570	3	**NBC—Affiliate feeds** Occasional live/taped network feeds.
571	4	**Occasional Transmissions** Sporting events, news and network feeds.
569	5	**NBC—Pacific** Live/taped network programming. Daily, 24 hours.
522	6	**NBC—Central** Live/taped network programming. Daily, 24 hours.
523	7	**Conus/Hubcom** Remote newsfeed access channel.
525	8	**PSN (Private Satellite Network)** (s) Occasional teleconferencing and private video network feeds.
526	9	**NBC—News feeds** Occasional live/taped news feeds and network programming.
527	10	**Occasional Transmissions** Sporting events, news, and network feeds.

SATCOM K2 (K2) 81 W

No Active Video

SPACENET 2 (S2) 69 W

ID#	TR.#	SERVICE
529	20	**Occasional Transmissions** Sporting events, news, and network feeds.
532	22	**Occasional Transmissions** Sporting events, news, and network feeds.

Reprinted by Permission. Copyright 1986, CommTek Publishing.

THE BIRDS AT A GLANCE

USE SPACE BELOW TO LIST YOUR ACTUATOR NUMBERS

TRANSPONDER	AURORA F5 143° W	SATCOM F1R 139°	GALAXY 1 134°	SATCOM F3R 131° W	TELSTAR 303 125°	WESTAR 5 122.5° W	SPACENET 1 120°
1			Occasional	Nickelodeon – East	CMTV (Country Music Television	Occasional	
2			TNN (The Nashville Network)	PTL Satellite – The Inspirational Network	SelecTV	The University Network (The Unchannel)	
3			WGN – Chicago	TBN (Trinity Broadcasting Network)			Occasional
4			The Disney Channel – East	FNN (Financial News Network) /SCORE			
5			Showtime – East	Viewer's Choice(s)		Occasional International Transmissions	Occasional
6	Occasional		SIN (Spanish International Network)	SPN (Satellite Program Network)			
7		Occasional	CNN (Cable News Network)	ESPN Blackout Network		CBS – Contract Channel/ Occasional	Occasional
8		NBC – East	CNN Headline News	CBN Cable Network	Occasional	PASS (Pro Am Sports System) Occasional	
9		Occasional	ESPN	USA Network – West			Occasional
10			TMC (The Movie Channel) – East	Showtime – West			
11		Ocassional	CBN Cable Network	MTV: Music Television			HSN (Hospital Satellite Network)(s)
12		Ocassional	Request TV(s)	EWTN (Eternal Word TV)			
13		NASA – Contract Channel	C – SPAN	HBO (Home Box Office) West (s)			Occasional
14			TMC (The Movie Channel) West	Open-net/ Occasional			
15		Yonkers Horse Racing (s)/ Occasional	WOR – New York	VH1/Video Hits One		Occasional	ACTS (American Christian Television System)
16			Occasional	HTN (Home Theater Net.)/TLC (Learning Channel)		Occasional	
17			PTL Satellite – The Inspirational Network	Lifetime			
18		Occasional	WTBS Atlanta	NJT/ Reuter(s)	American Extasy	Occasional	Occasional
19	Occasional		Cinemax – East (s)	Occasional			
20	Learn/Alaska Television Network		GalaVision	BET (Black Entertainment Television)		Telebet Racing(s), Occasional	
21	Occasional		USA Network – East	The Weather Channel		Occasional	BTN (Baptist)(s)/ VMT (Vanderbilt)(s) /Occasional
22		Occasional	The Discovery Channel/AVN (Alternate View Network)	MSN/Home Shopping Network/ Occasional	KTVT – Dallas	The Meadows Racing Network/ Occasional	
23			HBO (Home Box Office) – East (s)	Cinemax – West (s)			Occasional
24	Alaska Satellite Television Project	Occasional	The Disney Channel – West	A&E (Arts & Entertainment)		FUN Channel(s)/ PPV—Pay Per View Channel(s)/o/v	

For more detailed information about the satellite services, see pages B12-B14.

For information about the audio/test subcarrier services, see pages B15-B16.

Reprinted by Permission. Copyright 1986, CommTek Publishing.

MORELOS 1 113.5° W	ANIK B 109° W	ANIK D1 104.5° W	WESTAR 4 99° W	TELSTAR 301 96° W	GALAXY 3 93.5° W
		Occasional		Occasional	
		TSN (The Sports Network)		CBS - Central	Occasional
				ABC incoming feeds Occasional	
		Global TV Occasional		ABC - Contract Channel Occasional	Occasional
		Occasional		Occasional	Occasional
		MuchMusic	Atlantic Horse Racing(s) Word of Faith Network o/v	Occasional	Occasional
		CBC Occasional		CBS Occasional Feeds	Occasional
		CHCH Hamilton(s)			Senate Republican Conf. News Occasional
		WDIV - Detroit(s)		WOLD Communications	Occasional
		WXYZ Detroit(s)	Occasional	ABC Central	Occasional
		CBC - North Pacific	CTNA (Catholic Telecommunications Net.)(s)o/v	Occasional	Occasional
XEW - Mexico City				ABC Contract	Occasional
		Occasional		Occasional	
XHITM - Mexico City		TCTV (Telemedia Communications Television)(s)		Occasional	
		CBC French - Eastern	PBS (Public Broadcasting Service)	Occasional	
		Can. Parliamentary Network Occasional	CNN Contract Channel Occasional	Occasional	
		Occasional	PBS (Public Broadcasting Service)	Occasional	
		CITV Edmonton(s)	Occasional	Occasional	
		CBC North - Eastern	Occasional	Occasional	
Occasional		CBC - Montreal	Occasional	Occasional	
		WTVS - Detroit(s)	PBS (Public Broadcasting Service)		
XHDF - Mexico City		BCTV - British Columbia(s)			
		WJBK - Detroit(s)	PBS (Public Broadcasting Service)	WOLD Communications	
XETV - San Diego /Tijuana		Can. Parliamentary Network/CBC North - Eastern	Occasional	Occasional	

o/v = Occasional Video

THE BIRDS AT A GLANCE

USE SPACE BELOW TO LIST YOUR ACTUATOR NUMBERS

Reprinted from Satellite ORBIT magazine

WESTAR 3 91° W	TELSTAR 302 86° W	SATCOM F4 84° W	WESTAR 2 79° W	GALAXY 2 74° W	SATCOM F2R 72 W	SPACE[illegible] 69 W	
	Occasional	HSN (Home Shopping Network)					
		Bravo; C.O.M.B. Value Net			Occasional		2
		Occasional					3
		Nickelodeon West					4
CNN Contract Channel		ABC Affiliate Feeds Occasional					5
		BIZNET/MSG (Madison Square Garden)					6
		NCN					7
		Occasional					8
	Occasional	Sportsvision					9
	ABC - West	AMC (American Movie Classics), Occasional					10
	Occasional	HSE (Home Sports Entertainment)					11
		The Playboy Channel/ Heartbeat Network					12
		NESN (New England Sports Network)			NASA Contract Channel		13
	Occasional	Private Ticket/ HSE Affil./ Occasional					14
	Occasional	SNL (Success N Life Network)		Occasional			15
	CBS - West	The Silent Network		Occasional			16
	Occasional	Occasional		Occasional			17
		Hit Video U.S.A.		Occasional		Occasional	18
Occasional	Occasional/ CBS Encryption Test Channel	WPIX New York					19
	CBS - Central	Prime Ticket/ Occasional			AFRTS (American Forces Radio and TV)	Occasional	20
	Occasional	The Nostalgia Channel					21
	Occasional	HTS (Home Team Sports)/ Occasional					22
Brightstar feeds	Occasional	Occasional					23
	Occasional	Nightline TV Network/ Occasional					24

VERTICAL HORIZONTAL

Chapter 7

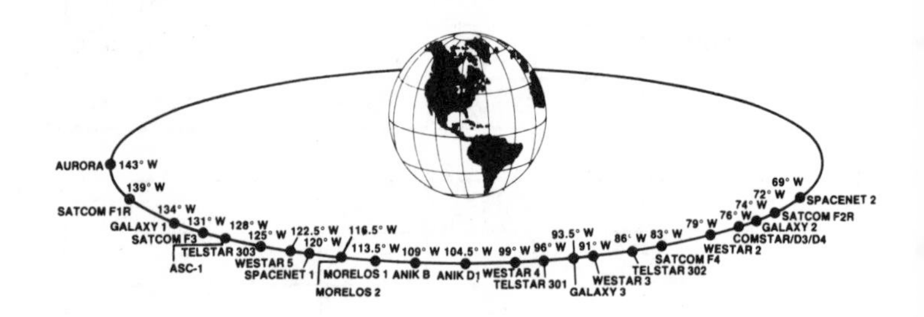

TVRO Considerations and Design

THERE ARE A THOUSAND AND ONE QUESTIONS THAT WILL OCcur as you begin consideration of your first (or last) television receive only (TVRO) earth station. What size antenna do you need, made of what material, what temperature low-noise amplifier (LNA), how good should the receiver be, and is there need for two downconversion frequency stages instead of one? Who are the outstanding manufacturers or suppliers, are they really qualified, what about cost, are remote-controlled aiming devices worth the extra expenditure, is there mounting room, and what about the site? How about wind velocity and survivability, what are the sunrays effects, will the reflector surface peel and lose gain, how about a secure mounting? After all these you should consider the possibility of microwave interference from any surrounding sources, county and state placement regulations, permits, and possible neighborhood curiosity and complaints.

True, you're beginning with the 4/6 GHz C-band right now, but how about the K and Ku frequencies among the direct broadcast (DBS) satellites to come? Will you be satisfied with current SATCOMS, WESTARS, AND TELSTARS for the foreseeable future, or will you subscribe to pay TV service as it begins in 1986?

If you're serious about quality signals, then investigate quality equipment. The signal-to-noise for broadcast-quality intelligence is 54 dB or better, and this takes a large dish, perhaps as large as 7 meters. But if "reasonable" quality signals are sufficient, then a minimum of 12-foot parabolic or 13-foot torus-type antennas are required in northern and southern latitudes and 10- or 12-foot dishes elsewhere. Of course, the larger the dish, the greater the gain, probably fewer offending sidelobes, but a somewhat more restricted pointing angle for the parabolic, and a larger reception area for torus. The latter, however, is not in large quantity production, although we're betting on its eventual emergence because of multisatellite receiving abilities. In the jargon of sales pitch, "where there's a need, there are things to fill that need." It would seem that torus has a place for general, if somewhat more expensive installations. At the moment, of course, the parabolic is everything. Among TVRO makers, we understand there is only one consumer-type spherical on the market, and because of this there'll be no discussion other than to say that in sphericals it's customary to move the feed assembly for satellite reception rather than the entire dish. Apparently most of the industry feels there are better overall results with the methods at least for the time being. Later, there will undoubtedly be changes as the market expands and the popularity of satellite reception becomes even greater than now. Motorized polarity changes, for instance, may become all-electronic even before this book is published. There are lots more in the wind that manufacturers won't even talk about because of ever pressing competition. With all sorts of R & D underway to improve both antenna dish strength, efficiency, sidelobes, gain, and feed systems, you can bet the day of the high efficiency, microprocessor-controlled consumer unit is at hand; so much so that during mid-summer of 1986 it was all we could do to collect most significant design changes. The industry in the 1980s is moving that fast, and prices are dropping! As little as two to three thousand dollars may get you some real goodies by 1987—so be prepared!

Before bringing you all the way up to date, however, let's look at a little history by way of explanation and develop some usable methods in this desirable and profitable (for the professionals) endeavor of madness. Not that these are intended for those who *must* build their own baskets of wire and struts, since these, nowadays, are often too inexpensive and jury-rigged compared to many excellent and fairly reasonable dishes and their electronics that are available for surprisingly small sums which, we suspect, will be constantly decreasing as commerce supports considerably greater volume. What follows constitutes only background and a base from which to understand what is sure to become one of the largest industries in the U.S.—that of satellite reception and all the remarkable varieties of communications that come with it. We're very confident that satellite backyard and rooftop receiving dishes will, in time, become just as common as television antennas are today, and at least 100% more effective since professional installers will be required.

THE PARABOLA

For this one, let's return to analytic geometry (bless it) and see why some things have developed in certain ways. Any parabola has both a curved surface and a corresponding straight line, with a point-of-origin called a focus (Fig. 7-1). The parabola, then, is defined as the locus of all points equidistant from fixed point (P) and the fixed line (L). As point P moves, with FP and LP remaining numerically equal, point P will automatically trace out a parabola of the form $y^2 = 2px$ or $x^2 = 2py$, depending on the parabola's direction and coordinate axes. From such equations you may plot the entire curve with various values of X and Y.

Let's not become bogged down in serious math technicalities before we've really started since we would probably lose a good many of you before the game becomes interesting. So once again we'll draw a parabola (Fig. 7-2), but this time we'll add a diameter, focal length, and note the depth of the parabola at its center. Further, we'll draw perpendicular Y, which is both parallel to the direc-

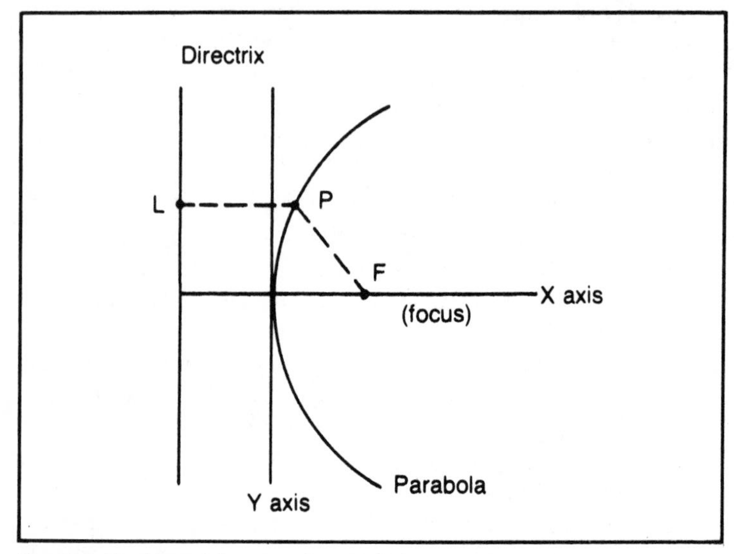

Fig. 7-1. Fundamental curve of a parabola.

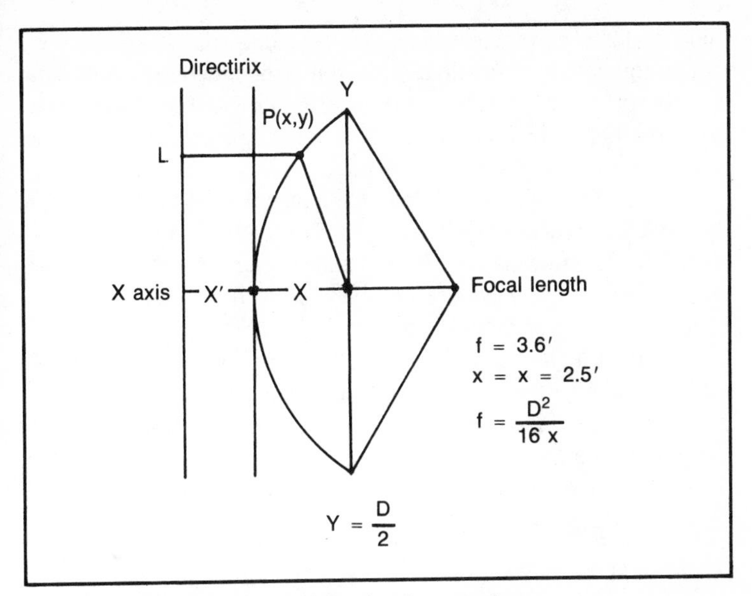

Fig. 7-2. Modified parabola with pertinent parameters.

trix and seemingly equal to diameter D: note we said "seemingly," which will be explained shortly. In addition, note the small letters px between large P and the directrix. Now, the *standard* form of a parabola equation as it relates to these antennas becomes:

$$Y^2 = 4px$$

This equation, incidentally, is only applicable where the parabola's vertex is at the origin and its axis coincides with the X axis; therefore, solutions aren't all that difficult. However, there are some parameter equation rules that *must* be followed: $Y = D/2$, and the ratio between the focal length f and diameter D of the dish can vary between 0.3 and 0.4. Usually 0.3 is selected. Consequently, in a 12-foot dish f would equal:

$$f = 0.3D = 0.3 \times 12 = 3.6'$$

Consequently, if $Y^2 = 4FX$

then, $(D/2)^2 = 4 \times 3.6 \times X = 36/14.4 = X$
and $X = 2.5'$

Naturally, in any equation with three parameters, you have to know or develop in some way two of the three so the other unknown can be equated. You should now be able to handle these initial conditions without any problems at all. Just keep all measurements in either meters or feet and don't attempt to mix the two or the results can become strange indeed.

You might also like to know that every time the size of a dish is doubled its power gain *increases* by 6 dB and is quadrupled. Furthermore, the same dish receiving double its normal frequency again quadruples gain if there are no other losses or distortions. However, as the size of the antenna increases, the narrower becomes its beamwidth and the more difficult it is to point. Consequently, a 10-foot dish with adequate electronics, once more, is the recommended TVRO size considering what we see at present, and all the goodies coming down the road, to look at the sky. But, with 65° to 80° LNAs at mid-CONUS, size *could* drop to 9 feet and still be adequate in high signal areas.

MATERIALS

Work out all your basic math angles, but be sure you know the ins and outs of satisfactory construction. Reflectors can be made of wire mesh, fiberglass, aluminum, and even steel. Most of the larger units, and certainly many of the best small antennas are made of precision, die-stamped panels backed by a frame of steel. The high thermal conductivity of aluminum—almost three times that of steel—helps reduce differential thermal distortion when heated by sun and cooled by rain or wind, and will offer substantial resistance to high wind conditions without bending and distorting signals. Fiberglass, with proper support, also withstands the elements very well and, if anything, is preferable.

Nonetheless, all such antennas are designed to concentrate the very small mount of rf energy emanating from space to some focal point for electronic pickup and amplification. Their shape, surfaces, feed design, and low-noise amplification all contribute to overall performance even before signal selection and demodulation. Even solar energy diffusive paint inside the dishes to prevent undue sun concentration on the feed assembly becomes a prime consideration. Of course, wind effects on the overall assembly and mount, plus manual or electronic orientation are additional factors requiring greater than merely offhand study.

In building metal receptors, such material is bent from its original flat surface and stressed to conform to whatever shape the antenna has been designed. The same conclusion can be reached of wire mesh since it must be stretched and secured to a retaining frame. Aluminum, of course, will develop an oxide coating when attacked by salt air or spray and even acid or smokey conditions unless properly prepared beforehand; and aluminum oxide is non-conductive.

Fiberglass, on the other hand, won't oxidize and is normally element-resistive and, if formed over a precision form, produces an excellent mold. However, since the dish must reflect rf signals, it has to have some sort of metal or metallic surface. Therefore, metal needs to be laid over either the form or surface of the dish, producing good surface tolerance so this type of receptor may be used at both 4 and 12 GHz (with inclusion of the proper electronics). Normally, wire mesh-type dishes are not that flexible since they are usually designed for a single band of frequencies of approximately 4 GHz. They may also warp and bend under ice or snow loads, and especially with inadequate mounts. Perforated dishes are better.

Fiber and resin normally constitute fiberglass rigs so there is no risk of delamination, and ultraviolet stabilized resins, with the addition of gel-coat, ordinarily provide sufficient ultraviolet radiation protection. Some also believe that metal antennas are easier to repair than fiberglass since the panels are easily replaced. Fiberglass may be patched with kits, although the metallic surface must be added also, and there are fiberglass-paneled antennas, too. Of course, fiberglass can burn, but should be treated to resist all but the hottest pyrotechnics which probably would melt metal anyway.

Mounts and Dishes

Briefly, and in sum and substance, that's the story of light industrial and consumer-type antennas, most of which are now made of fiberglass for the reasons given. There are specialty types, of course, but we're speaking of general offerings by the major manufacturers advertising diameters up to 15 or 16 feet max. For the larger commercial and military installations aluminum and steel are almost always used for precision alignments and structural strengths. Large antennas also have heating elements connected to their structures to remove ice and snow loading that becomes a hazard if permitted to accumulate. Consumer-type dishes may usually be cleared by hand in northern and western areas when required. In the tropics where rainfall is heavy, wire mesh installations are often quite successful in handling communications traffic, avoiding excessive moisture accumulations and reducing hurricane-force wind loading. Mesh materials may include galvanized and tinned iron, aluminum, and copper, which can be formed in shaped panels like solid metal. Although it is said they don't have to make electrical contact, overlaps between pieces are recommended for increased efficiency.

A group of five standard TVRO antennas by Microwave Associates is shown in Fig. 7-3. All have 24-channel tuning, threshold extension, TV modulator receiver plug-in, optional LNA temperature inserts, durable non-corrosive fiberglass construction, dual linear polarization, and high efficiency feed assemblies.

If you're using an AZ-EL mount, especially if looking at only one direction, then mounting is much simpler since there is no special fixed position required. Angles of elevation and azimuth adjustments may be cranked in for whatever specific point is needed and more or less permanently adjusted. If a polar mount is specified, then either the base or the mount will have to be pointed true north or south, depending on the manufacturer's instructions.

Location Instructions

Since true north is by no means magnetic north, a Topol (or topographical) map can be procured from any appropriate government office to pinpoint your exact location and indicate necessary variation; or you may call the nearest airport for general north magnetic correction; or buy a copy of the Farmer's Almanac, find your time for high noon, drive a stick in the ground, and mark the exact shadow cast precisely upon the hour. Any of these procedures should do the location job without problems, although the Topol method is obviously preferred even though just a little calculation is involved.

Fig. 7-3. Group of five TVRO earth stations (courtesy M/A-COM).

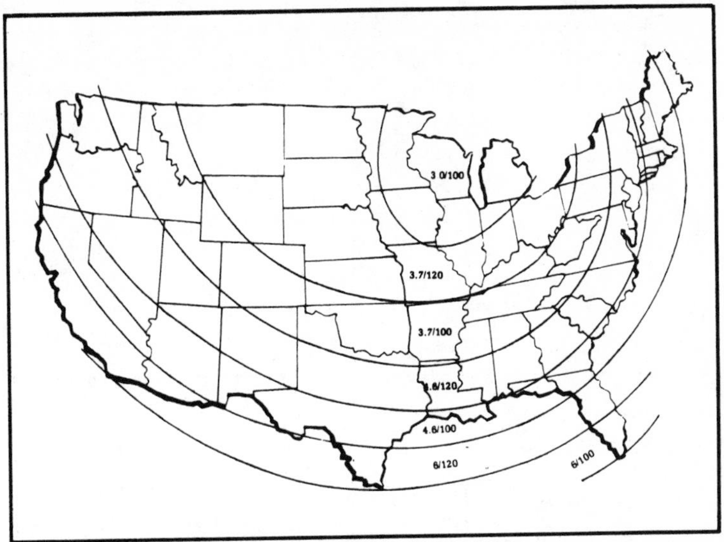

Fig. 7-4. Typical dish contours with recommended LNA temperatures in °Kelvin (courtesy of M/A-COM).

As an example of how to choose an array for your general location, accompanied by a suitable low-noise amplifier, see Fig. 7-4 that covers the entire U.S. for SATCOM. This is an excellent illustration that may be used as a general guide for other satellites having more or less the same longitudes and general elevation angles. Observe that LNA temperature in °K varies only between 100 and 120. In specific locations with difficult azimuths or angles of elevation, tighter tolerances for both antennas and LNAs may be required, so don't overlook exceptions to the general rule. If you want to convert antenna size recommendations from meters to feet, simply multiply by 3.281. Meters are nice, but feet are often more familiar, especially on pay day!

THE SYSTEM

At the slight risk of getting ahead too fast, let's use an excellent block diagram (Fig. 7-5) to show a combined satellite and broadcast TVRO installation with dual polarization, channel selector, downconverter, power supply, motor drives, selector switch, and audio/video demodulators. Observe there are both rf and baseband outputs for standard television receivers as well as monitors, with antenna remote control optional. This is a very realistic diagram and offers excellent insight into what and how this type of installa-

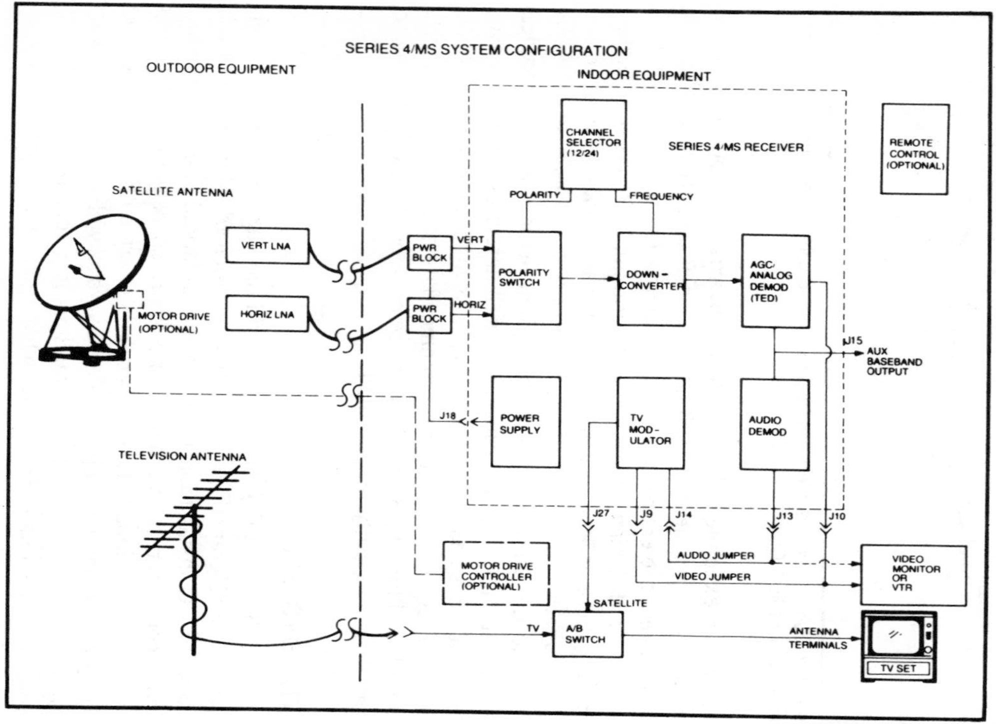

Fig. 7-5. Typical block diagram of satellite and broadcast TVRO receive installation (courtesy of Microwave Associates Communications).

tion should be affected. With polar mount permitting antenna azimuth traverse between 70° and 140° to cover at least existing geostationary satellites assigned to North America.

Antennas, of course, must have feeds, and both 3.7-meter prime focus with buttonhook and 5-meter with Cassegrain and three supporting struts are illustrated in Fig. 7-6. Polarization for the 12-foot dish is single or dual, and that for the 16.4-foot array is circular. Gains are 40.1 and 44.5 dBi, with the first sidelobe at 18 dB (typical) for the small and 12 dB for the larger. Nominal beamwidth measures 1.5° and 1.1°, respectively. Note how beamwidth *decreases* with dish diameter while weight increases from 930 lbs. to 1,945 lbs.

Prime Focus Versus Cassegrain Feed

Here, something of a wrangle in space communications grazing land begins. Cassegrain configurations use hyperbola subreflectors mounted at the focal point to reflect rf to the feedhorn, often offering more gain and easier polarization change/alignment, but suggests somewhat poorer sidelobe measurements. Prime-focus feed, on the other hand, has good sidelobe characteristics but possibly less gain because of waveguide losses with feedhorn at the focus. The Cassegrain assembly must also be covered with some type of radome or window as a moisture seal and is susceptible to rock throwing or BB gun vandals who can damage the feed and necessitate expensive repairs. Prime-focus types with LNA protective covers usually are safe. The Cassegrain feedhorn will also degrade in performance with excessive rainfall or a covering of

Fig. 7-6. Prime-focus and Cassegrain-fed antennas with polar and AZ-EL mounts, respectively (courtesy M/A-COM).

snow or ice. Some Cassegrain units are also difficult to service because of spars and other equipment interference at the rear of the antenna.

Obversely, prime-focus feeds often have lower noise temperatures, are easier to maintain on TVRO antennas and have no subreflectors to become misaligned, in addition they are usually less expensive. On large antennas, where money has to be spent on a variety of specialized equipment, Cassegrain feeds are desirable and most commercial units—at least the older ones—are so equipped. Whether future developments change such design concepts remains to be seen but, as of now, prime focus, parabolics, and LNAs with temperature ranges between 70° and 100° are destined to predominate in consumer-type products and smaller commercial TVRO dishes. As the market for these earth stations expands at accelerated rates, however, there's no telling what can happen and how fast. A number of changes are expected even before the summer of 1987-1988.

Antenna Gain

A simplified equation for calculating antenna gain in a parabolic array could be:

$$G = 4\pi A\eta/\lambda^2$$

A being the reflector area, and efficiency η being a number less than 1 (perhaps 0.5 to 0.8), depending on your feed. If your dish has a circular aperature (diameter) D, then

$$A = \pi(D/2)^2$$

Lambda (λ), naturally has to be expressed in feet, just like the parabola's diameter; so let's try a basic example with a 12-foot dish and efficiency of 0.55 equal to η:

$$A = 3.1416 \times 12^2/4 = 113.1 \text{ sq. feet}$$

You would think that just using lambda's wavelength of 984×10^6 feet/second would just do fine, but that isn't the case at all. Remember that 4 GHz frequency also enters into the equation. Lambda (λ), therefore equals $9.84 \times 10^8/4 \times 10^9$, with this, of course, being squared. There's another factor also, since we're talking about power gain, and this is expressed always in decibels.

So the real equation for the gain of *any* aperture antenna in dB is:

$$G = 10 \log 4\pi A\eta/\lambda^2 = 10 \log 4\pi \times 113.1 \times 0.55/0.246^2$$
$$= 10 \log 13028.18 = 41.1 \text{ dB}$$

However, the specific equation for a *parabolic* dish is:

$$G = 10 \log \eta\, (\pi D/\lambda)^2 = \text{which this time we'll do in meters (m).}$$
$$G = 10 \log (4 \times 3.66/3 \times 18/4 \times 10^9)^2 = 10 \log (11.498/.075)^2 \times .55$$
$$= 10 \log .55\, (23502.93)$$
$$G = 41.1 \text{ dB}$$

So if you like all this, how about trying an even easier way that *does* take into account both frequency and dish diameter (aperture)?

$$G = 20 \log D + 20 \log f - 52.5$$
$$= 20 \times 1.08 + 20 \times 4 \times 10^3 - 52.5$$
$$G = 41.1 \text{ dB}$$

where log f is designated in megahertz (here 4,000 MHz) with the log of 4,000 becoming 3.60 times 20 equal 72 dB added to 21.6 and subtracted from 52.5, for a resultant of 41.1 dB once more. Of course the latter is considerably easier to use, and this useful equation may also be written:

$$G = 20 \log f \text{ (in GHz)} + 20 \log D \text{ (in ft.)} + 7.5.$$

Beamwidth

Since we're in business of making life a little more bearable, how about a beamwidth equation that won't further destroy cranial incubation? Normally expressed at the half power point:

$$\text{Beamwidth} = 68{,}700/f \text{ (in MHz)} \times D \text{ (ft.)}$$
$$\text{at} = 68.700/4 \times 10^3 \times 12$$
$$-3 \text{ dB} = 1.43125°$$

Close enough for government work? Let's hope there's enough substance for all.

True, the last two gain and beamwidth equations are simplifications, but they're in the ballpark, nonetheless, and with considerable accuracy considering all the limitations. If, later, you require especially definitive and exact equations for particular arrays, then you'll need to backpack all the different parameters to a computer—and they're a bunch! For day-to-day involvement, you have all the working equations necessary among those already given.

Low-Noise Amplifiers (LNAs)

Now, we're moving into the nitty-gritty of these earth stations electronics with the one piece of equipment that can affect all the rest. Its noise temperature and amplification over a large range of conditions really spells success or failure for your TVRO or any other receiving array. For no matter how good the antenna, receiver, demodulator, and downconverter, initial signal handling and good temperature characteristics will determine what everything else does in the entire system. If you want to see how your dish and downlink signal are going to interact satisfactorily or not, just measure operations of the various satellite transponders through the LNA with a spectrum analyzer, such as the Tektronix 21 GHz model 492, and you'll find out in a hurry. Noise and nonlinear amplification will be evident immediately. Even if you have to go to the downconverter output following the LNA, there will still be plenty of prime evidence if faults are already in the system. At the output of an LNA, of course, there's also 50 dB of gain which even the most sensitive spectrum analyzer will cordially appreciate. With its ability to execute fast Fourier transforms and display waveform amplitude versus frequency, the analyzer should become your prime tester both for linearity and relative gain. Just remember that dc at 50 ohms should never enter this instrument's swept front end; so be careful and always provide a dc block. At GHz, almost any capacitor looks like an ac short to incoming signals.

Modern low-noise amplifiers usually appear in two categories. Parametric types use varactor diodes for good gain and very low noise temperatures. These are favored in most all large earth stations and are either cryogenically cooled or uncooled. But even if uncooled, they must be temperature stabilized. As you surmised, they *are* expensive! The second best—but perhaps not for long—are GaAsFETs (Fig. 7-7), or gallium arsenide field effect transistors. Their power-handling and noise rejection abilities are increasing almost every day. Furthermore they're relatively cheap! So in much of the lesser-cost commercial and consumer-type world, GaAsFETs are used almost entirely and will become more so as

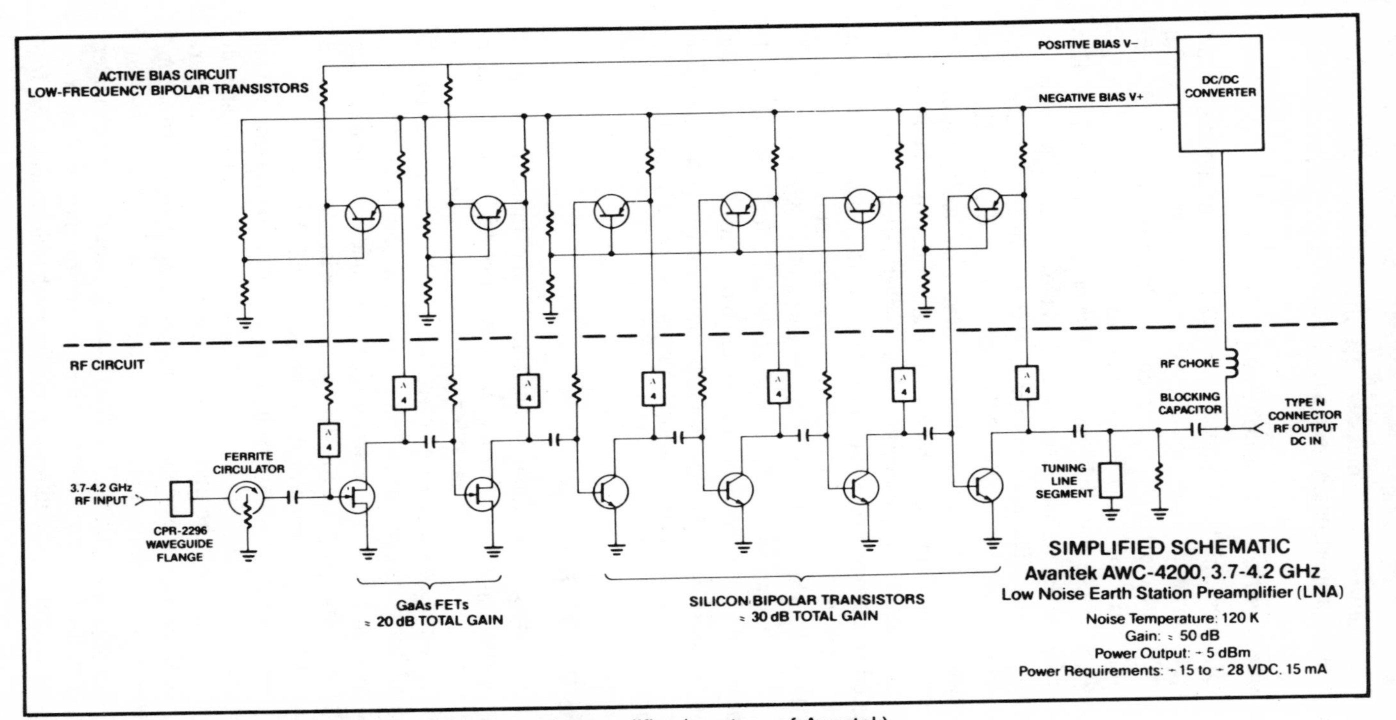

Fig. 7-7. Avantek's block diagram of a modern low-noise amplifier (courtesy of Avantek).

their metallized gates measure less than 0.5 micron and noise decreases proportionally (see Fig. 7-8).

Downconverter and Receiver

Of course, all these signals from 12 to 24 transponders must be set to certain frequencies for uplinks as well as downlinks with a 500 MHz band spectrum and channel bandpass of 36 MHz. In the illustration of Fig. 7-9, however, such a satellite has 24 transponders, whose downlinks range in frequency between 3,720 MHz and 4,180 MHz, with an offset of 20 MHz to accommodate horizontal polarization. The uplink is similarly structured, having frequencies that begin with 5,945 MHz and end at 6,405 MHz. This particular spectrum plan would serve CONUS, Alaska, Hawaii, and Puerto Rico. Therefore, when we casually refer to 4/6 GHz satellite downlink/uplink operation, this is what's meant since all assigned frequencies must be clustered generally in this spectrum for C-band operation. Guardbands between channels measure 4 MHz each, whether horizontally or vertically polarized.

As these signals reach your parabola, spherical, or torus earth station, they are reflected in a Cassegrain or prime-focus feed, then into the LNA/LNB and so on to the downconverter and satellite receiver for further processing, which is described below (see Fig. 7-10).

The VR-4XS Receiver

Polarized vertical (odd) or horizontal (even) inputs permit passage through the rf switch, into the downconverter section through a 3.7 to 4.2 GHz bandpass filter and into the anti-paralleled diode mixer. Here, the signals are heterodyned with the local oscillator's second harmonic, producing an i-f output of 550 MHz. This second harmonic develops in the anti-parallel diode mixer from inputs between 1,575 and 1,825 MHz as synthesized by channel and thumbwheel switches and the push-pull oscillator.

Oscillator generation results from a pair of voltage-controlled oscillators 180° out-of-phase delivering what's known as push-pull oscillation to both the mixer and as feedback to the synthesizer. Synthesized frequency control, at this point, accepts oscillator fundamental frequencies of from 792.5 to 907.5 MHz by way of a pad, divides them down through both a fixed and programmable divider and compares to a crystal-derived reference frequency. Any error voltage is integrated and returned to the push-pull oscillator for fre-

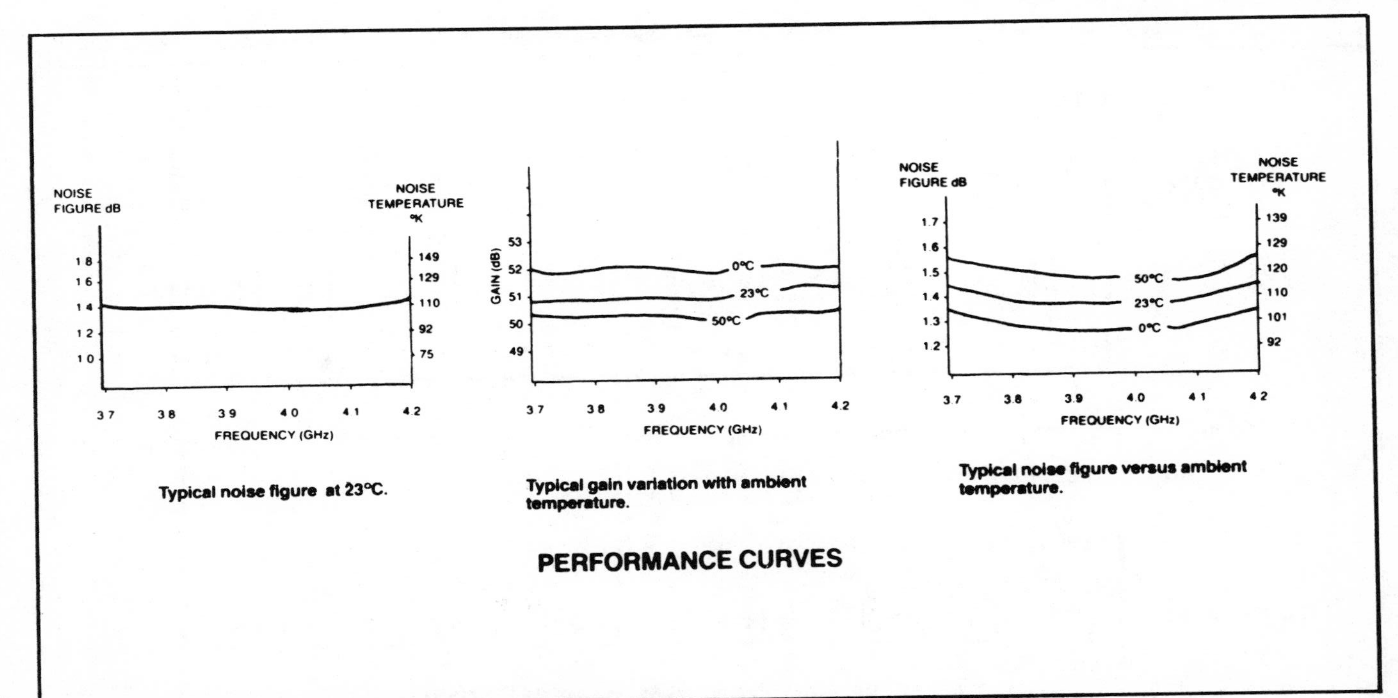

Fig. 7-8. Good sets of performance curves for 110° GaAsFET LNAs supplied by M/A-COM Video Systems.

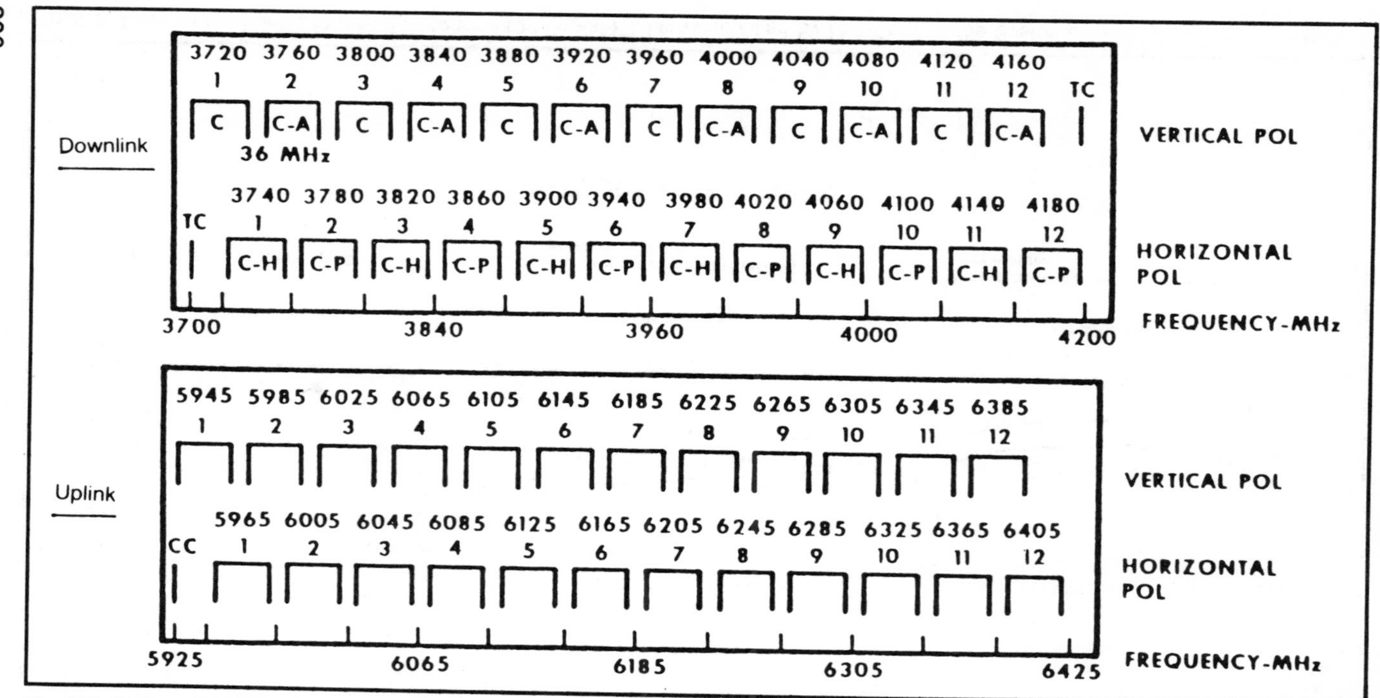

Fig. 7-9. Recent AT&T request to the FCC for satellite stationing and transponder frequency approval (courtesy AT&T).

quency control. BCD inputs from thumbwheel and channel switches operate on a total of 36 memory locations—12 for a single-polarity and 24 for dual-polarity subsystems.

The 550 MHz first mixer output next passes through an amplifier and then a second bandpass filter before reaching the second mixer and its local oscillator. This one is fixed at 480 MHz so that the difference frequency between 550 and 480 MHz develops 70 MHz as the final i-f output at a bandpass of 30 MHz. A variable center frequency in an all-pass equalizer with inverse phase response couples this intelligence through broadband integrated circuit amplifiers, compensating for filter losses and producing an overall rf-to-i-f gain of some 23 dB.

I-f amplifier PN 1808340-1 contains four, fixed gain IC amplifiers, each developing gains of 14 dB, the first two of which are AGC-controlled, maintaining an essentially constant output of +2 dBm over a 40 dB input swing. There is also a threshold extension demodulator, using digital phase-locked loop techniques to detect modulation in the 70 MHz carrier. The PLL tunes series resonant filters to track FM, limiting noise bandwidth and reducing the i-f frequency swing to 15 MHz instead of 30 MHz. Power supply operating voltages have positive and negative IC regulators, protected by 1.5 A fuses.

The video demodulator circuit board (Fig. 7-11) receives its 70 MHz i-f at J1 through an input level adjust pad and impedance matching transformer to a limiting amplifier and divide-by-two circuit, which reduces 70 MHz to 35 MHz. This carrier and its intelligence passes to both a digital demodulator and digital AFC, which includes two divider circuits, a level translator and phase detector whose reference originates from a 1.4 MHz crystal oscillator and divider. This AFC voltage is amplified by U10 and U11 and serves as frequency control for the local oscillator (but is not operative in the synthesized receiver version).

The U3-U4 demodulator recovers baseband and all subcarriers on the satellite signal (including dispersal) from the 35 MHz information, which are amplified and then distributed to various connecting configurations for final processing at 1 and 2-volt outputs, respectively. Additional operating voltage regulators supply two pairs of extra dc voltages. Subcarrier processing (Fig. 7-12) passes either 6.2 or 6.8 MHz (whichever is specified), while attenuating other subcarriers and video. A double balanced mixer and another local oscillator heterodyne this signal, permitting a 13.5 MHz difference signal to pass, where a T^2 L counter divides by three,

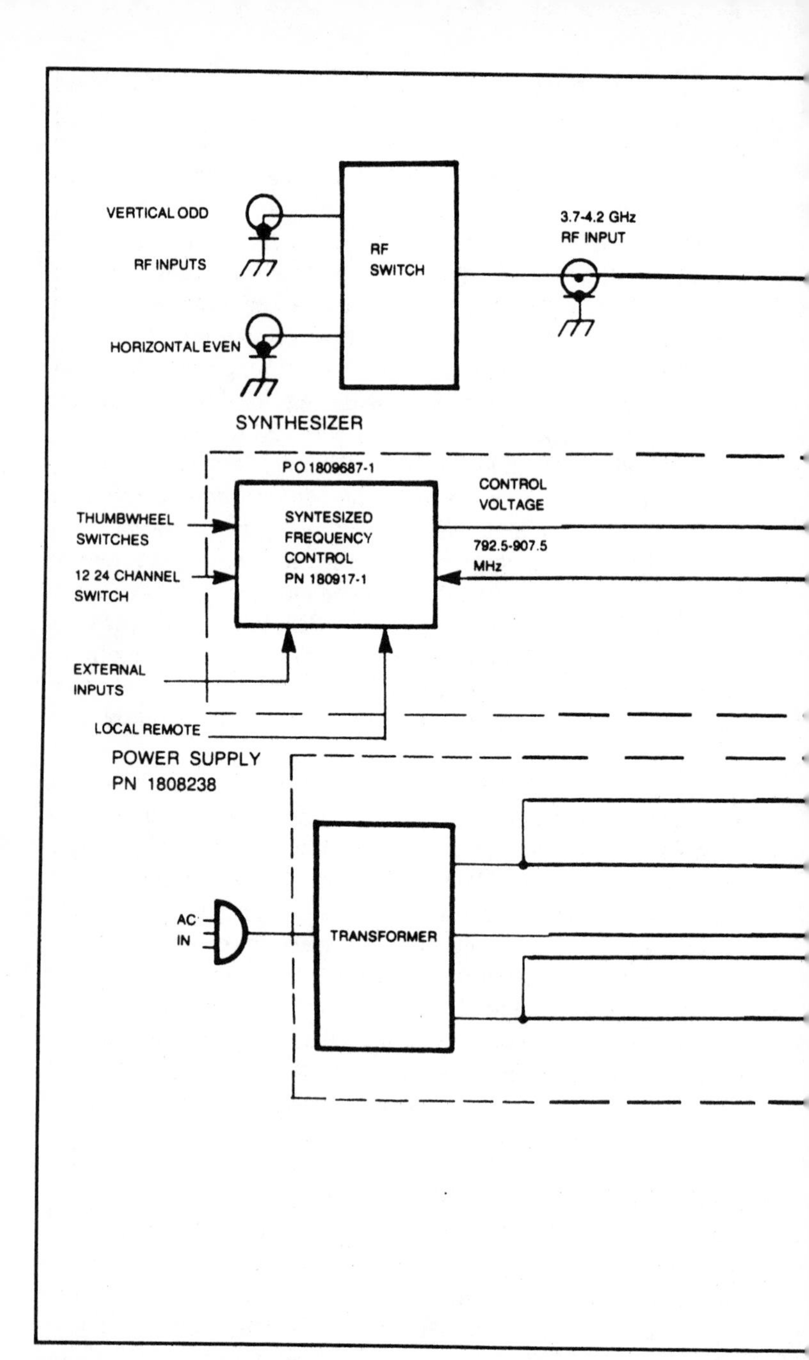

Fig. 7-10. Block diagram of M/A-COM VR-4XS satellite receiver (courtesy of M/A-COM).

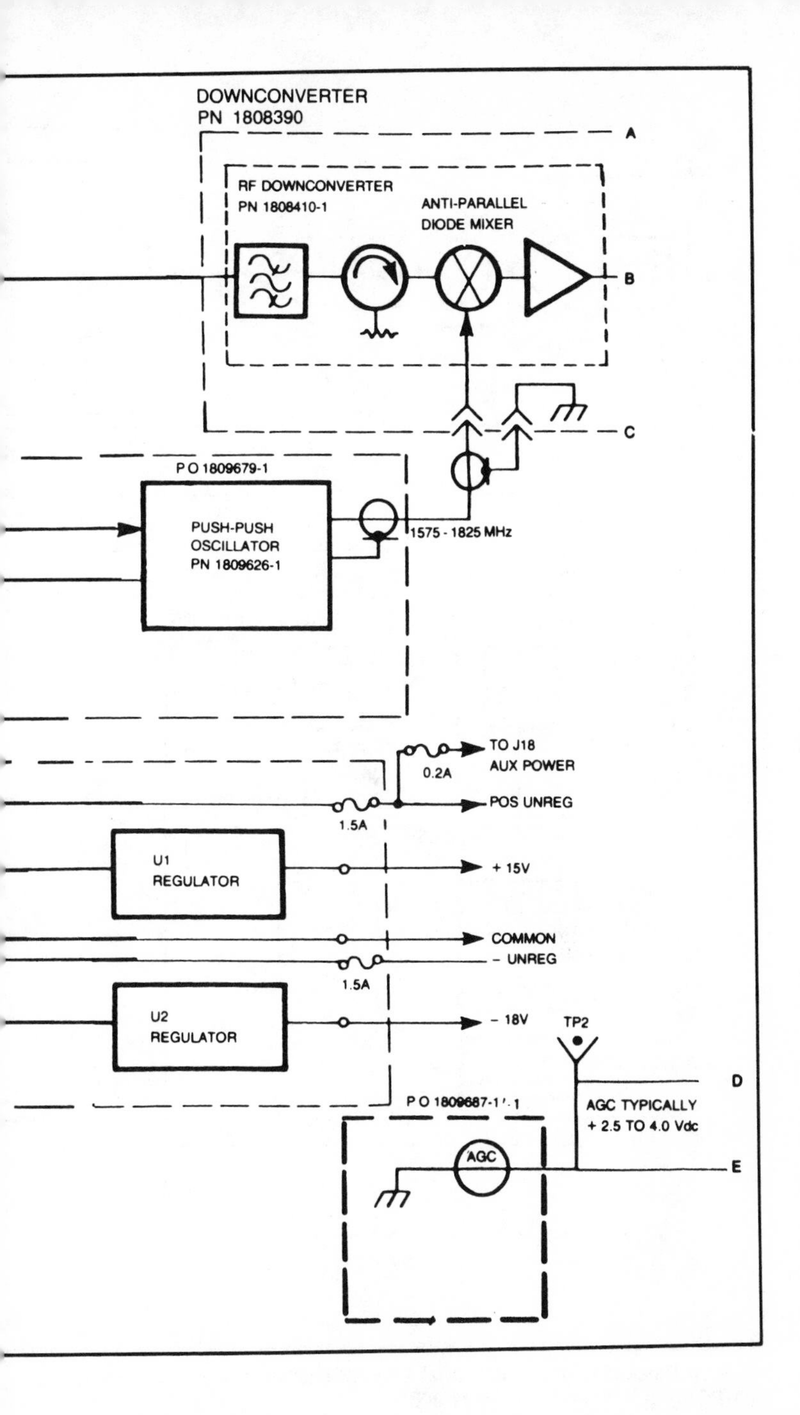

DOWNCONVERTER
PN 1808390
A
RF DOWNCONVERTER
PN 1808410-1
ANTI-PARALLEL
DIODE MIXER
B
C
P O 1809679-1
PUSH-PUSH
OSCILLATOR
PN 1809626-1
1575 - 1825 MHz
TO J18
AUX POWER
0.2A
POS UNREG
1.5A
U1
REGULATOR
+ 15V
COMMON
– UNREG
1.5A
U2
REGULATOR
– 18V
TP2
D
AGC TYPICALLY
+ 2.5 TO 4.0 Vdc
AGC
E

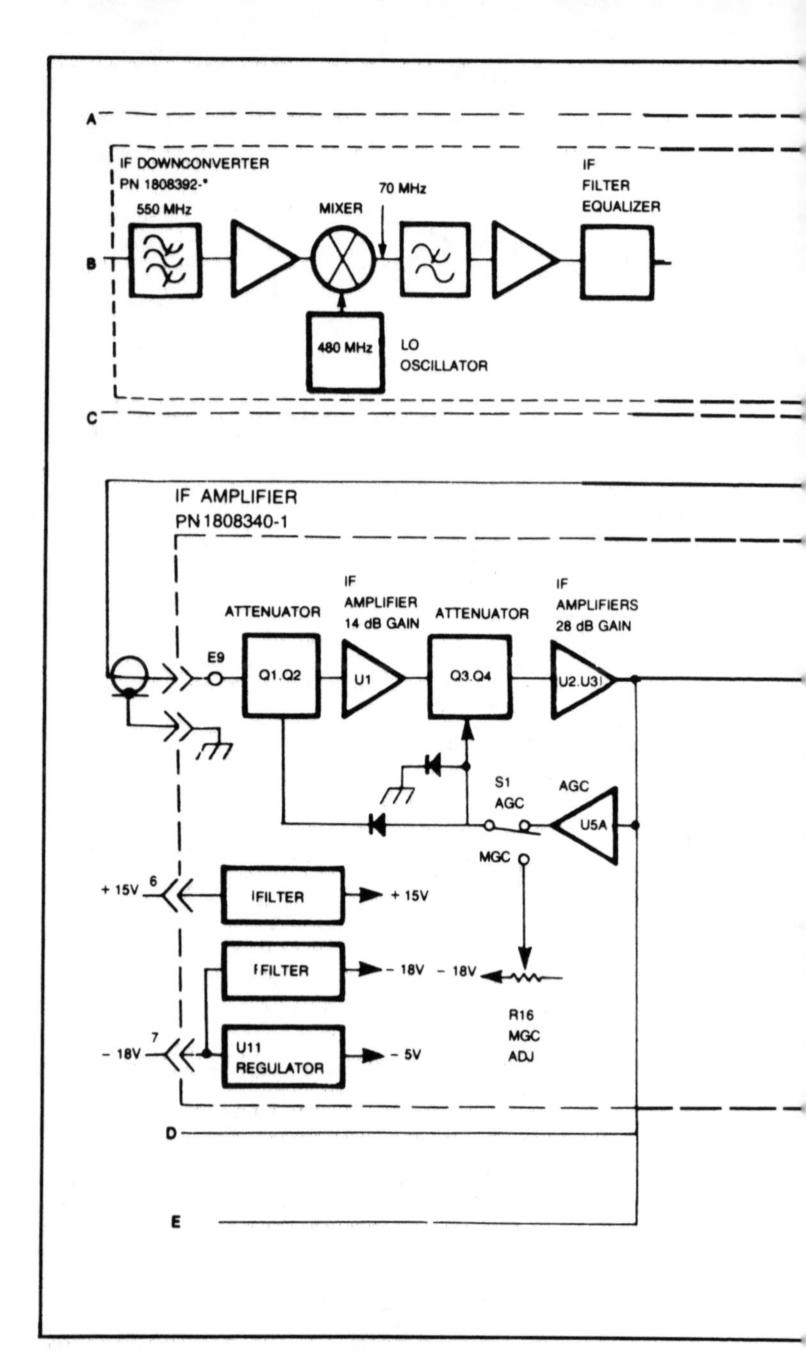

Fig. 7-10. Block diagram of M/A-COM VR-4XS satellite receiver (courtesy of M/A-COM). (Continued from page 208.)

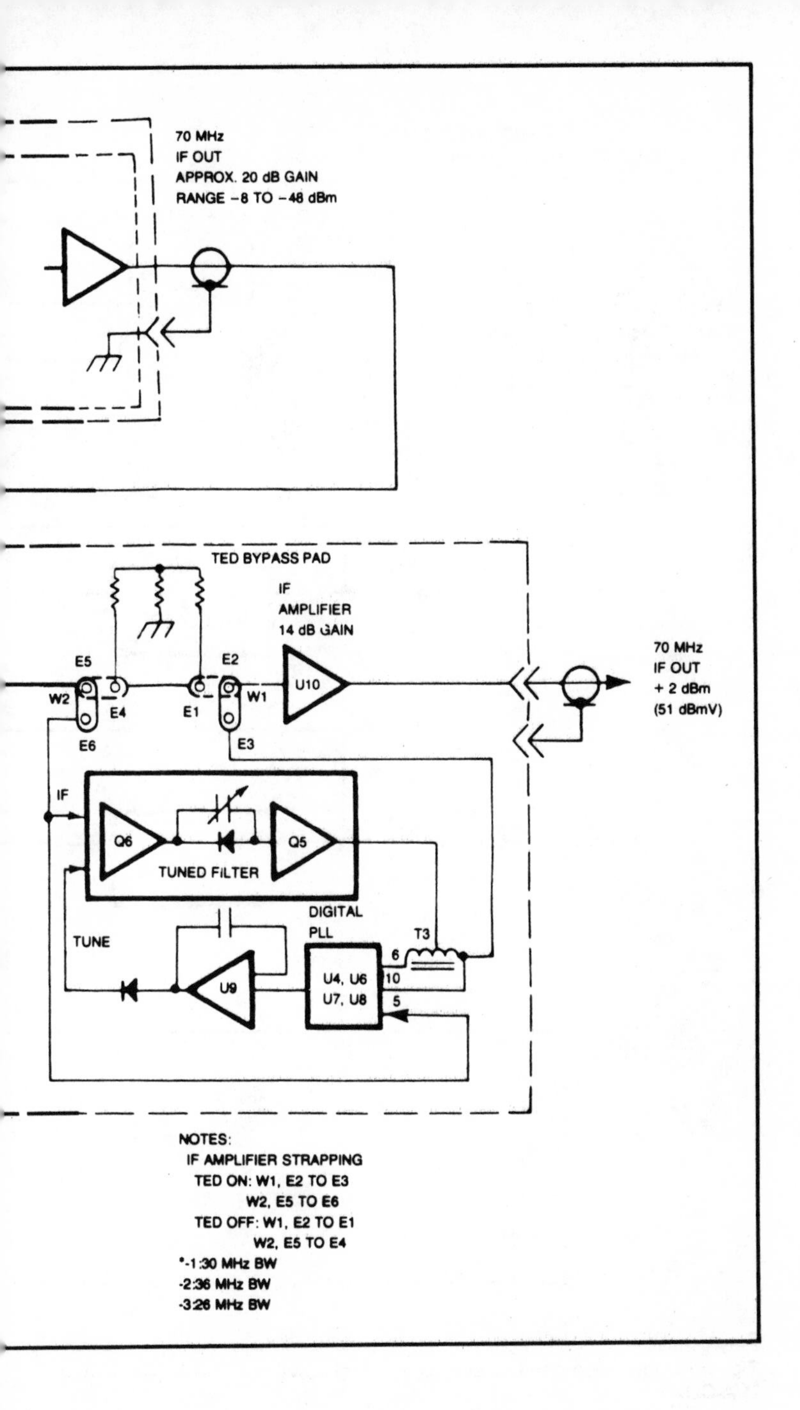

70 MHz
IF OUT
APPROX. 20 dB GAIN
RANGE −8 TO −48 dBm
TED BYPASS PAD
IF
AMPLIFIER
14 dB GAIN
E5
E2
W2
E4
E1
W1
U10
E6
E3
70 MHz
IF OUT
+ 2 dBm
(51 dBmV)
IF
Q6
Q5
TUNED FILTER
DIGITAL
PLL
T3
TUNE
6
U4, U6
U7, U8
10
5
U9
NOTES:
IF AMPLIFIER STRAPPING
TED ON: W1, E2 TO E3
W2, E5 TO E6
TED OFF: W1, E2 TO E1
W2, E5 TO E4
*-1:30 MHz BW
-2:36 MHz BW
-3:26 MHz BW

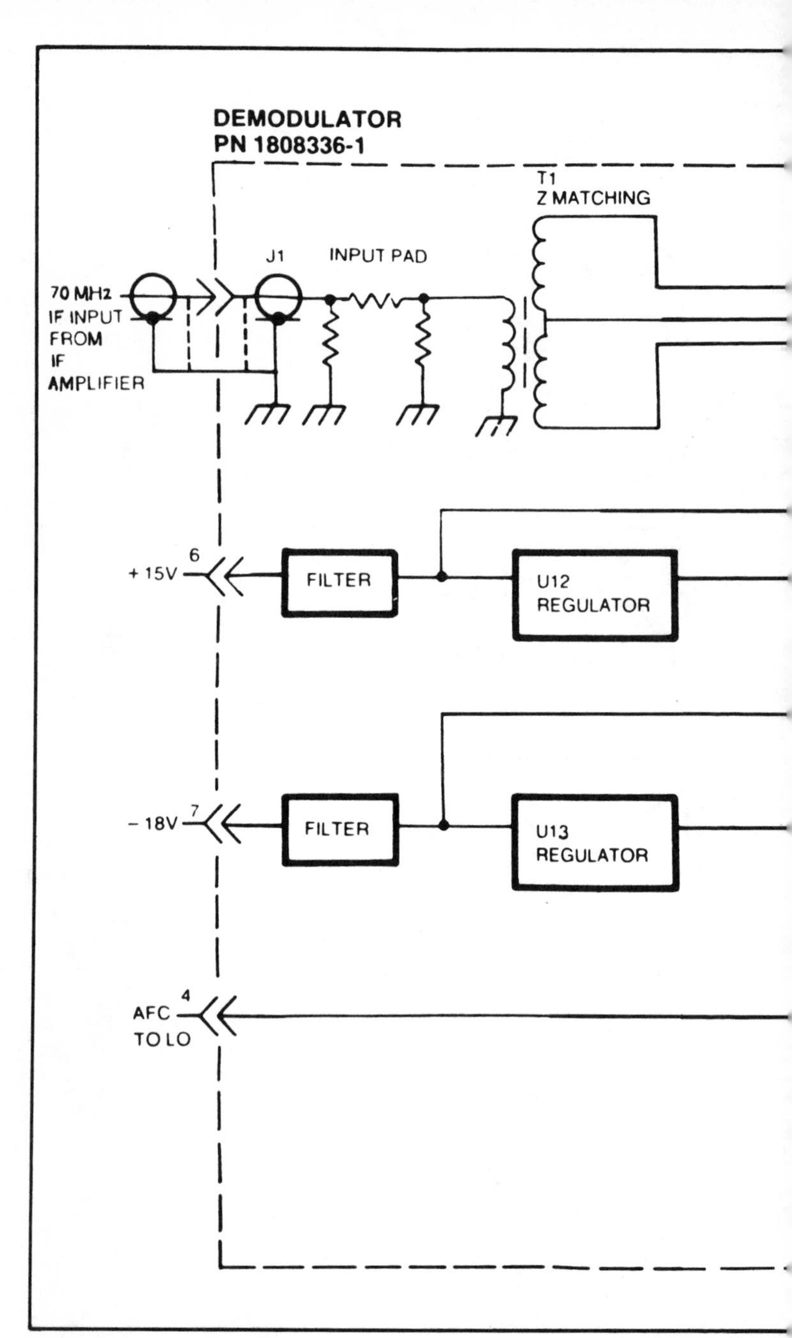

Fig. 7-11. FM video demodulator for AR-4XS satellite receiver (courtesy of M/A-COM).

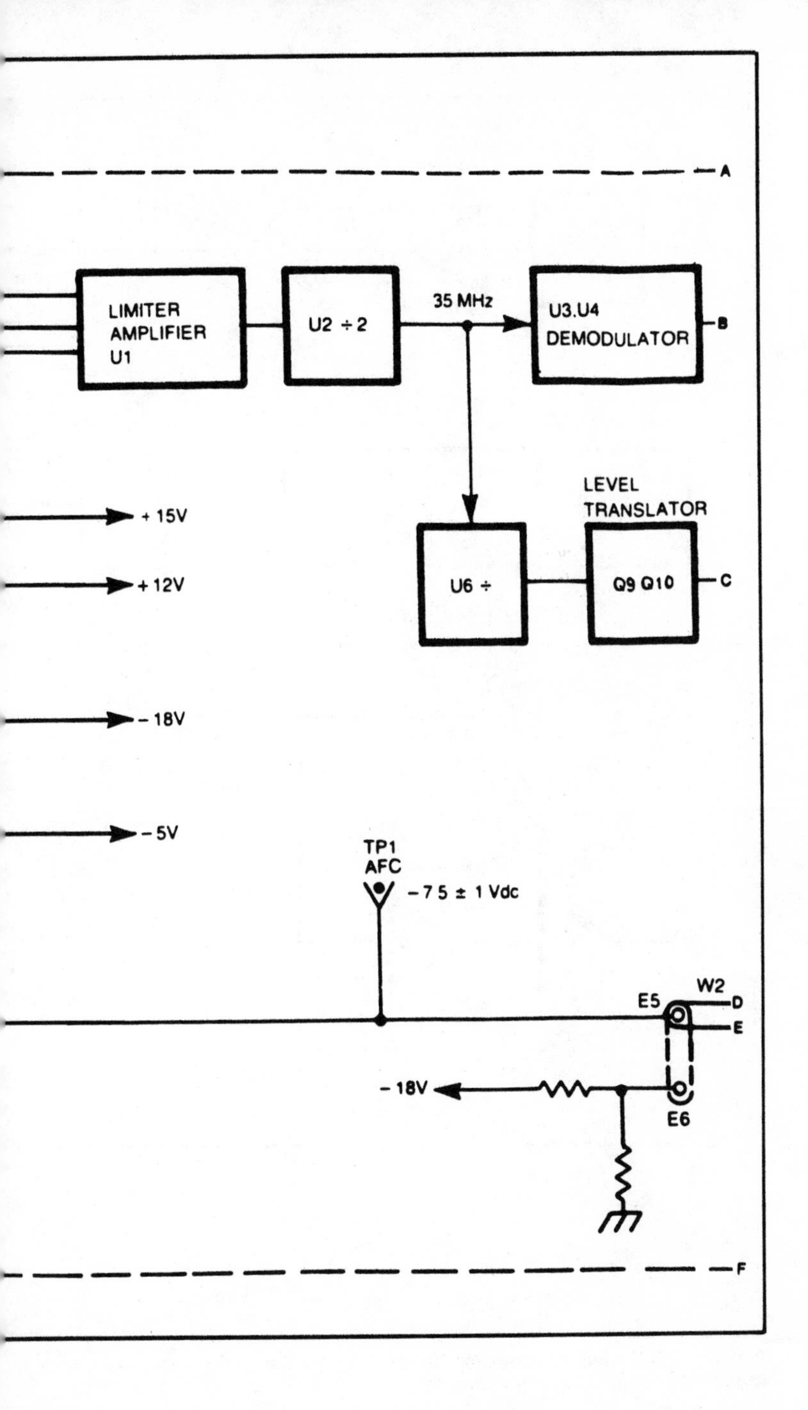
A
LIMITER
AMPLIFIER
U1
U2 ÷ 2
35 MHz
U3.U4
DEMODULATOR
B
+ 15V
+ 12V
LEVEL
TRANSLATOR
U6 ÷
Q9 Q10
C
− 18V
− 5V
TP1
AFC
− 7 5 ± 1 Vdc
W2
E5
D
E
− 18V
E6
F

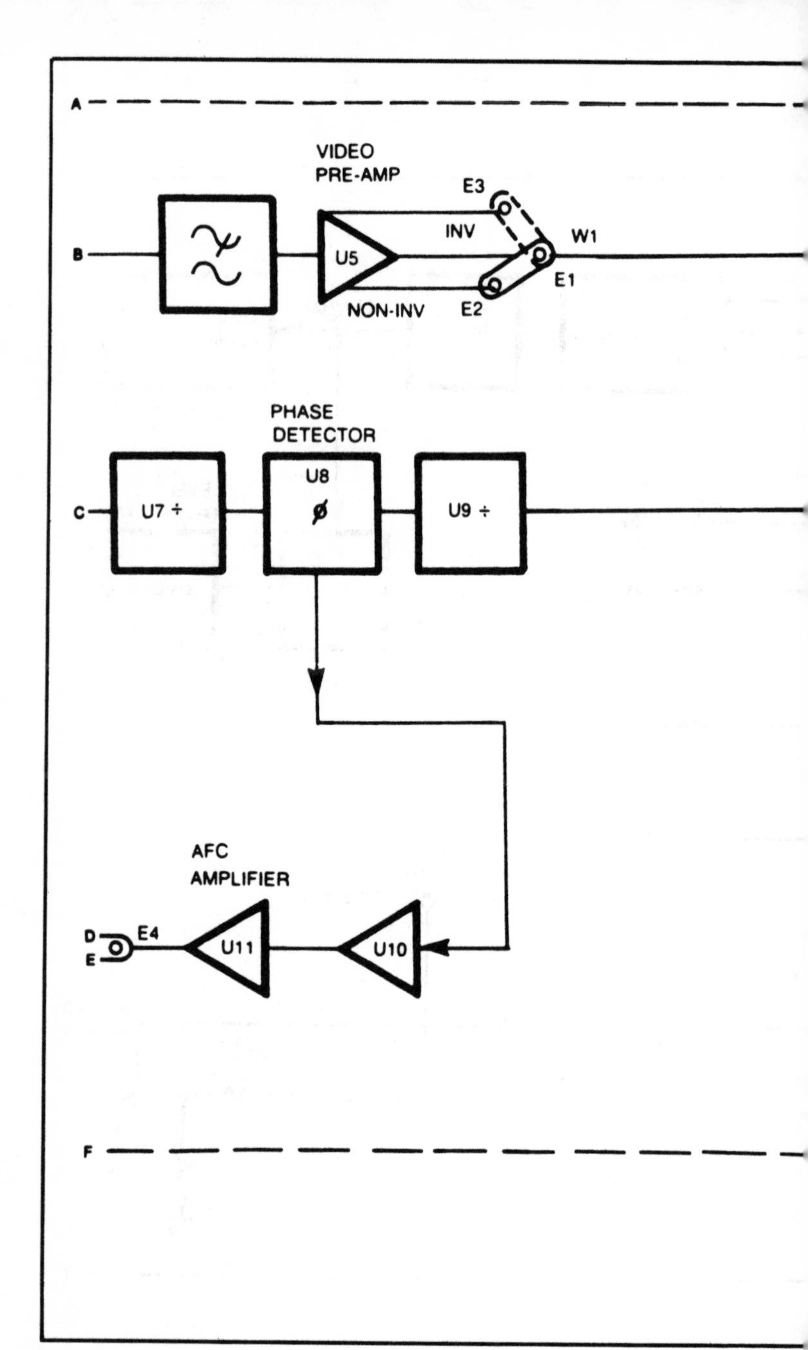

Fig. 7-11. FM video demodulator for AR-4XS satellite receiver (courtesy of M/A-COM). (Continued from page 212.)

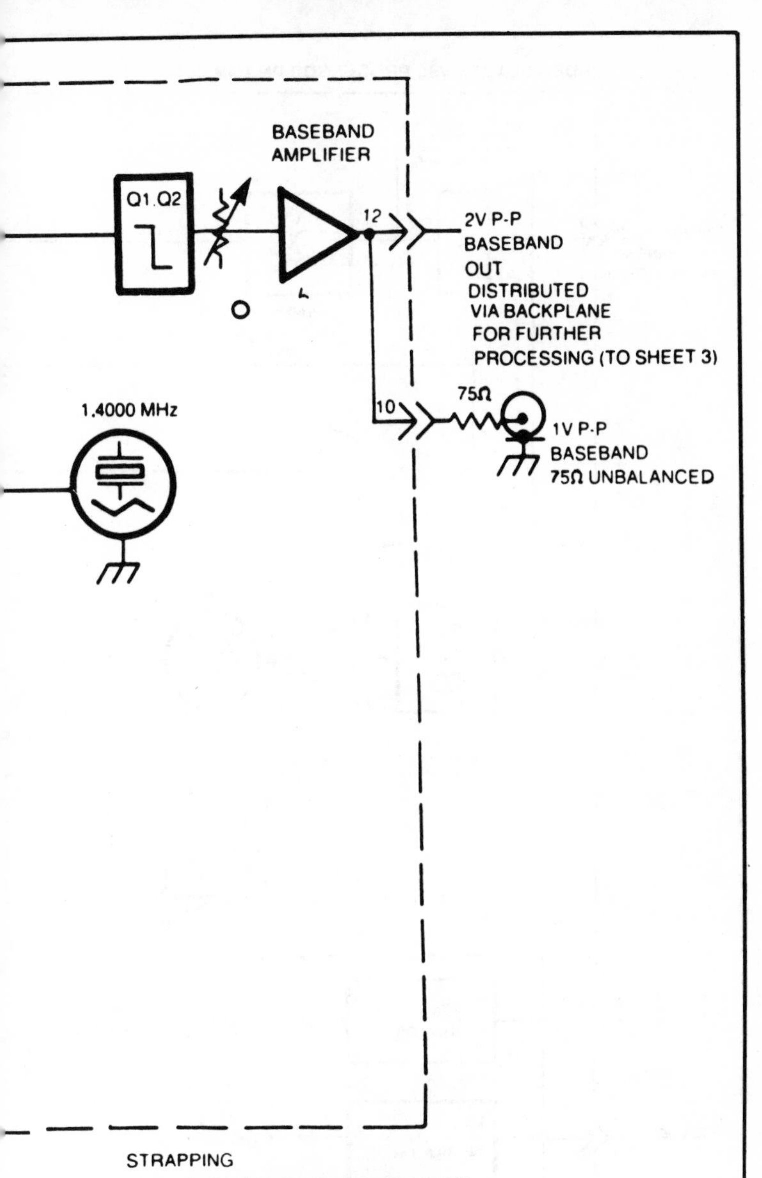

STRAPPING

W1: E1 TO E2 NON-INVERTED VIDEO

E1 TO E3 INVERTED VIDEO

W2: E4 TO E5 AFC VOLTAGE ON

E4 TO E6 REFERENCE VOLTAGE REPLACES AFC VOLTAGE

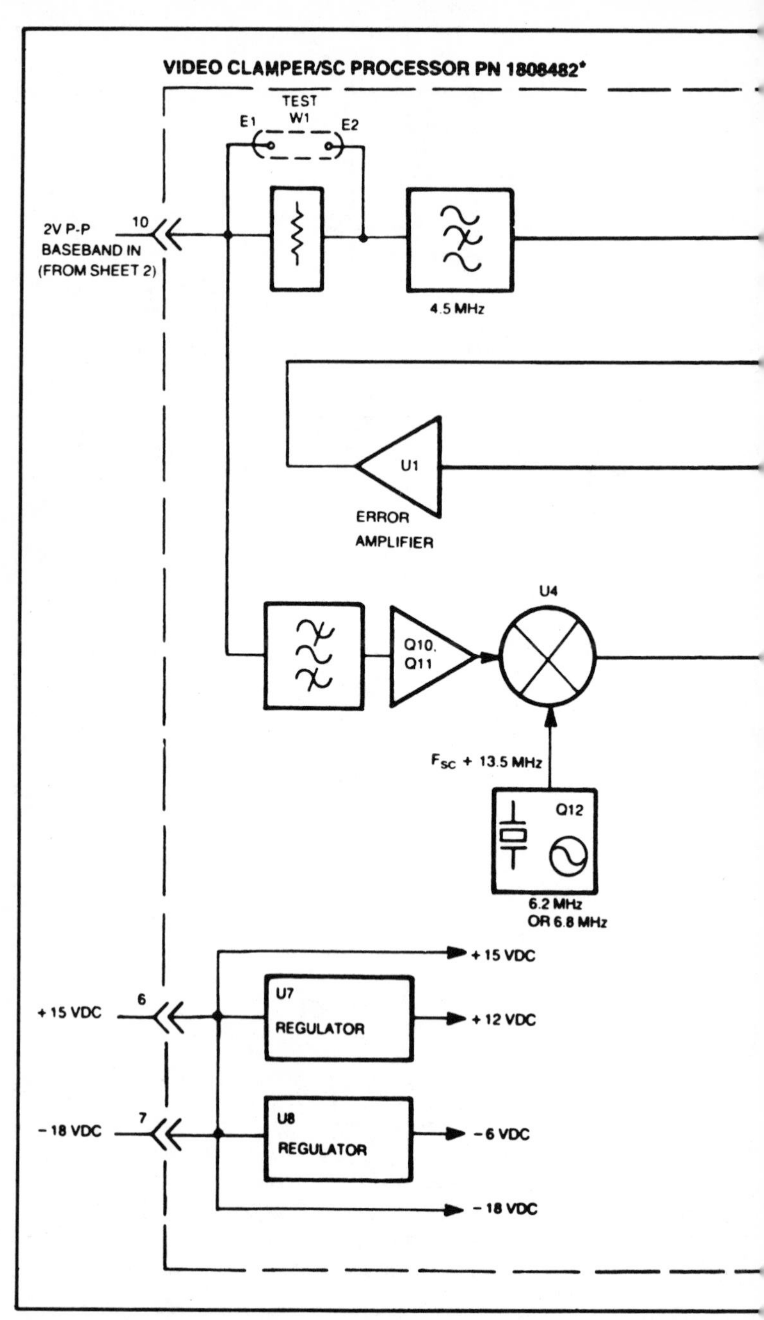

Fig. 7-12. Baseband processor and audio demodulator for the VR-4XS (courtesy of M/A-COM).

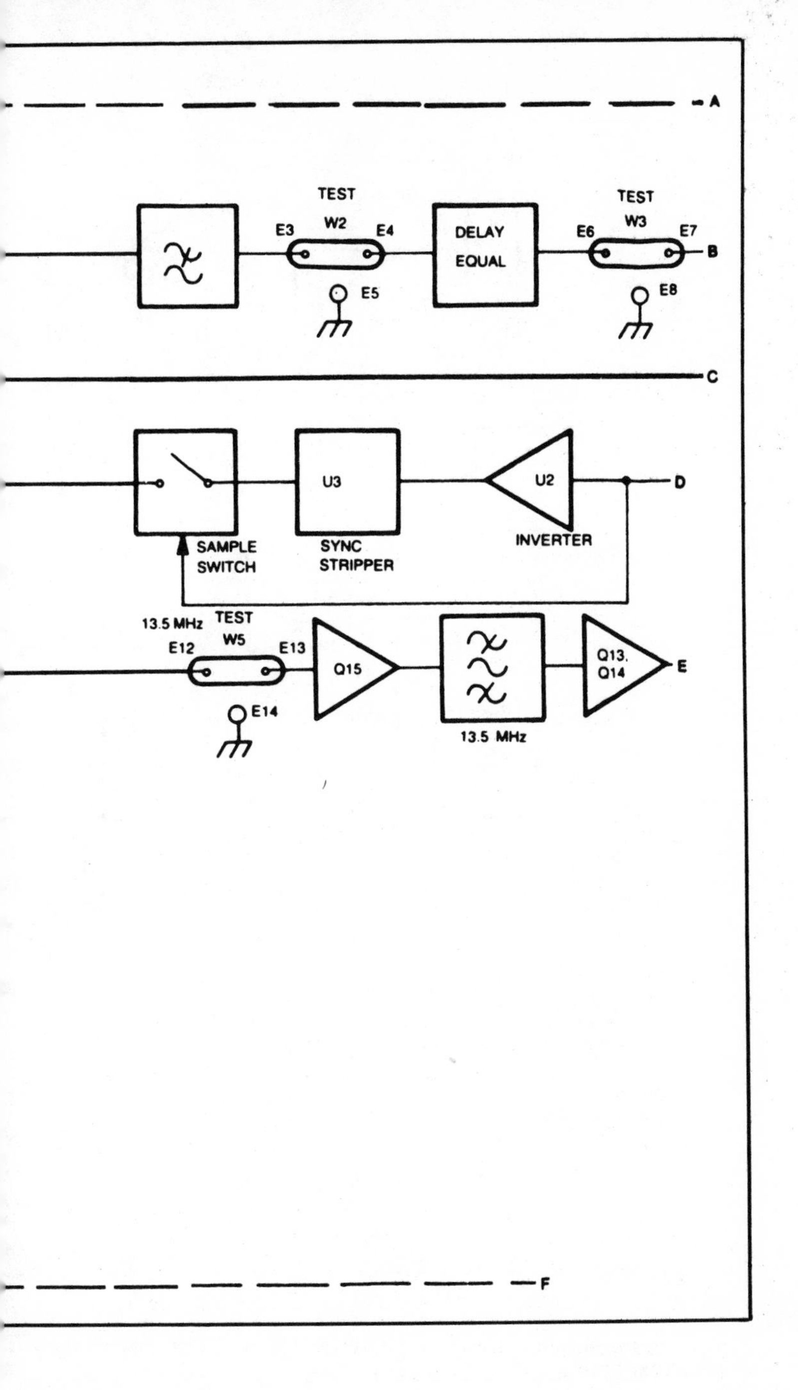
A
TEST
W2
E3
E4
E5
DELAY
EQUAL
TEST
W3
E6
E7
E8
B
C
U3
U2
D
SAMPLE
SWITCH
SYNC
STRIPPER
INVERTER
13.5 MHz
TEST
W5
E12
E13
E14
Q15
Q13,
Q14
E
13.5 MHz
F

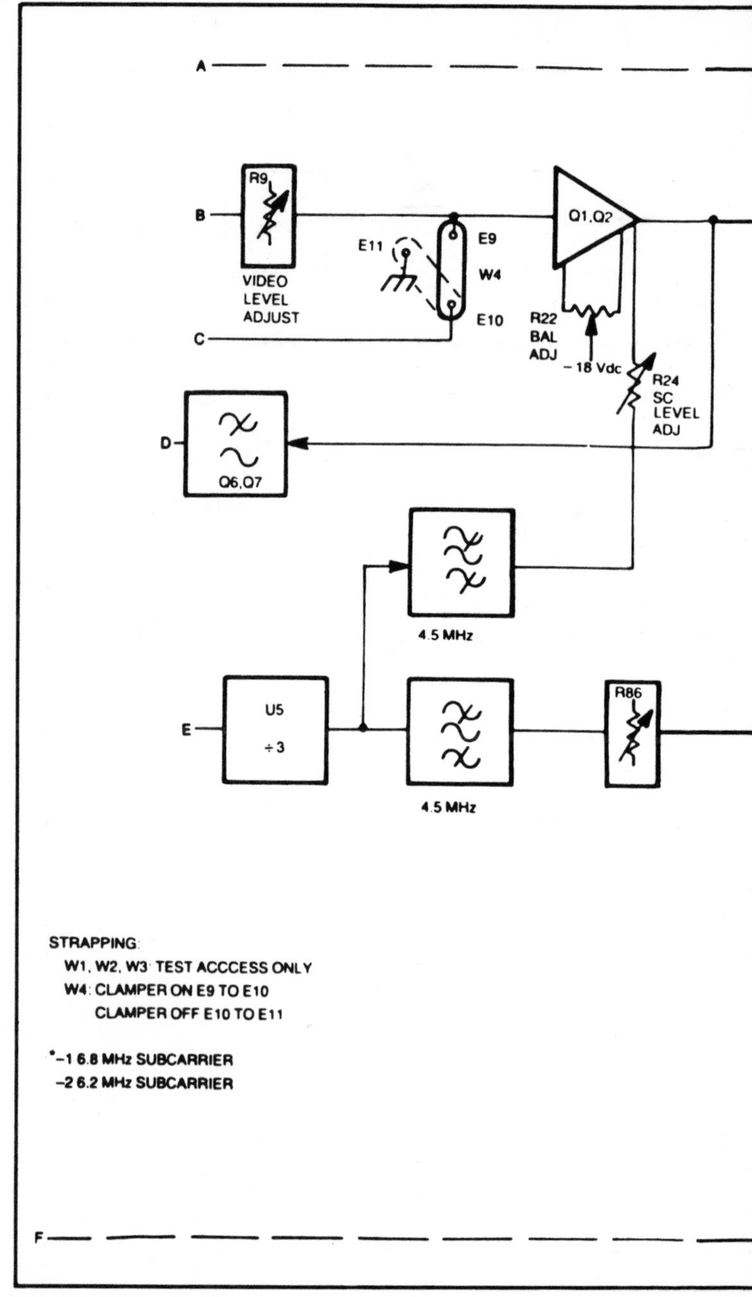

Fig. 7-12. Baseband processor and audio demodulator for the VR-4XS (courtesy of M/A-COM). (Continued from page 216.)

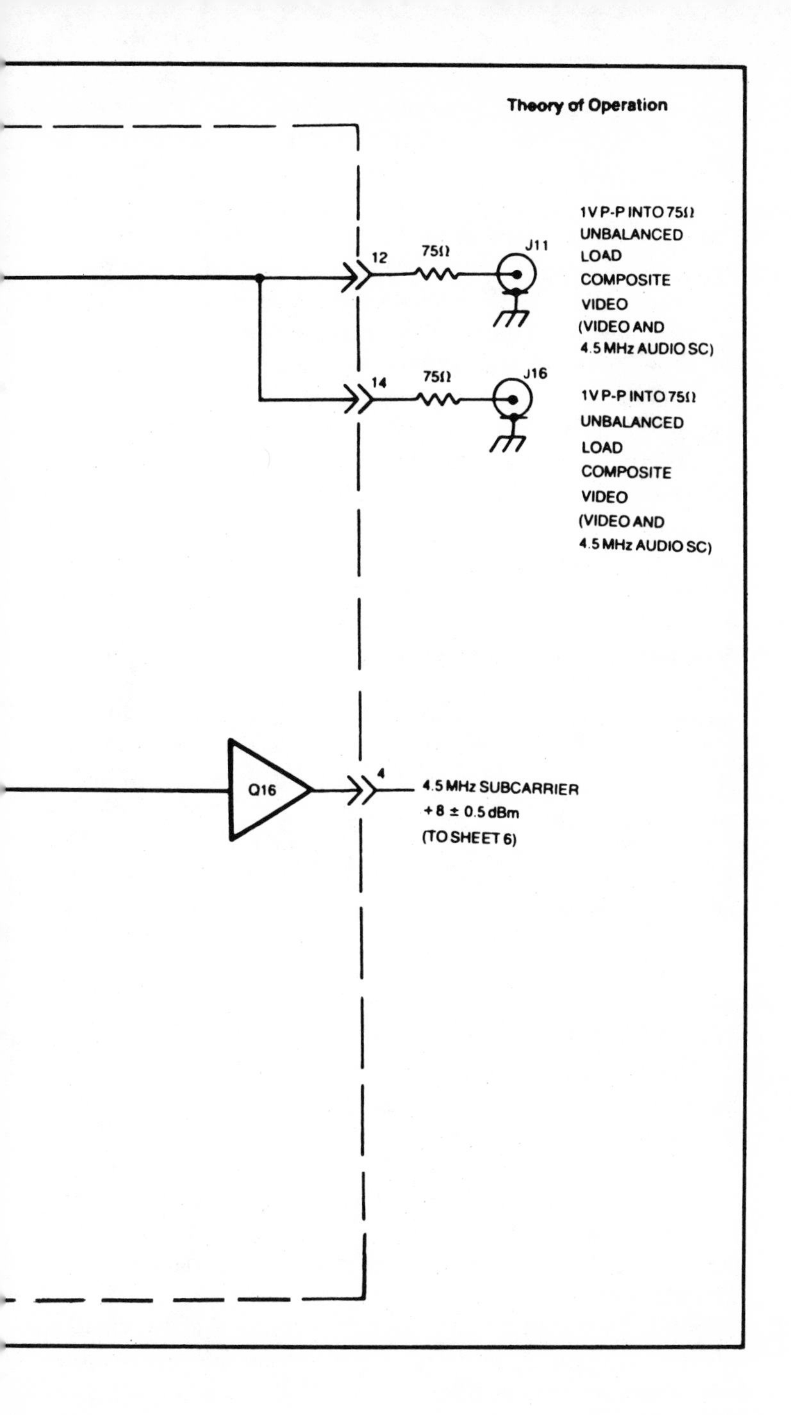
12
75Ω
J11
1V P-P INTO 75Ω
UNBALANCED
LOAD
COMPOSITE
VIDEO
(VIDEO AND
4.5 MHz AUDIO SC)
14
75Ω
J16
1V P-P INTO 75Ω
UNBALANCED
LOAD
COMPOSITE
VIDEO
(VIDEO AND
4.5 MHz AUDIO SC)
Q16
4
4.5 MHz SUBCARRIER
+8 ± 0.5 dBm
(TO SHEET 6)

producing the 4.5 MHz audio intercarrier. Here 75 kHz deviation is reduced to ±25 kHz for standard TV sound deviation and the subcarrier output is split. Half of it goes to the video clamper amplifier for addition to the video signal and the other portion passes through pin 4 for the cable modulator option. Afterwards, the subcarrier demodulator recovers audio signals from the FM-modulated rf subcarrier in baseband for balanced 600-ohm audio outputs.

M/A-COM's VR-4XS is a pretty good example of how receivers and their downconverters and audio-video demodulators operate. While we did not include all the diagrams, by any means, those referenced are the major units and should offer prime suggestions of what to look for in well-designed consumer or commercial components. Of course, if outputs are destined for TV/rf inputs, then both sound-sight basebands will have to be remodulated—the video on AM carrier—before a TV can handle inputs on either conventional channels 3 or 4 (60 to 72 MHz). Audio/video baseband, of course, is already available for display monitors which readily accept sound between 30 Hz and 15 kHz, and video from dc to 4.2 MHz. Later, as high definition 1125-line TV enters closed circuit and direct broadcast satellite pictures, bandwidth will have to be expanded to something like 20 MHz (even companded), while all audio should be received dual-channel, or stereo, even though its ±25 kHz is not expected to be necessarily increased to FM radio's full deviation of ±75 kHz—but it could happen. Leaders in HDTV at the moment are Japan's Ikegami, Panasonic, and Sony. It could be successfully used in CATV, movie theaters, teleconferencing, and all types of satellite transmissions, with picture quality equivalent to 35 mm transparencies and superior to 35 mm motion picture film.

OTHER CONSIDERATIONS

This heading is a little misleading in that further contents of this chapter may have been touched upon previously in other parts of the book. Here we are dealing with television receive only antennas and the information applies directly and *only* to TVRO. Therefore, liberty is taken to work out a few more equations and show several more antennas which are especially popular at this writing, or have very recently been introduced to the market for the 1983 sales year. We think both offerings will prove beneficial since a little repetition and practical extension of information covered previously should only aid in increasing knowledge rather than truncating it. In addition, some of the downlink information will be relatively new, including computerized extra knowledge which considers specific space factors. Most or all of this valuable information on TVRO equations was generously supplied by our friends at M/A-COM, who have obviously done innumerable, worthwhile studies to arrive at the various conclusions and derivations plus the very handy charts shown under the next heading.

Carrier/Noise and G/T

Downlink carrier-noise involves transponder isotropic radiated power, signal output, G/T (net system gain over noise temperature), receiver bandwidth, and Boltzmann's constant as a conversion factor. EIRP typical values range between 30 and 35 dBw (dB in watts).

$$C/N \text{ downlink} = EIRP - S^* + G/T - 10 \log BW_{Hz} + 228.6$$

Now, satellite transmission from 22,300 miles in space does, most certainly, involve more than a little attenuation, about 196 dB. Without considering bad weather and some other extra losses which vary with frequency, here is the clear-view space loss value:

$$^*S = 123.5 + 20 \log F_{MHz} + 10 \log$$
$$1.42 - 0.42 \text{ TVRO latitude} \times \cos$$
$$(\text{Satellite longitude} - \text{TVRO longitude})$$

When using these equations, pay particular attention to frequency and how it is defined. Hz does *not* mean MHz, and vice versa!

There is also a useful equation for G/T which you may or may not want to use for one reason or another but it's presented nonetheless for reference as the spirit moves. You will have to know a little more than just the equation information to put it to full use, especially any loss between LNA and antenna cable and couplings. On the other hand, the M/A-COM chart that follows pretty well does all this for you without the necessity of some complications, using an elevation of 25° to remove the pointing angle from looking at the 290°K earth. Even though approximate, you should find it very useful (Fig. 7-13). Antenna sizes are 3 to 10 meters.

$$G/T = G_a \text{ (ant. gain)} - 10 \log (T_a \text{ ant. noise temp} + L_1 - 1)\ T_0$$
$$290\,°K + L_1 (T_{LNA} \text{ noise temperature}).$$

The L_1 item is defined at the antilog of the loss between the an-

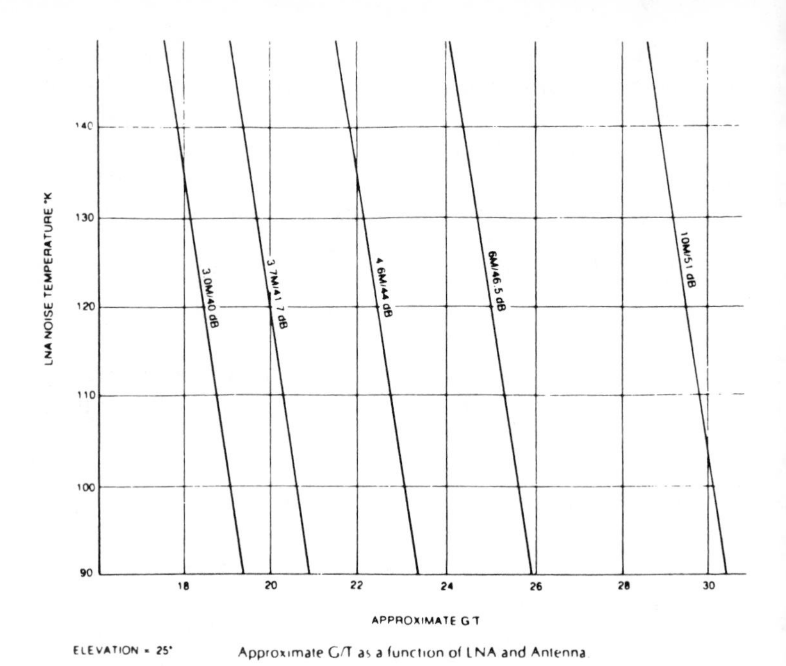

Fig. 7-13. Three- to 10-meter dishes and their gains in terms of G/T (courtesy of M/A-COM).

tenna and LNA in dB divided by 10. If you'd like to work this through for either sheer delight or experience, please do. But, the chart that follows may be somewhat easier to comprehend. LNA temperatures are shown vertically on the chart's left side, antenna size and gain register on the black slant lines, and the approximate G/T you select at the bottom. For in between antenna sizes you should find it easy to interpolate with reasonable accuracy.

According to M/A-COM, you may also use a chart or calculations to find the approximation of G/T with respect to EIRP and V_S/n. The chart is illustrated in Fig. 7-14, the equations are as follows:

$$G/T + EIRP \cong 55.7$$

and $$C/N_t \cong 10 \log \frac{1}{\dfrac{1}{\text{antilog } \dfrac{C/Nd}{10}} + \dfrac{1}{\text{antilog } \dfrac{C/I}{10}}}$$

where C/I is the ratio of carrier to interference, which are terrestrial microwaves at 4 GHz, antenna received adjacent satellite signals, and satellite distortion, itself. Typically these values are 18 to 25 dB and rise with larger antennas. Combining downlink C/Nd with C/I gives a total carrier-to-noise C/N_t. Now, having determined total C/N_t, the video signal-to-noise above receiver threshold isn't at all difficult:

$$V_s/n = C/N_t + 37.5$$

TVRO system G/T losses following the LNA are usually insignificant since a 50 dB LNA gain can nullify cable and power divider loss and receiver noise figure. Consequently, we won't quote M/A-COM's suggested equation for these factors since your spec sheets will probably give you enough information to work these out with simple arithmetic.

We would, however, like to add a noise temperature/dB conversion chart that will come in quite handy (Fig. 7-15). The noise figure is in dB and the noise temperature, of course, in °Kelvin. Equations from which these derive are:

$$T = \left[\left(\text{antilog}\ \frac{dB}{10}\right) - 1\right] 290^\circ$$

$$dB = 10 \log (T/290^\circ + 1)$$

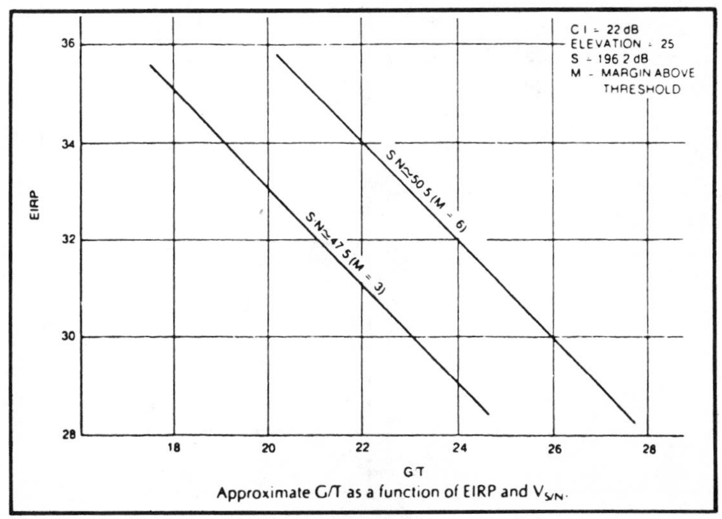

Fig. 7-14. EIRP and S/N also approximate G/T (courtesy of M/A-COM).

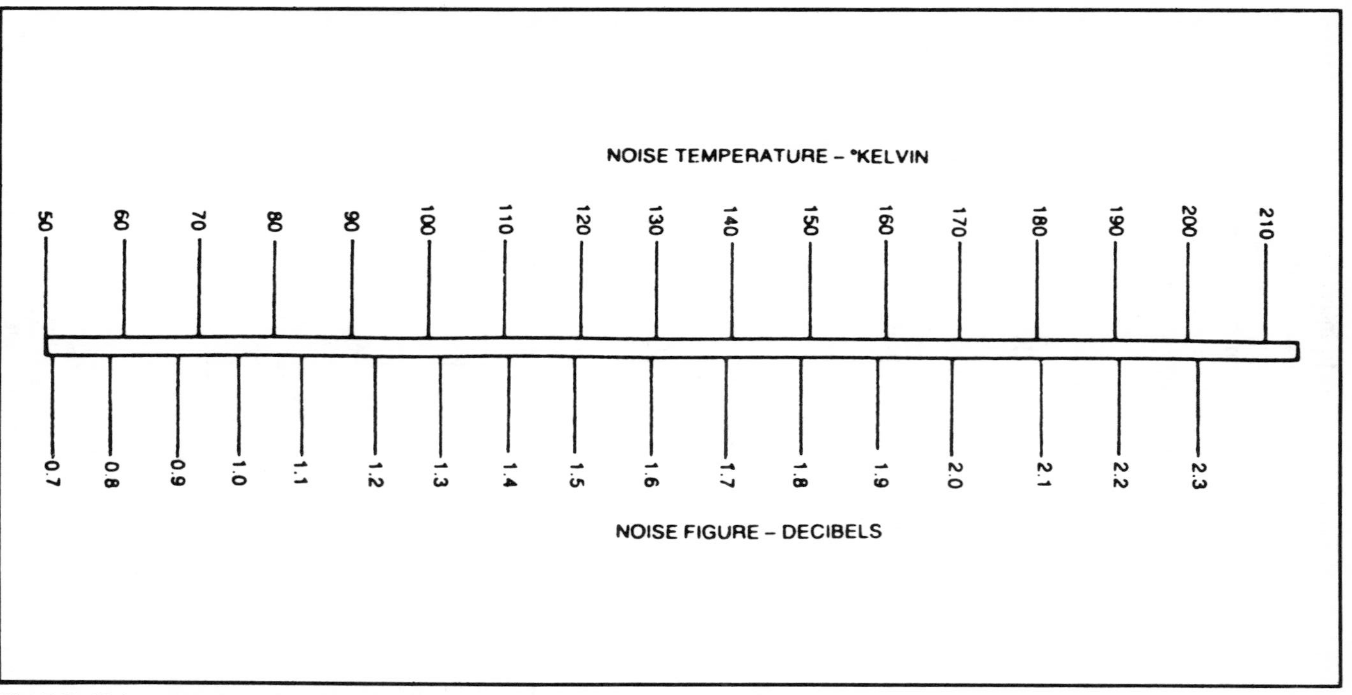

Fig. 7-15. Noise temperature in °K versus noise in dB (courtesy of M/A-COM).

The 290 ° K, if you remember, represents the earth's temperature. The chart, however pretty well does all the interpolating for you. The antilog symbol on almost any reasonable calculator is 10^X, if you're hesitant, and is usually written $\log^{-1}$. But if you're dealing with negative dB numbers (−30 dB, for instance), your calculator must have this negative sign before −30 dB to change the mantissa and crank out the right answer.

POINTING THE EARTH STATION

Surmising you've already produced a Topol map from some nearby government agency, or had someone kindly run off a computer location for you, it's time to find out how and where to point your dish for the various satellites. As usual, there are several very good methods of doing this: one with direct calculations, another by handy charts, and a third with a calculator. As usual, we'll try and give all three so you may use whichever suits best. Although some of these instructions apply to products of specific manufacturers—as will installation instructions in another chapter most or parts of such information universally applies to all earth stations in one way or another. As you will find in establishing your own, there will be individual instructions for your unit, but you may want another one or two perspectives that could well prove useful in dealing with unexplained situations or unusual conditions not normally encountered.

In determining satellite dish pointing information you are still required to know certain things about either your specific location and certainly the satellite's precise location. You can't, unfortunately, pop a wet finger toward the sky and determine anything but the direction of some breeze, if there is one. However, many companies selling earth stations, and all professional installers will calculate pertinent information for you, so this is no problem. But, when you begin looking for additional satellites with AZ-EL mounts or have difficulties with true north in a polar mount, then azimuth (horizontal traverse) and elevation (the angle between earth and the satellite) can make a world of difference. Just by knowing these two key parameters, you might be able to reset your polar mount quite conveniently by just the angle of elevation and swinging the dish through the range of 70 ° to 140 °, which are nominal eastern and western limits of U.S. geosynchronous coverage. Then, by simply knowing the general locations of the various satellites in question, their identities and longitudinal positions measured from east to west relative to the earth's surface can be established. This means, of course, your satellite dish must have relatively clear field areas east and west, and most certainly south, at least within all pointing angles you wish your dish to swing. If this isn't possible, then you have a distinct problem with 4 GHz downlinks, and may have to wait for 12 GHz dBs downlinks where dish diameters and mounting constraints will be much less because of antenna size and possible/probable rooftop mountings. Unfortunately, urban use of the fixed satellite service—at least for average consumers—will be limited due to both microwave and line-of-sight interference prevalent in many of the larger cities.

Now for both calculations and handy charts:

$$\text{Azimuth}\quad \text{Az} = \cos^{-1}\left[\frac{-\tan(\text{LAT})}{\tan \text{Y}}\right]$$

$$\text{Elevation}\quad \text{EL} = \tan^{-1}\left[\frac{\cos \text{Y} - 0.15116}{\sin \text{Y}}\right]$$

where $\text{Y} = \cos^{-1}[\cos(\text{LONG} - \text{SAT})]\,[\cos(\text{LAT})]$ and north/west degrees are positive.

Latitude and longitude charts for the same results follow in Figs. 7-16 and 7-17. One is in terms of azimuth in degrees from true north and the other deals with the earth station's angle of elevation. Observe in both charts that Δ L becomes the absolute value difference in degrees longitude between the earth station and satellite, establishing both tilt and E/W direction.

MICROWAVE DIRECTIONS

M/A-COM has yet another powerful equation that will enable you to calculate bearing and distance between microwave locations and another point such as your antenna installation. True, this does involve calculator programming, but with a little ingenuity (engineers are supposed to be ingenious), you should be able to work this out on any worthwhile scientific or engineering calculator.

$$\text{Bearing A to B} = \cos^{-1}\frac{\sin(\text{LAT B}) - \sin(\text{LAT A}) \times \text{C}}{\cos(\text{LAT A}) \times \sin(\cos^{-1}\text{C})}$$

$$\text{Distance A to B} = \cos^{-1}\text{C} \times 60 \times 1.150779$$

where $\text{C} = \sin(\text{LAT A} \times \sin(\text{LAT B}) + \cos(\text{LAT A}) \times \cos(\text{LAT B}) \times \cos(\text{LONG B} - \text{LONG A})$

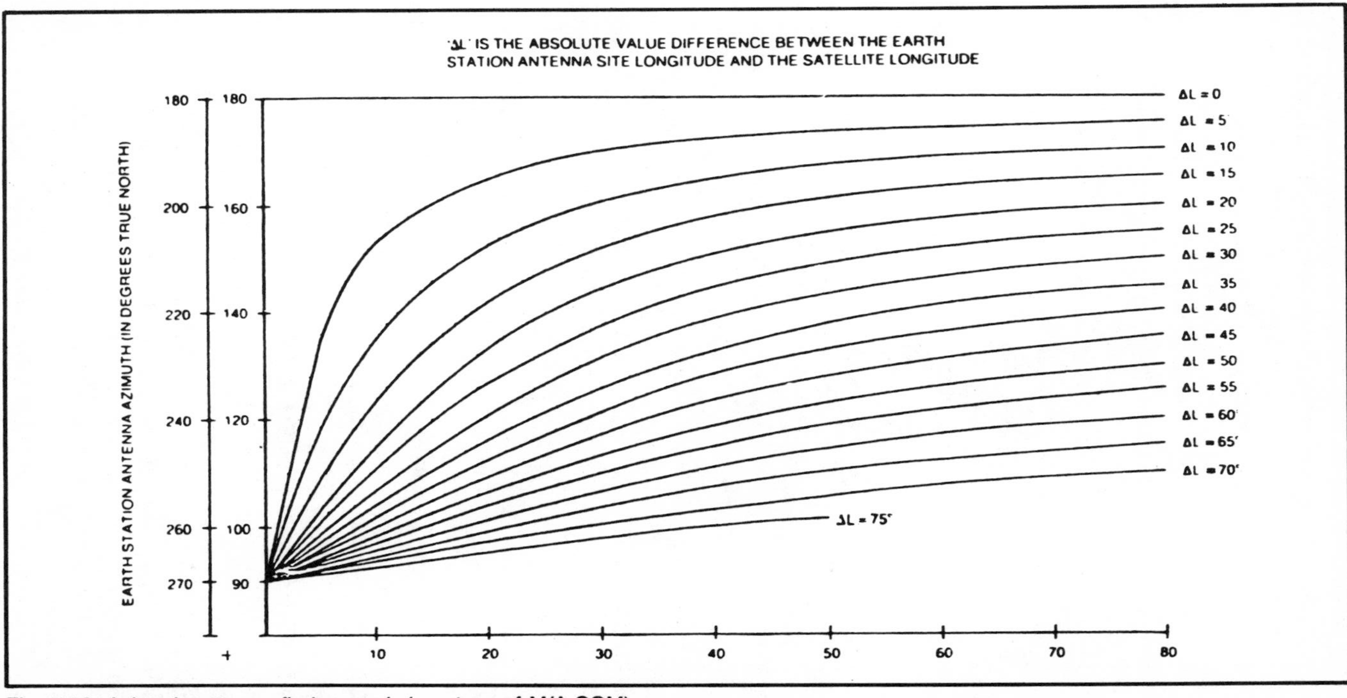

Fig. 7-16. Azimuth antenna fix by graph (courtesy of M/A-COM).

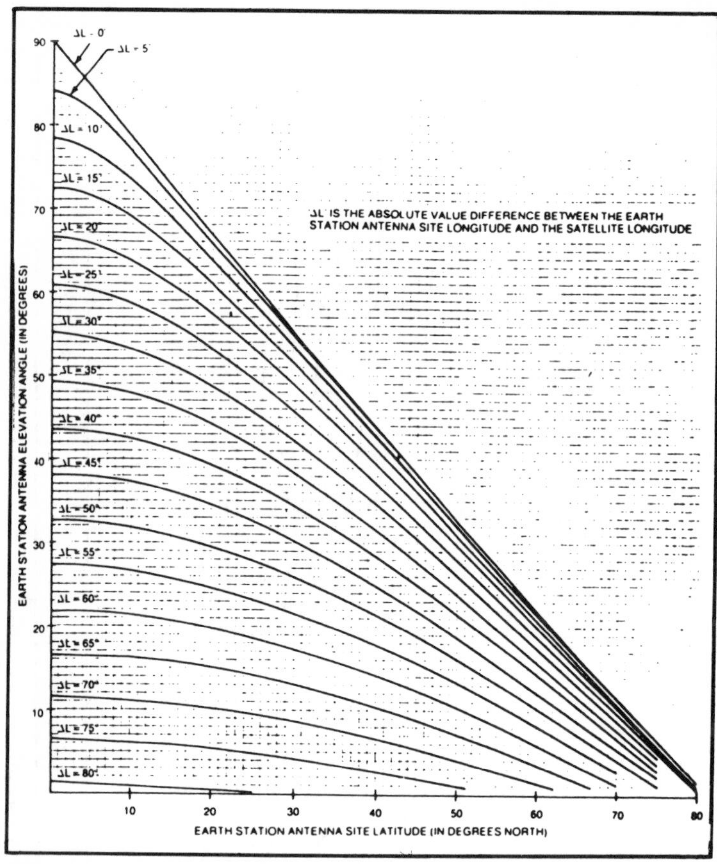

Fig. 7-17. Elevation of antenna fixed by graph (courtesy of M/A-COM).

If you do have a handy dandy programmable, put in the figures at site A for LAT A and LONG A, and at site B for LAT B and LONG B, then run down the rest.

HARRIS HAS AN EASIER WAY

Although much of the foregoing will get you well within the "ballpark," Harris Corporation Satellite Communications Division has now developed an extremely useful *single* nomogram that will virtually pinpoint satellite locations almost precisely from your own geographical location. See Fig. 7-18. All you need to know is your *own* latitude and longitude which is always available from local airports, nearby military installations, or the usual global lat/long projections. Afterwards, it's only a matter of simple arithmetic and a straightedge to find the "bird" of your choice parked at whatever West Longitude it's assigned by the Federal Communications Commission. See Fig. 7-19. Of course, if you must use equations to satisfy your mathematical urge, go back a few pages and you'll find them. Otherwise, here's how this one works—and we'll use Harris' example since that's already printed on the diagram for easy reference.

SATCOM 1, for instance, is positioned at 135° WL. You, let's say, are at 95.5° W and 29.5°N. Now all you have to do is subtract the satellite's longitude from your own:

$$135° \text{ (sat.)} - 95.5° \text{ (you)} = 39.5° \text{ difference in longitude.}$$

Plot this on the difference between satellite and site longitudes horizontal scale, then plot the site latitude (you) at 29.5° N on the vertical scale. Take your straightedge from the two plot points and place a dot where the two meet. Here you will find two sets of curves, one moving horizontally and to the left denoting the elevation (pointing) angle, one other designating azimuth in a vertical

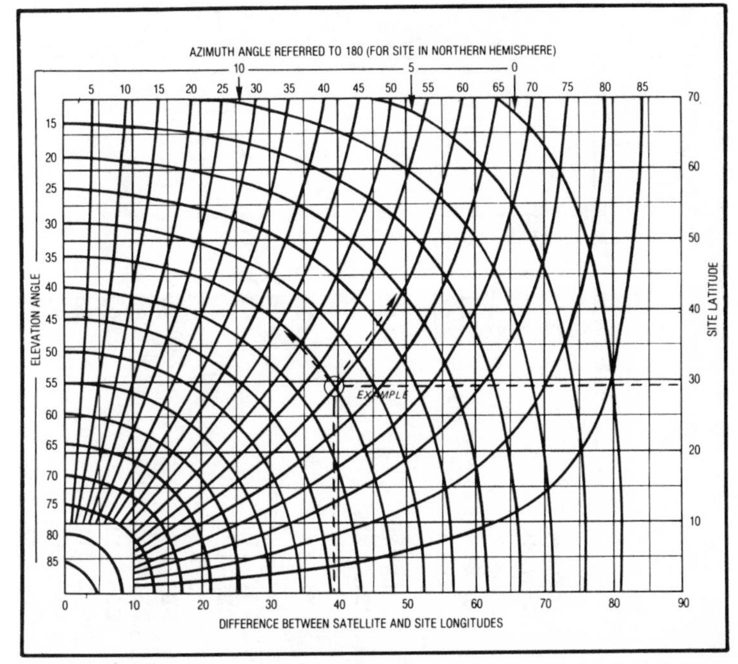

Fig. 7-18. A simple nomogram for estimating both satellite azimuth and elevation from *your* U.S. location (courtesy Harris Corp.).

UNITED STATES DOMESTIC SATELLITE SYSTEMS

Satellite	Orbit Locations West Longitude	Frequency Band (GHz)	Date Launched	# of Xpdrs/ BW (MHz)
SATCOM V	143°	4/6	Oct. 1982	24/36
SATCOM I-R	139°	4/6	Apr. 1983	24/36
GALAXY I	134°	4/6	June 1983	24/36
SATCOM III-R	131°	4/6	Nov. 1981	24/36
ASC-1	128°	4/6; 12/14	Aug. 1985	12/36 & 6/72 6/72
TELSTAR 303	125°	4/6	June 1985	24/36
SPACENET I	120°	4/6; 12/14	May 1984	12/36 & 6/72; 6/72
WESTAR V	122.5°	4/6	June 1982	24/36
GSTAR I	103°	12/14	May 1985	16/54
SBS IV	101° (temporary)	12/14	Sept. 1984	10/43
SBS I	99°	12/14	Nov. 1980	10/43
WESTAR IV	99°	4/6	Feb. 1982	24/36
SBS II	97°	12/14	Oct. 1981	10/43
TELSTAR 301	96°	4/6	July 1983	24/36
SBS III	95°	12/14	Nov. 1982	10/43
GALAXY III	93.5°	4/6	Sept. 1984	24/36
WESTAR III	91°	4/6	Aug. 1979	12/36
TELSTAR 302	86°	4/6	Sept. 1984	24/36

SATCOM IV	84°	4/6	Jan. 1982	24/36
SATCOM Ku-2	81°	12/14	Nov. 1985	16/54
WESTAR II	79°	4/6	June 1974	12/36
COMSTAR D_2 & D_4	76°	4/6	Sept. 1976 & Feb. 1981	24/36 24/36
GALAXY II	74°	4/6	Sept. 1983	24/36
SATCOM II-R	72°	4/6	Sept. 1983	24/36
SPACENET II	69°	4/6; 12/14	Nov. 1984	12/36 & 6/72; 6/72
WESTAR I	retired (8/83)	4/6	April 1974	12/36
SATCOM I	retired (5/84)	4/6	Dec. 1975	24/36
COMSTAR D_1	retired (9/84)	4/6	July 1976	24/36
COMSTAR D_3	retired (8/85)	4/6	Sept. 1978	24/36
SATCOM II	retired (2/85)	4/6	March 1976	24/36

In Orbit as of January 24, 1986

SATCOM - RCA American Communications, Inc.
WESTAR - Western Union Telegraph Company
SBS - Satellite Business Systems
GALAXY - Hughes Communications Galaxy, Inc.
COMSTAR - owned - Comsat General Corporation
- operated - AT&T Co.
TELSTAR - AT&T Co.
SPACENET - GTE Spacenet Corporation
GSTAR - GTE Satellite Corporation
ASC - American Satellite Company

Planned Launches in 1986

Shuttle	**Ariane**
Westar IV-S	GSTAR-2
GSTAR 3	Spacenet F3
	SBS-5

Fig. 7-19. North American satellite systems in domestic satellite services.

direction. Following the arrows, you read an elevation angle of 35° and an azimuth of some 60°.

Now you know that 60° azimuth can't be the final figure since you're only working between approximately 270° West and 180° South. Therefore, add 180° to the 60°, and you have an azimuth location of 240°, give or take one or two degrees. Then, taking into account local magnetic variation, add or subtract this for west or east, and take a true compass bearing from your position to that of the satellite. You'll find it works every time! Do, however, allow just a little for the usual do-all graphics, and also remember that offset feeds are often 20 to 22 degrees less (subtracted) than normal prime focus pointings. (We had a difficult time with this one ourselves.) For example, if you were looking at SATCOM 1 with a 20° offset, then the actual pointing angle would be 15°. This is especially true for Ku-band antennas where the feed horn and supporting structure have been repositioned for better gain and sidelobe characteristics. Such offsets, although subject to larger noise temperatures due to reduced elevation angles, are normally mounted on structures *above* the earth and, therefore, not abnormally affected.

We would also point out that block downconverters in the 12 GHz downlink Ku spectrum have considerably higher noise temperatures than those at C band (100° versus 200° (round figures) and any noise saving at Ku becomes system-noise appreciated in the total compilation. Reflected noise through mesh and perforated reflectors in normal prime-focus feeds may also be a contributing factor to extra sparklies at C band, especially since amplification must be considerably greater to accommodate only 5-W to 10-W transponder outputs.

Therefore, considering all these factors in addition to added carrier-to-interference problems occasioned by 2° spacing and poorly designed antenna beamwidths of the less expensive and smaller reflectors becoming quite popular for C band, precise antenna pointing now becomes more of a necessity than ever. Fortunately, however, as reflector sizes increase, noise temperatures decrease, so there is more than minor compensation between 1.2-meter and 3-meter reflectors . . . Proving that it never hurts to think *big* rather than small in TVRO.

One further (small) note, if you have a Ku band choice between 1.2 and 1.8 meters, and there are no mounting or location problems, choose the larger for better gain *and* approximately 3° *less* noise temperature. For 10° increment angles between 10° and 40°

elevation, the 1.8-meter reflector has measured temperatures of 46° to 29°. Just thought you'd like to know. A tip such as this is often helpful. You will also find the pointing error somewhat less.

MOUNT AND REFLECTOR MAINTENANCE

There's no way to contrive a set of rules and procedures that cover every situation, but we can work with our own experiences and those of others to offer useful applications for mount and reflector maintenance. Because the LNA/LNB, reflector, feed, mount, actuator, cable and feed support are subject to weather year 'round, you might well start with these and regress to positioner, receiver, any i-f switches and/or other attached electronics. The logic here is that if space electromagnetic signals aren't being received accurately, then the receiver, et. al, is no better than the inputs. At the moment, of course, we're speaking of a *complete* system checkout after, say, a year or two of full operation. Incidentals and breakdowns we'll tackle at an appropriate time following what you may perceive as a fairly good argument for annual TVRO service either on an hourly or service policy basis. If handled discreetly, you could sell a considerable amount of new equipment such as full bandwidth monitors, more fully automated positioners, orthogonal LNBs to eliminate skew, new cable, and even another reflector when the old ones (especially metal are damaged, corroded, bent or rusted. Further, with Ku-band showing such active promise, a dual C- and K-band feed might be very appropriate for those with block down conversion receivers, or they may even want an entirely new system for stereo sound and additional satellite transponders. Certainly if you can prove that satellite reception is vastly superior to cable and even better than local TV broadcasts, many with *good* TV receiver/monitors can appreciate the difference. If they haven't got a monitor with 4- to 8-MHz bandpass, that might be a worthwhile ticket item, too. In short, we see many, many possibilities for some sort of annual TVRO service and sales combination that could be a boon to both customer and the industry if handled forthrightly, honestly, and with discretion. With Ku-band initially unscrambled, an initial or second TVRO 12 GHz system with a 1.2- or 1.8-meter reflector might just be the ticket for those in the city or near-city areas where larger reflectors aren't always possible.

Hardware

Hardware begins with the mounting post and includes whatever remains out in the weather during the four seasons. If it's metal, there's guaranteed maintenance; if hand-layup fiberglass, you probably have cracking and peeling to contend with; if metal mesh, there may even be bent or missing panels to replace and at least an eyeball check for parabolic accuracy. If the reflector's design characteristics have changed, so will the resulting picture. In aggravated cases, the entire reflector should be substituted with a newer and stronger unit that will *not* repeat its predecessor's problems. Ripples in the metal or mesh are telltale signs of parabolic abnormality, and dents or holes must be repaired (accurately) or restructured and replaced before you're sure of satisfactory reflector operation. For *any* distortion that upsets design curvature will certainly divert some or many of those tiny dBm signals from outer space and either weaken or produce unpleasant interference in the picture. Sound, too, with its 6.2 and 6.8 MHz carriers (with perhaps more to come) may be affected as well. Parabolic or rectangular accuracy, therefore, is essential to worthwhile audio/video reproduction.

The Mounting Post

Usually these are (or should be) sunk in 40-odd inches of reinforced concrete in a hole of some 18 to 20 inches in diameter for normal installations. For accurate polar tracking, or even AZ/EL pointing, such posts have to be precisely perpendicular with *no* discernible movement permitted in any weather conditions. Normally of at least 30 (40 preferred) formulation, such posts are usually either supplied by the reflector factory or are picked up from some junk yard. In either instance, they must not only be perfectly straight pieces of metal but maintain that true and tested shape throughout their TVRO lifetime. If not, your mount (as we know it today) will not track correctly and some transponders and even a number of satellites will be lost.

Metal, of course, weathers and rusts. To help prevent this and extend its usefulness, we would suggest a covering ranging from old Navy zinc chromate to Rustoleum® be applied *before* installation over the full pipe length. This will not only preserve at least the exterior but also relieve its eventual owner of an obviously rusting eyesore and give the appearance of a more finished installation—all of which aids your eventually welcome return. If the post has already rusted, then it's advisable to remove as much as possible with sandpaper, steel wool or possibly plumbers crocus cloth before applying such rust retardant coating. Rustoleum, of course, comes in colors and might help in partially disguising some of the larger installations. Our own preference is for silver or aluminum colors that add a bit of sparkle and freshness, especially where the mount, itself, is black-coated steel. For metal reflectors, usually charcoal or darker, a matching post would probably be more suitable. But with a solid fiberglass, compression molded reflector, the effect becomes rather professional looking and much in keeping with the rest of the installation. In addition, what looked like a rusty old iron tube on the junk pile, now becomes almost a thing of beauty depending, of course, on the beholder.

Should the mounting post have deteriorated considerably or warped out of shape, you might cut off the offending upper portion and bolt a new section to it if you have a very snug fit, and if the old mount will accommodate the substitution. Otherwise, a smaller pipe will have to be placed *inside* the old one, bolted to it very securely in several separate positions, and probably a new and smaller mount procured. Or, perhaps, you'll have to supply an entirely new installation. This will occur, anyway, if the surrounding concrete is cracked, powdered, or insufficiently reinforced to take seasonal strains. Frankly, it may be better to do the latter rather than try a makeshift arrangement that might not retain its perpendicularity when blasted with wind, rain, and snow. Customers usually don't like to pay for the same service three or four times. Moral of the story being: always sell the best installation you can—it usually makes friends.

You might also remember that nothing takes the place of good, firm cement to anchor the mounting post. Then, following curing time, do put a bead of good quality caulking around the post's perimeter where it enters the cement to keep moisture from entering the foundation. Then be sure the post's top is capped, also, to resist element entry there. In time, we would imagine that some entrepreneur will devise a suitably universal collar that can be adapted to most normal-size posts to firmly support either larger or smaller reflectors that may eventually occupy original installations.

The Mount

You'll have to depend on some manufacturer's reputation for this one. And with Ku-band requiring closer and closer azimuthal tolerances, it might well be wise to pay a little more for this essential hardware and be especially careful of its north-south positions when undertaking and installation. Despite past assurances, we suspect that there are many reflector mounts today that won't accurately declinate nor accurately scan even a full 90 degrees, let alone from horizon to horizon. Therefore, instead of attempting to repair some of the older mounts, you may be able to adapt one of the newer units successfully and access even more satellites with even better elevation and azimuth angles. Our preference would be not only a new mount but a new reflector to go with it, since a single source design usually works better than mating complete strangers. There could be stress points or angles that foreign components can't compensate.

In addition to the usual predators of corrosion and rust, you also have a lubrication factor plus wear and tear on plastic or nylon "bearings" or washers. All these have to be attended to with suitable rust retardants and greases to maintain satisfactory operation. You may also want to recalibrate azimuth and elevation for maximum arc tracking, and as the mount and reflector swing, look for any electromagnetic interference or RFI that may be floating about. An occasional trap or filter can clean up many a mild case of wavy lines or picture ripples if applied at the offending source. But do use a color monitor rather than simple black and white because of the various frequencies and chroma/luma interleaving involved.

Should the mount appear loose or shaky, the culprit will have to be found and remedied, otherwise it will cause the reflector to "flutter" in the wind and access only a limited number of satellites. Possibly new washers or bolts will rectify such conditions, but extended usage and poor original design may require a new unit altogether. Patching up a poor quality mount may be difficult, or totally impossible, and *you* must be both judge and salesman in any solution. Around the 1990s we'll probably see a fair amount of these replacements because of inadequate equipment originally installed in addition to the C- and Ku-band combinations that will most certainly abound. By then, too, polar Ku-band mounts will be readily available, and it may be that separate C and Ku installations can be more economical rather than an entirely n-w unit for both. Prices and applications, naturally, will determine marketing positions and appropriate sales.

The Actuator

This arm of the TVRO earth station only does exactly what its positioner tells it to. Strictly an electromechanical device with Hall, pulse, reed, or mechanical sensors, its gears and/or arm completely determine the pointing accuracy of your feed and reflector. If water, ice, or congealing grease is allowed to accumulate on its gears or extending arm, it cannot execute commands and, therefore won't move the reflector when told. Further, excess moisture may easily cause a short circuit in the motor or rear electrical connections, rendering the entire assembly unusable.

This is another part that actually requires at least an annual inspection for water tightness, gear wear, lubrication and ease of operation. Early spring just after the first rains might be a good time to conduct such examinations in preparation for increased summer activity. You may also observe if the actuator has or needs a bellows-type weather sleeve and also if that sleeve and the gear housing is permitting any accumulated moisture to readily drain away. The same is generally true for horizon-to-horizon gear trains and their coverings, which may or may not drain themselves adequately. Waterproof grease for these larger arc gears, of course, should be relatively mandatory subject to the various manufacturers' recommendations Because special metals and other materials are often used in such equipment, we won't attempt to recommend anything other than the greases and oils specified. Dirt and rust need to be removed and their entrance ways countered immediately. Electromechanical equipment always encounters severe problems if dirt is permitted to collect in any amount.

As more and more C- and Ku-band satellites are placed in orbit, the linear, 90-degree actuators may not have sufficient range to access all available satellites—some probably don't even now. You may want to point this out in talking with existing or prospective TVRO owners, especially those who want maximum use from their receive only terminals. The subject of mounts and actuators will be a relatively hot topic for sometime until one can direct-dial both satellite and desired transponder from a multi-hundred storage memory. At that juncture, we may well revert to an AZ/EL type arrangement that will obsolete all true north-south mount installations. But for the moment, this is just something to think about. Next year . . . who knows?

Bent shafts, worn gears, intermittent motors, excess rust or corrosion, naturally, warrant new actuators and these should be installed as rapidly as possible. Be sure the new unit has the same command-sensing mechanism as the one it replaces, or you have double troubles with both your time and the TVRO owner. Fuse blowing in the positioner just might tell you something—bad!

Well-designed and serviced actuators should last for a number of years without major overhauls. But those that are neglected will certainly fail in only a matter of months following any extended period of continuous operation. Said another way, after the first year watch out! And after two or more years of non-service, small or major repairs are guaranteed!

The Reflector

Often called a *dish* if parabolic, these reflectors are made of aluminum, steel, fiberglass and even plastic. The metal ones may dent, warp, corrode, rust, lose small or large sections, and generally disintegrate due to weathering. Compression molded fiberglass, however, has a pretty good history of resisting the elements and may be repaneled under most circumstances without disturbing its all-important curvature. Hand layup units, on the other hand, can peel, warp, and result in all sorts of unpleasant problems if produced in haphazard "backyard" operations by the cheapies. Of the metal units, solid, good quality aluminum or stainless steel are preferred to most others. Generally, they hold their shapes and produce adequate reflectance for any and all satellite signals. In extremely windy or high rainfall areas, however, perforated, may be preferable if manufactured under strict standards and by reliable manufacturers. In addition, if close tolerances are met, such solid metal and fiberglass reflectors may be entirely suitable for combined C- and Ku-band feeds now making their market appearances. To avoid undue carrier-to-interference problems, however, such reflectors should measure at least 9 feet in diameter and have beamwidths of 1.8 degrees or less for 2° satellite spacing separations now mandated by the Federal Communications Commission. Most available authorities say even 8-foot reflectors will not be adequate under the new regulations.

Solid reflectors, however, do require substantial mounting posts and heavy-duty mounts in windy regions, and considerable care is needed in their installations. A 50-inch deep by 20-inch diameter post hole is not unreasonable for ordinary soils and blustery weather, accompanied by steel reinforcing bar (rebar) support. The same may be true for many of the finely perforated antennas now offered for sale because their 80-thousandths (or less) holes are going to permit little wind passage after about 20 mph. Any further loss of strength due to additional perforations is bound to become detrimental.

In working with the reflectors, you'll find the considerations are many, and anything beyond superficial and easily-repaired damage certainly justifies immediate replacement. Just be careful the new reflector is more substantial than its predecessor.

Feeds and Feed Supports

Feeds may be Cassegrain, Gregorian, or focal-fed, the latter being the most popular and placed directly at the focal point of the antenna. Another version of this is the offset feed, usually found in the smaller Ku-band only antennas.

As you are well aware, such feeds or their reflectors (if Cassegrain or Gregorian) must be positioned fairly precisely so that rays from the reflector converge at that particular point. What is *not* generally understood is that phase position (skew) is extremely important and has a distinct bearing on cross-polar interference rejection as well as maximum signal input. In other words, you must keep horizontal and vertical transponders from interfering with one another even though this may mean slight signal amplitude loss or diversion. Without instrumentation you naturally *look* for best signal, but with a spectrum analyzer you can *see* both amplitude and cross-pole effects as well as any interference present. A *clean* signal always means best picture and sound—an absolute necessity in hostile environments, especially EMI and RFI.

Lightning damage and moisture intrusion are the prime failure modes in these scalar prime focus feeds, most of which are made of metal since metal-coated plastic is usually subject to flaking. With a direct lightning strike, the polar probe and its motor often take the brunt of the charge and require replacement. Water or snow ingress over time will also result in failures that freeze probe rotation.

The best precaution we can suggest for feeds is grounding its metal scalar portion to a common point with the antenna mount. Later, several manufacturers will have some MOV (metal-oxide varistor) protection that will help to a certain extent in normal instances. These MOVs are zinc oxide varistors that operate similarly to back-to-back Zener diodes. In standby, they offer high impedance, but when energized by high voltage transients, they conduct at a very low impedance, clamping such abnormal voltages to a relatively safe level. Their I/V graph is much like that of a very sharp-knee diode and will dissipate considerable power, depending on individual ratings. G.E. first introduced MOVs in 1972.

As for supports, these should remain rigid, free from corrosion and rust, and feed centering should be checked from time to time, especially if there is any question of feed-arm movement or instability. Usually you have little problem with the prime focus variety.

Cassegrain feeds, on the other hand, have a second reflector about the focal point and so must have their horns pointed upward. This means the the second reflector has to be precisely adjusted and the horn portion carefully protected from the weather; sometimes on larger reflectors even being filled with a suitable gas under pressure. Any disturbance or weather problems that upset reflector alignment and puncture feed horn protective coverings will, of course, affect reception to a greater or lesser degree depending on the damage or deterioration.

There is also an O-ring at the back of the feedhorn flange that may be replaced in the field, but O-rings between window and feed are factory installed. Once in a while the 1/2 pound psi dry air or nitrogen pressure is exceeded and this will "blow" the feedhorn cover. That's about all you should encounter in 30 feet or less diameter reflector assemblies and their feeds, unless some sharp-clawed bird punctures the cover while attempting to land.

GLOSSARY OF TERMS

The following are terms commonly used in TVRO systems. One should become very familiar with each term and its meaning.

TVRO—Television Receive Only

TVRO Site Lattitude—latitude of the proposed TVRO site in degrees, minutes and seconds. Positive entries indicate north latitudes. Negative entries are used for south latitudes.

TVRO Site Longitude—longitude fo the proposed TVRO site in degrees, minutes and seconds. Positive entries indicate western hemisphere. Negative entries are used for eastern hemisphere.

Satellite Longitude—is given in degrees west of Greenwich, England, for positive entries. Negative entries indicate east of Greenwich, England.

AZIM—true azimuth bearing in degrees.

ELEV—elevation angle above the level reference.

ANTK—effective noise temperature, in degrees Kelvin, of the antenna to be used. This value will vary with elevation angle and antenna size. A chart has been provided to estimate this variable.

LNAK—effective noise temperature, in degrees Kelvin, of the LNA (low noise amplifier).

ANTG—gain, in decibels (dB), of the antenna to be used at 4 GHz. A chart has been provided to show this value for each size antenna.

EIRP—Effective Isotropic Radiated Power of a transponder or group of transponders of the satellite proposed for use with the TVRO station. This value is expressed in decibels above one watt (dBw). A map showing the average EIRP has been provided.

Nominal Performance values—are calculated assuming clear sky path losses.

C/N—i-f carrier-to-noise ratio in decibels (dB) which can be measured at the i-f test point of a receiver equipped with a 30 MHz bandwidth i-f amplifier.

MRGN—the difference in decibels (dB) of the C/N value, above or below, the threshold of the receiver discriminator, usually considered as 8 dB.

VS/N—video signal-to-noise ratio, in decibels (dB), at the video output of the receiver.

AS/N—audio signal-to-noise ratio, in decibels (dB), at the audio output of the receiver.

Worst Case Expected—will repeat the calculations with a degraded performance for the downlink which includes precipitation loss, the antenna pointing error such as is encountered with high winds and degraded transponder power. These losses add up to 1.71 dB for the purpose of this analysis.

Chapter 8

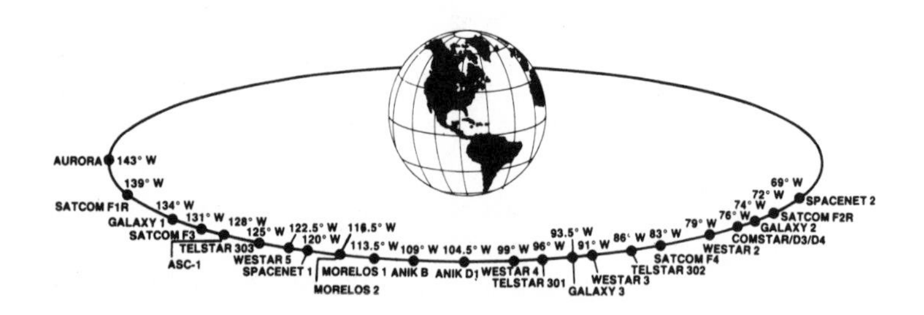

Interference and Mounting Constraints

OUT IN THE COUNTRY, SURROUNDED BY SLOPING HILLS, AND far away from terrestrial microwaves, almost anyone with experience can erect a satellite earth station and make it work. In the big city with its tall buildings and complex communications, simple installations turn into very difficult situations which require more than considerable expertise and not a little good fortune.

You're subject to video deviation of interfering carriers, excessive noise, data link overrides, law enforcement, multi-distribution systems, and even the Department of Transportation. Then there are the independent Bell systems busily constructing their own radiators and actually forming a communications grid. Add to this some of the land mobile base stations and their wheeled counterparts, and you can readily see the extent of this very real communications problem.

There's another difficulty that's just now rearing its unpleasant little head. With large stations, the dish aperture (front-looking beamwidth) was relatively narrow and largely unaffected by some of the surrounding electromagnetic clutter. But with smaller dishes and the coming of 12 GHz downlinks, these 3-meter or less receivers may have relatively broad apertures and could be subject to a great deal more pickup which, while not anticipated, could come from anywhere.

For instance, a CATV head-end requires a good tall tower for TV signal reception. An earth station trying to look past such a tower not only absorbs associated radiated carriers which are reflected down, but may also have to look through power lines and guy wires—and this is an almost impossible situation, not to mention probable feedhorn or Cassegrain damage from falling tower ice.

INTERFERENCE

Cities may be fine places for people to work and live, but they are undeniably nightmares for many forms of broadcast communications. The building next door, for instance, instead of becoming a friendly bulwark between your antenna and the rest of the city, often reflects unwanted signals and requires that you relocate the entire earth station away from its intended spot. As you can readily see, some of these problems are pretty rough stuff, and must be solved by exceptional engineers with both imagination and considerable experience. Once again, we suspect that the 3-meter or larger dish will prove invaluable both in cities and the country for both single and multi-satellite coverage. High gain and narrow beamwidths can mean the difference between usable and non-usable signals. Antenna shrouds and special passive trapping may become even more important as metro areas fill up with satellite earth station paraphernalia.

Since TVRO stations do *not* require FCC licensing or approval, anybody's dish and associated electronics won't necessarily guarantee enjoyable reception. Have the prospective vendor trot out a 10- or 12-foot dish on platform or wheels, hook it up to your receiving equipment and try the rig *both* day and night before you buy. A seven-day trial during the week and on weekends ought to do the proving. Location adjustments may then be made *before* final installation and happiness should descend upon all. Play it safe and keep interfering carriers better than 20 dB below what you're receiving and/or transmitting. If not, you're going to have trouble sooner or later and their presence may be even more difficult to eradicate later as additional microwave users creep into the spectrum.

A pretty good example of such possible interference can be seen in Fig. 8-1 (with a little imagination, of course). As indicated by Tektronix 7L12's convenient digital readouts, the vertical attenuator is set at −50 dBm, with measurements thereafter at 10 dB per division. Resolution for these signals measures 30 kHz, and there is a 0.5 MHz (500 kHz) frequency difference between each division. To be honest, these multi-signals were taken directly off the air at a spectrum analyzer setting just about the top of the 108

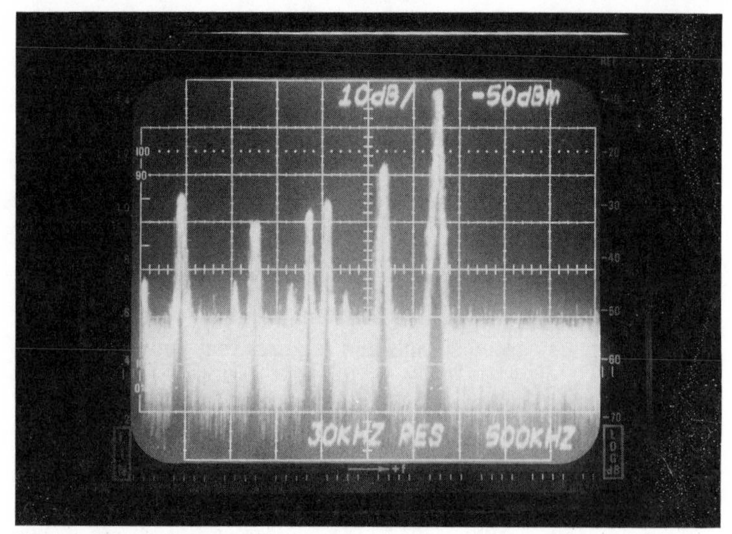

Fig. 8-1. Bunched FM and other carriers could be devastating in the TV band.

MHz FM band. Notice that the separation here is about 200 kHz, which is adequate, although one carrier seems to be only about 13 dB down from its neighbors. Were these TV signals with their 6 MHz bandwidths, you can easily understand surprising amounts of interference that could have resulted. The same identical situation confronts satellite signals and electronics, and too much of a good thing can get us all into trouble. This is also why the industry right now opposes 2° satellite spacing; electronics and dishes will have to be *better* to separate and accommodate any such closely bunched signals.

Just for the record, let's look at a conventional uhf television station (Fig. 8-2) that's operating just below 700 MHz and show you exactly what its three carriers should look like. Naturally various measuring parameters have been shifted to accommodate the much broader band TV signals, and you can see a 4.5 MHz difference between the left-hand video carrier and audio carrier on the right. Next to the audio carrier is the "suppressed" 3.58 MHz chroma subcarrier which is always in evidence despite the transmitter's balanced modulator's best efforts. In TV, these same signals appear everywhere that adequate video rf systems are operating. Baseband (demodulated video) is another matter entirely.

WEATHER PLAYS A PART TOO

Then, you've got to take into account the weather. When those "sparkles" and popping noises appear in video and audio, how do you know their source, and do they disappear during changing humidity conditions even before a quick check? How about differences between summer and winter, especially tree leaves, and local construction? Remember, too, that different programs on individual downlink channels generate changing signal-to-noise ratios and these, too, become good or bad factors. Snow and ice may clog your feedhorns and place undue strain or weight on the dish. A deluge of rain often reduces or even cancels received signals completely if the torrent can be measured in inches per hour. This will be especially true at the 12 GHz K- and Ku-band downlinks and the 17 GHz K band uplinks as these frequencies gradually come into common use—all the more reason for justifying both lower-noise LNAs and high-gain parabolic or torus-type dishes. Sufficient signal strength can usually nullify most marginal conditions and you'd never know even if they have a tendency to occur.

Wind factors are another consideration that's important, especially if your dish isn't shielded by either terrain or buildings, and the area is subject to north and west blows in the winter and strong southerly breezes during summer. Hurricane-prone Florida should offer an excellent testing ground, and many antenna companies make their homes there for this and other allied reasons.

Salt corrosion and smog coatings may also play just as much

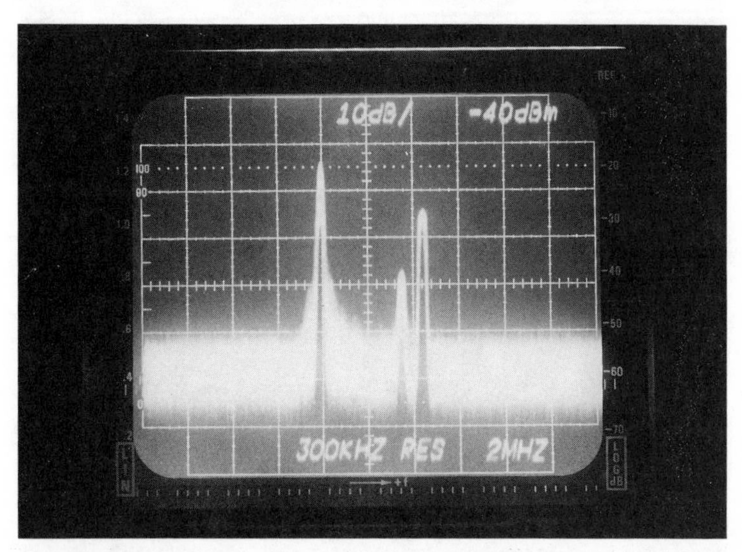

Fig. 8-2. Audio and video carriers, and chroma subcarriers occupy a TV channel's allotted 6 MHz.

of a part in satellite receiving equipment as they do on ordinary television antennas. At least an annual cleaning and inspection should be in order to guard against such possible or probable effects. As Gertrude Stein might have said; "an antenna is an antenna is an antenna." What else could it be?

The answer is "several things," depending on your LNA, downconverter arrangement and an adequate receiver. Manufacturer Harris suggests the following LNA temperature ranges for the U.S. that's worth a moment of consideration (Fig. 8-3). Notice how the low LNA curves swing through New England and the Northwest. As we suggested, Florida is the toughest, and an 80° K LNA does cost a mite of cash now—less later.

PRELIMINARIES

Once a customer has signified his intention to install an earth station, all preliminaries should be examined before actual construction commences. You'll have to know if it will become a transceiver, what data or video will be transmitted or received, community, county, city, and state ordinances or requirements, possible microwave and/or other stray or transmitted types of interference, site ownership, proximity to other adjacent services, accessibility, probable equipment needed, estimated time for construction, cost, distance from transmitter/receiver, clearview area to accessed satellite(s), alternate sites, if available, and necessary permits. Long,

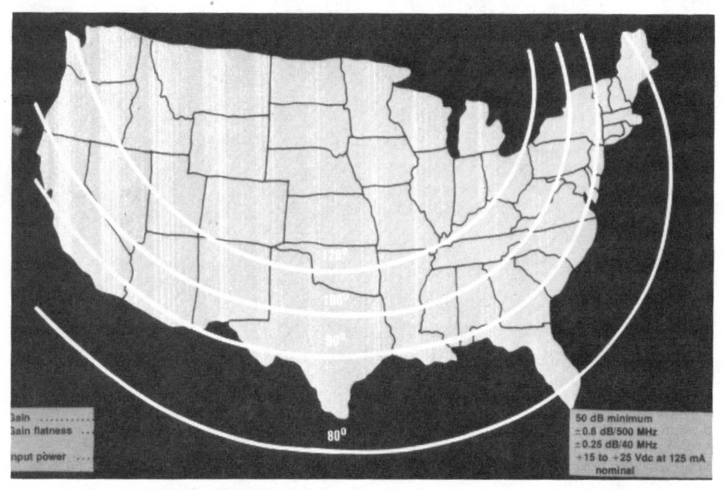

Fig. 8-3. Harris' suggested LNA temperature range for the U.S. (courtesy of Harris Satellite Communications Div.).

expensive cable runs between the dish and its outboard electronics are decidedly not advisable, and adequate station security should be considered whenever and wherever possible.

LOADS

When installing antennas, you must take into consideration the *dead* load, *ice* load, and *wind* load. Charts are available for 50-year wind conditions at a particular location, and annual weather bureau reports can give some idea of what ice and snow loads may be encountered. The dead load, as you might imagine, is just the antenna weight itself without the other two factors. The Electronic Industries Association can furnish an RS-222-C standard for structural code antenna support structures. In warm climates only the dead load and wind load need normally be considered; but in temperate and colder sections, the *dead* load, *ice* load, plus the *wind* load should be included. Horizontal force, as an example, varies directly as the square of wind speed, so be guided accordingly.

In addition to clear lines of sight to the satellites of interest, the mounting area needs to be fairly level, no underground obstructions, recent landfill, or sinkholes, and should be well drained with satisfactory porosity. If not, you may have to install drainage facilities to keep the mounting area dry.

Electrical facilities and property ownership are two other factors and a civil engineering report may be needed to verify your initial findings. If this is an especially large and costly installation, you may want to select a competent general contractor who's familiar with the area to do part or all of the work. Telephone facilities are another matter of importance, plus the usual permits associated with *both* large and small stations. Finally, to comply with FCC regulations in the commercial installations, you'll have to send out a frequency coordination questionnaire to all affected terrestrial station licensees in the area, and then allow 30 days for responses *before* filing your FCC application in the case of common carrier earth stations.

Things become tougher with huge installations such as wind survival with 0.5-inch radial ice (87.5 mph wind), or 125 mph wind with no ice at 40° elevation. They must withstand rainfall of 4 inches per hour, and be able to survive earthquakes of 0.25G horizontal and vertical, or comply with local codes if they are more stringent. Foundations must not show differential settlements of more than 0.03 inch/year across the foundation. Maximum levels of nonionizing radiation are also limited to 10 mW/cm^2 over

6-minute periods at normal operating areas about the earth station.

FOUNDATIONS

Foundations are of two general types (Fig. 8-4): the first is a single mat reinforced concrete slab, capable of handling all three loading factors with the antenna firmly installed; and the second consists of reinforced concrete cylinders that snugly fill either augur or hand excavated holes. Usually, the cylinders are bolted to a triangular metal frame so they will react in unison to antenna loads above. Naturally this results in concrete savings and becomes an excellent method of earth station anchoring when soil conditions permit. Variations in such mountings are required under unusual conditions or when the installation is extra large or overly complex. A large Andrew Corporation antenna is shown in Fig. 8-5.

Soil Conditions and Drags

Dense sands and stiff clay have a bearing capacity of some 8,000 lbs/ft^2 and average soils only 3,000 lbs/ft^2. Under stringent conditions, antenna patterns are to meet those required by the FCC, the first sidelobe gain has to be at least 14 dB below the main beam and symmetrical within 1 dB, and the antenna must point to *any* geostationary satellite without structural changes or relocation of jackscrews. Wind load factors to be considered in these installations are drag, lift, and elevation.

Calculations show that in survival winds of 120 mph, a 3-meter antenna at 60° elevation has a drag force of 3,200 lbs; a 4.6-meter rig at 110 mph, a drag force of 9,300 lbs; and a 5-meter dish at

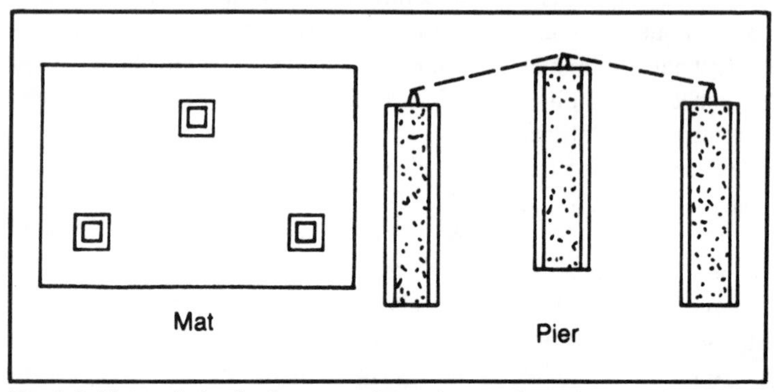

Fig. 8-4. Mat and pier footings usually anchor earth stations.

Fig. 8-5. A large 10-meter earth station with piers, equipment house and protective fence (courtesy Andrew Corp.).

125 mph, a drag force of 14,000 lbs. All these maximum forces actually occur when striking the antenna from the *rear.* Therefore, when pouring your foundations, take into account these various drag conditions and be guided accordingly. TVROs, normally, have specific instructions for consumer or commercial mounting and these should be strictly adhered to, and even expanded upon if there are weather or soil reservations for your particular location. Business and home insurance policies should also cover any damage if one of these earth stations tears loose. Aggravated neighbors are

sometimes tough on the pocketbook in or out of court.

By the way, there are both state and county agencies that can supply soil condition reports with, perhaps a few calculations to help you along. Clay soils, of course, are best, and some may have cohesive strengths of between 1,000 and 2,000 pounds-per-square-foot. Therefore, the side area or any mounting pier multiplied by the soil cohesion amounts to its skin friction resistance; but halve this number for a safety factor. Sandy soils obviously require deeper holes and more concrete, sometimes in the shape of pyramids, and even earth anchors are needed in extreme conditions where cohesion is suspect.

You will also need to establish accurate north/south compass headings, taking into account the difference between true and magnetic north. Magnetic declination maps are available from the U.S. Geological Survey and possibly other sources in the area. Pointing the antenna accurately means accessing the satellites with the same inclination angle and only varying azimuths. Check end-of-arc coverage between 60° and 140+° *before* locking your dish down. Otherwise, polar mounts will never be able to use a motorized aiming arrangement when other satellites become attractive.

Other Foundations

Mat foundations (or pads) also have earth excavations with reinforcing bars and anchor bolts supported by templates. The concrete is poured, permitted to set, and then the base plates and mounts are installed. Depth has to be at least equal to or greater than maximum frost penetration, in addition to any and all loading factors involved. Occasionally, rock foundations are suitable when conditions and equipment are available for this type of installation. Top-of-building installations are very tricky and should only be done by highly experienced contractors who know stresses, sway factors, load frames, and all the rest. We'll not attempt an explanation of what's involved since there are so many different circumstances extant in such an undertaking. It may be that these "simple" DBS home installations to come are not as simple as glibly depicted unless the antenna is often shielded from prevailing winds by one of the roof gables. True, we're considering a 1-meter receiving dish only, but insecurely mounted this, too, can become a slicing disk or boomerang.

CABLE

There are opinions galore on various types of cables and how they can or must be used. The good ones are expensive and the poor ones pop up with miserable high frequency losses, poor standing wave ratios, improper shielding, slow propagation, and no quality-sweep tests before they leave the factory. Just like an ordinary broadcast, CATV, or basic television antenna, if you have good equipment and installation, it pays to put in good cable. Either outgoing or incoming signals only travel where they're sent, and the less they're interfered with, the better all can be seen. Here, of course, we're primarily limiting our discussion to television cable since this is more vulnerable than lesser level transmissions. Frequencies involved range between megahertz for broadcasters to gigahertz for satellite up and downlinks. Very different cables are required, depending on frequency, power, and construction constraints. Weather, of course, is a factor, too, as well as above and below ground installations. We won't attempt to describe all such circumstances, but the usual types will be considered.

HELIAX®

This is the Andrew Corporation's registered trademark cable designation and their transmission lines will be used in this section of the chapter because of both their broad and specialized applications in microwave transmissions and receptions. Figure 8-6 shows a total of five types of transmission lines for frequencies from megahertz or less to 26 GHz. That's about as broad a range of frequencies as you could wish for. Reading from top to bottom they're identified as air dielectric, foam dielectric, elliptical waveguide, rectangular waveguide, and circular waveguide.

☐ *Air dielectric* is recommended for antennas with feeds for 2.7 GHz and below. It's available in long lengths, easy to install, and quite maintenance free, according to the company.

☐ *Foam dielectric* is designed for systems with nonpressurized coaxial feeds, and some types have very low VSWR measurements.

☐ *Elliptical waveguide* cable earns the recommendation for most microwave antenna systems from 3.54 to 15.35 GHz. Assemblies are cut to specified lengths and terminated by connectors.

☐ *Rectangular waveguide* can be used for elbows, twists, etc., in elliptical and circular waveguide systems and connections between antennas and radio equipment. Ranges are from 3.3 to 26.5 GHz.

☐ *Circular waveguide* tends to minimize feed attenuation and

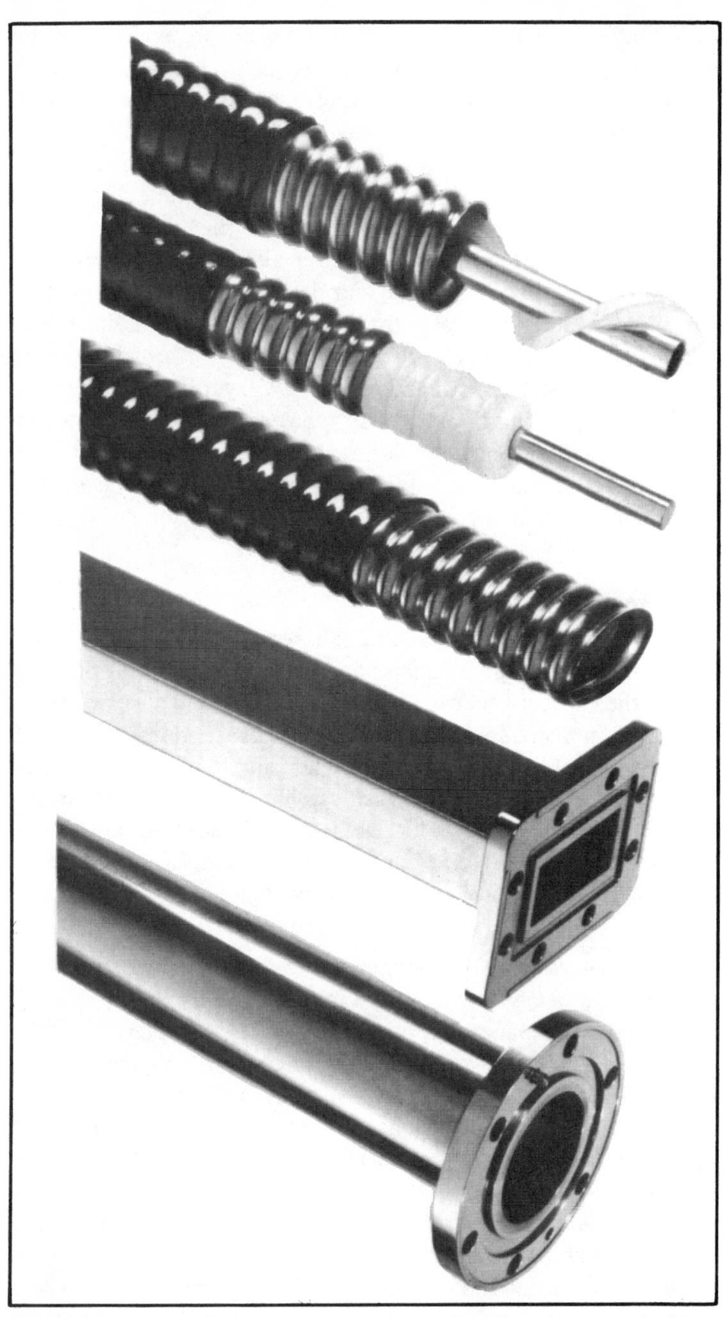

Fig. 8-6. Five types of microwave transmission lines from hertz to gigahertz (courtesy Andrew Corp.).

1/2" 75-OHM, LDF SERIES FOAM-DIELECTRIC HELIAX® CABLE

Type **LDF4-75** foam-dielectric HELIAX coaxial cable provides a combination of strength, flexibility and efficiency not available in other cables. It is ideal for CCTV security systems, 5 – 450 MHz CATV distribution, 70 MHz IF earth station antenna systems and military data links. An annularly corrugated outer conductor, in conjunction with the "O"-ring seals, provides a longitudinal moisture block. Differential expansion is eliminated by mechanically locking the outer conductor and bonding the inner conductor to the dielectric.

A fire-retardant, Rulan*-jacketed version is also available. Order Type **41690-17.**

Shipping Information. See pages 187 – 189.

DESCRIPTION

Nominal Size	1/2"
Outer Conductor	Copper
Inner Conductor	Copper-Clad Aluminum
Impedance, ohms	75
Type number	**LDF4-75**

CHARACTERISTICS

Impedance, ohms	75
Maximum Frequency, GHz	10
Velocity, percent	88
Peak Power Rating, kW	13
Attenuation and Power Rating	See graph below and pages 178, 179
Diameter over Jacket, in (mm)	0.64(16)
Minimum Bending Radius, in (mm)	5(125)
Cable Weight, lb/ft (kg/m)	0.14(0.21)

Fig. 8-7. A very good cable for general microwave use (courtesy Andrew Corp.).

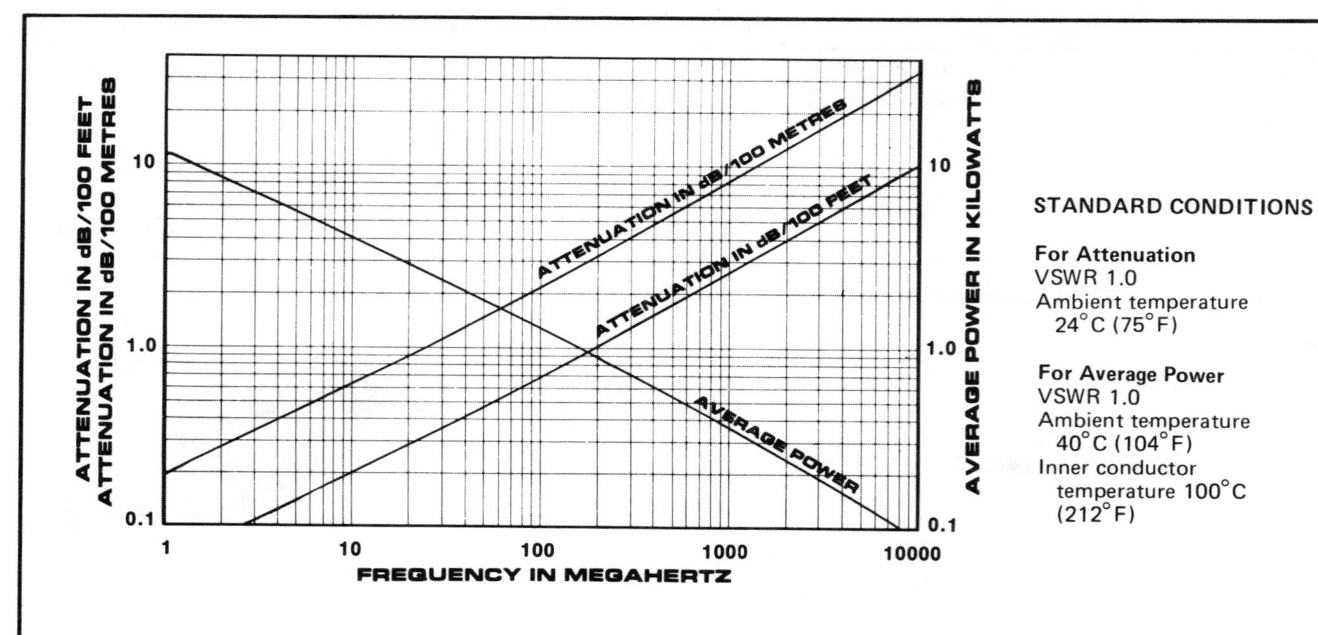

*Trademark of du Pont.

Fig. 8-7. A very good cable for general microwave use (courtesy Andrew Corp.) (Continued from page 253.)

is especially suitable for long vertical runs to tower-mounted antennas. For our TVRO purposes, type LDF4-75 foam dielectric HELIAX® with peak power rating of 13 kW, good high frequency response, and minimum attenuation in dB/100 feet looks like a winner. In conjunction with Andrew's F type connector, O-ring seals and sturdy outer conductor offering good moisture resistance, the 88 percent velocity rating, and copper clad aluminum conductor in nominal 1/2-inch cable seems to offer a good many advantages over competition. Note the accompanying chart in the illustration (Fig. 8-7) as well as the very low VSWR rating at just 1. Naturally, these somewhat fancy cables are going to dig the exchequer a bit,

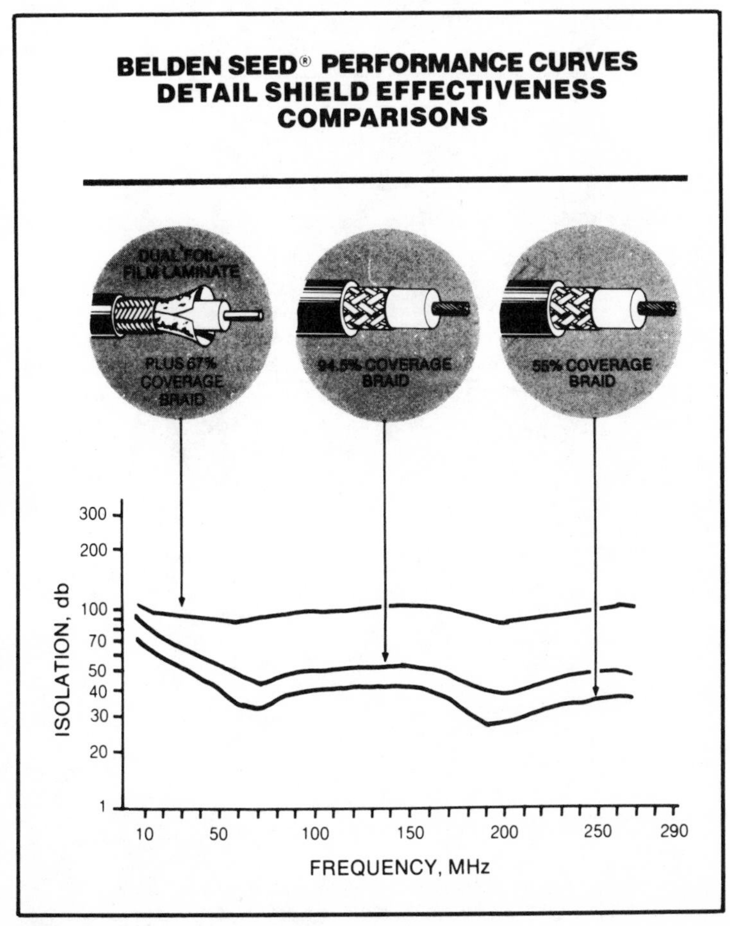

Fig. 8-8. Belden's excellent MATV and home installation TV cable (courtesy Belden Corp.).

but they'll bring all the signal that's fit to see at maximum resolution and minimum loss.

For regular MATV and home installations (Fig. 8-8), Belden, Corp. has a 9116 steel conductor, copper coated 6U Duobond II® plus braid that does well through 900 MHz. Losses range from 1.5 to 6.9 dB per hundred feet between 50 and 900 MHz. Propagation velocity amounts to 78% with 18 gauge center conductor at 75 ohms impedance. The 60+% shielding and aluminum foil offers 100 dB isolation from external interference. Such cable has been in personal use for the past several years and shows no deterioration or further signal attenuation.

Chapter 9

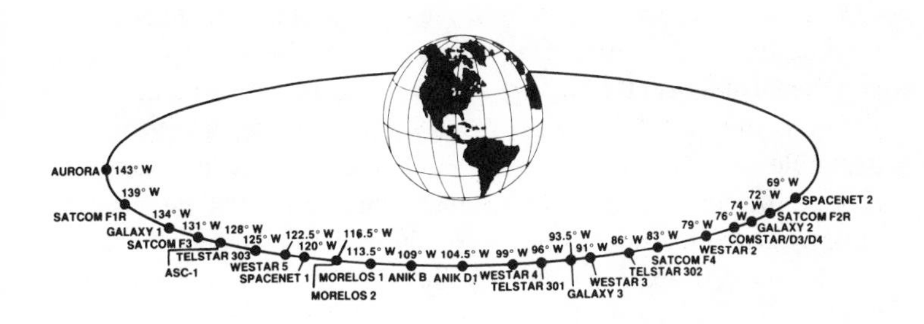

Security and Scrambling Devices

TO PROTECT REAL AND/OR IMAGINED "RIGHTS" OF ONE DEscription or another, many film and program originating sources are testing the murky world of scrambling and descrambling devices for both commercial and home use. The basic barrier remains, traditionally, what it's always been: COST!

THREE TYPES

Hard video signal security with good-to-excellent intelligence recovery is estimated to cost more than $10,000 for each "receive terminal." This, of course, could be done in digital techniques at a very high bit rate (90 megabits/sec in color), and probably with some question of long-term reliability considering the state of the art just now. Transmitting equipment could cost even more, depending on uplink systems, especially multichannel.

Medium video signal security—with programming denial to the majority of receivers—is attractive from the dollars standpoint ($300 to $1,000), but probably results in some degradation for cheaper systems. Smart electronics designers would, undoubtedly, find ways to break the electronics contortions, and some claim they already have, but with marginal proof.

Soft video signal security would constrain only those who have no desire to defeat the system but the cost could be nominal—about $100 per subscriber.

Obviously, the medium security classification would conceivably be most appealing to transmitters who would probably rent such decoders to the general public without tying up considerable funds in the process. At the same time, security would be reasonably well preserved and the customer served satisfactorily. As time goes on, of course, such systems will improve considerably, especially with further applications of digital logic and LSI analog devices which can perform a great many operations in relatively compact hardware. In the meantime, we'll print what information is generally available and whatever else can be gleaned from available contacts around the country. However, the best we can do will be block diagrams, since schematics on the really worthwhile systems would be too much to expect from anyone with an investment of more than casual engineering time. What can be done in some instances, however, is to establish either a trend or pattern, along with a few constraints that may be helpful in understanding both goals and parameters (See Fig. 9-1).

SOPHISTICATED DEVICES ARE COMING

For industrial or corporate teleconferencing, the existing and developing art of digitizing and some encryption seems to be the safest method momentarily, especially since color video isn't a prerequisite. Voice and slow scan TV apparently answer many needs in the early stages of this developing science/art and should

Voice/Data Protection Technology

TECHNOLOGY	VENDORS	PRODUCTS
Voice Scrambler-Analog (VS-A)	18	71
Voice Encryption-Narrowband (VE-N)	11	13
Voice Encryption-Wideband (VE-W)	11	23
Data Encryption (DE)	21	58
Facsimile Scrambler-Analog (Fax-A)	1	1
Facsimile Encryption-Digital (Fax-D)	3	6 (10)*

Total Vendors: 35 Total Products: 172 (10)

**Although there are 172 specific products listed, 10 of these are claimed to work with Fax-D applications. Therefore, if a product is used for both DE and Fax-D, the product is counted only as a single device.*

Fig. 9-1. Topical listing of scrambling vendors and their products (courtesy of Communications News).

satisfy most clients for the time being. Later, when competing systems begin translating the 12 and 14 GHz downlink and uplink primrose paths and patterns, much tighter encryption will necessarily become the order of the day because of the privacy these business firms must have. In another several years, workable schemes in *companding* (expansion and contraction) of these high bit-rate signals will probably be developed to the extent that living, moving color can be displayed with very reasonable fidelity. In the meantime, designers can concentrate on the analog systems, extracting the final drops of electronic scrambling or waveform alterations possible to still produce good quality signals but deny their access to the general public.

ANALOG AND DIGITAL TECHNIQUES

As most are aware, cable TV often scrambles enough video to discourage nonsubscribers using video inversion and even additive signals that require some ingenuity to defeat. Unfortunately, there are those who make a practice of finding usable solutions and then selling appropriate decoders to any who would cheat. Of course, these are usually entertainment systems, and there are break-even points to factors relating to total subscribers versus more or less expensive security systems. Then there are combinations of video and audio scrambling that do seem attractive to less secure systems, causing double effort for those who would overcome the dual obstacles.

Home Box Office has already made a somewhat similar decision and has gone past the bid process for results. HBO is interested in pseudo-random sequencing since sequence length or frequent changes makes code breaking more a problem than less complex methods of straight decoding. Various channels and systems could also have different coders and decoders, and certain components might not be identified or epoxied in a complete seal, for instance, just like some automobile ignition systems. HBO would like such design targets to be fulfilled as: audio and video security, addressable control, and prevention of system compromise. Operating principles include: fail safe operation, no operating restrictions on the user, and reliable operation in any receive/transmit mode environment.

Audio Signals

Analog scrambling could add to audio a constantly varying

mode with certain masking distortions inserted—but success or failure would depend on complete removal of such distortions and a truly decoded voice and music signal without intruding variations. Equipment complexity and reliability factors could well make such a task both undesirable and impractical. With baseband audio not exceeding 15 or 20 kilohertz, digitizing and/or encrypting these signals would seem considerably simpler. Binary levels would appear to be more appropriate since they could then be placed near the video carrier for dual transmission and probably no discernible crosstalk.

Video Signals

These can be scrambled both by analog and digital means, just like audio, but the bit rates are considerably higher, especially when coding and decoding NTSC color. Video can be inverted at about 55 IRE so that any 100-IRE video appears at some small amplitude with large signal inputs and large with small signal inputs. Sync may be interrupted or inverted also, or additional signals may be added to both.

Digitized video does cost considerably more money, and high bit rates, especially for color, can cause distortion unless an expensive system is used. Whether this much complexity is worth the price remains to be seen. If, however, picture companding becomes essentially successful, then both cable TV and satellite transmission may happily adopt the decreased bandwidth and digitize and encrypt whatever they wish at some nonprohibitive cost. On this, for the moment, there is little more than speculation and a great deal of work to be done before the trend becomes noteworthy. Another several years will tell.

Remote Decoder Control

There are lots of fancy words such as "addressability" that go with such electronics, but it all boils down to station or transmitter keying of customer decoders. All this has to be folded onto the video signal, including sync commands, and these also require encryption. That adds another level of complexity to an already difficult condition. Sure, there are many ways to "scramble" signals of every description, but the more pseudo-random, followed by sequencing, the more expensive all this becomes. The bottom line then adds up to "how much will the consumer pay?" Sometimes, when the competition is fierce, the payoff isn't always very much, especially if more than several dollars per month constitutes the difference. However, satellite pay TV and CATV could have their dedicated clientele, and one situation may not precisely mirror the other. Again, when theory becomes practice, then we'll have better hindsight—at the moment, a very depressed TVRO industry.

Scrambling Beginnings

Also called encryption, again seems to rely on this pseudo-random sequence that HBO has identified, along with something called Data Encryption Standard. Scrambling signals would be modulated within the video and audio envelopes and all programming could originate from the transmitter in automatic sequence.

According to R. Alongi and R. Fitting, of Comtech Telecommunications Corp., Satguard has already been developed with high security with good frequency range to at least 10 kHz and low S/N distortion. Video has medium security scrambling and meets EIA RS250B broadcast quality specifications. Existing satellite receivers would not have to be modified, and subscriber electronics could be automatically turned on and off and resynchronized periodically if and when there was power-line failure. Further, an additional data line of up to 9.6 kilobits/sec would be added.

Comtech selected a pseudo-random line inversion application for video because recovery of the original signal is not degraded. Random line inversion would also combine with suppressed line sync pulses, and audio would be multiplied with a complex digital waveform at line rate. With such methods, the picture would become thoroughly scrambled and audio would sound like noise.

The company says that a single LSI integrated circuit could handle decoding, each system would have a special sequence covering some 60 thousand subscribers, with signal recovery very good. There are, we are told, a minimum of 1,000 different codes available as each video line becomes either inverted or level shifted, or may *not* be inverted, depending on code bits. When there is inversion, it takes place in the middle of the horizontal sync pulse and horizontal sync pulse and horizontal sync pulses are then suppressed. Half of the picture information is automatically lost and the receiver may not sync at line or frame rates. Video can then be recovered by reinverting and back-porch clamping. But audio recovery is another matter, and only those with code knowledge could accomplish this mission. Standard earth stations in satellite broadcasting require no additional circuits to handle such scrambling.

Descramblers have two addresses, one unique and one common. The common address turns on all descramblers at the same moment, while the unique input requires a refresher signal about every two minutes. So, unless both addresses are known, reconstruction of the audio signal apparently becomes impossible.

If everyone could or would encrypt general traffic then there would be few problems. However, it's difficult to convince satellite or cable receive owners to spend more for a scrambling system than their own individual terminals are actually worth. Even the very rapid expansion of computer and voice interchange hasn't yet produced any considerable fear of electronic intercepts, even with the special vulnerability of dial-up systems. Many argue for bulk encryption of all satellite uplinks to substantially guard message confidentiality and also reduce costs to both transmitter and user. But, the idea seems to persist that such security is only the responsibility of the receiver and so program originators continue merrily on their way. Inadvertent crosstalk between channels (such as you occasionally hear on phone lines) hasn't become a werewolf yet and probably won't until both the satellite and cable industry have progressed considerably beyond their initial growing pains.

In the meantime, several million satellite dishes are to be erected by the late '80s, and many eyes and ears will happily "look" and listen to all sorts of results. Perhaps it's still a unique circumstance that most "airways" remain "free." At least many of us will obtain much needed additional education. Results will depend on how we use it! But Congress and the courts will, regrettably, have the last word. "Free," we're told is a relative thing anyway—so enjoy it while you can!

ORION

Oak Communications, Inc. has its own ORION (Oak Restricted Information and Operation Network) satellite encryption scheme which is rather similar to the general trend of relatively secure video and highly secure audio. It's useful at both C- and Ku-bands with standard TVROs and orbiting satellites, and should be very compatible with DBS systems arriving in the near future. According to Oak, it has already been tested on WESTARs I and III, SATCOMs I and II, and Canada's ANIK-B.

Fully addressable, it permits controlled delivery of information to specific decoders, each operating on a unique code which depends on digital directions from the uplink. There is time-varying

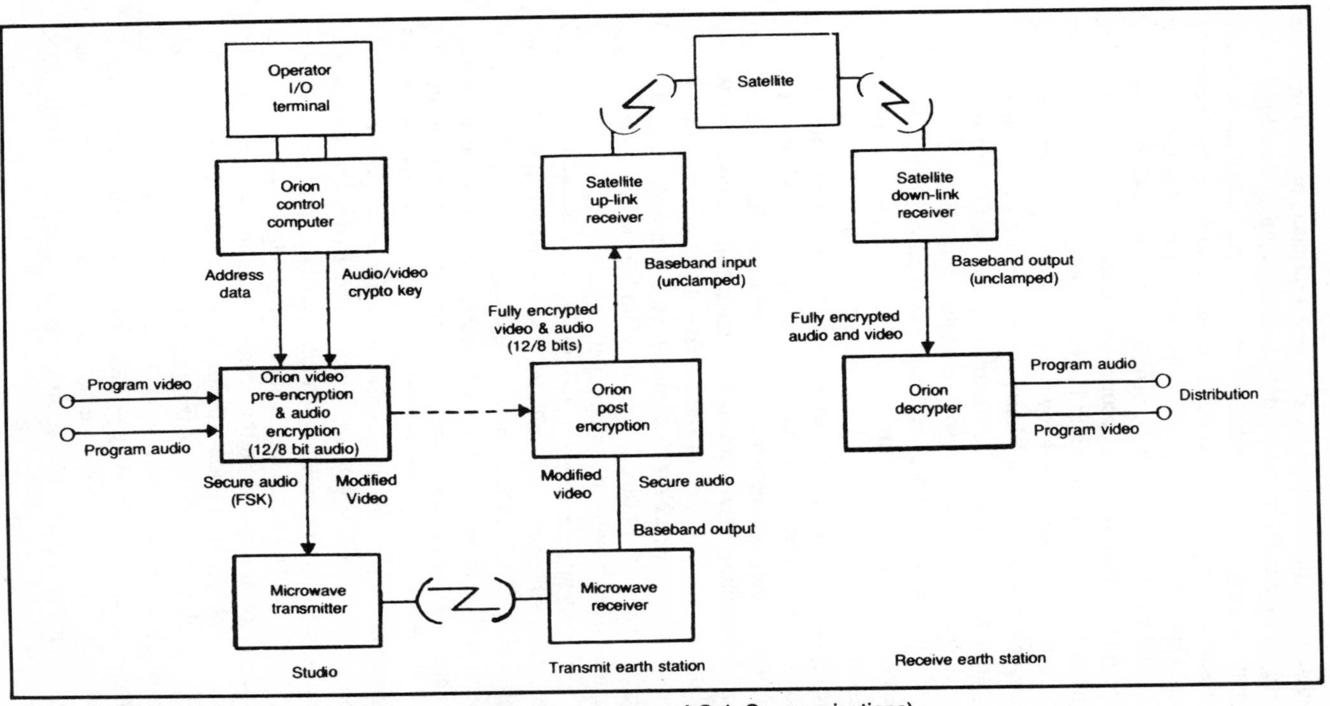

Fig. 9-2. Oak Communications encryption block diagram (courtesy of Oak Communications).

video and audio encryption, digitized and encrypted audio, standard baseband TV inputs and outputs for uplinks and downlinks, broadcast quality signal processing, tiered program control within the secure channel, and proven computer control of the entire system. Oak's diagram in Fig. 9-2 shows the setup for major phases from program origination to program recovery—all in baseband. There is also an optional second encrypted audio channel, along with another data channel and TVRO control units. Many levels of tiering are available, allowing time-sharing multiple satellite transponders by several "classes of authorized subscribers." Any decoder may be blocked from all tiers by master control.

An outsider attempting to view the signal sees severely scrambled video without sync, and hears audio only as white noise. At master control video is partially scrambled, baseband audio digitized and encrypted, and all addressing information formatted. At the uplink, video is further encrypted and its channel, masked audio and video are all combined for modulation and uplink transmission to the geosynchronous satellite. Here the intelligence becomes translated down in frequency and transmitted by TWT or solid-state transponder to the receive dish, its LNA, and downconverter-receiver. Now, the encrypted signal may be distributed "as is" to a series of receivers or unscrambled in the Orion Decrypter for direct baseband distribution.

Addressing rate allows 7,200/minute with the number of possible subscribers set at 2 million. It's compatible with Teletext, Teledon, VITS, VIRS, source ID and Captioning, and the video meets RS250-B specifications. A decoder is told what programs to decode by automatic computer control which also identifies the specific decoder. Signals are time varied pseudo-randomly, and decryption is virtually impossible even for most people in electronics. Standard TVRO earth stations are used without modification to process scrambled downlink signals.

While *not* Orion we also understand there is another system of secure communications that converts analog to digital line by line, scrambling each line with a pseudo-random code. This is converted back to analog *before* transmission, sent through the satellite, received and the process reversed by decoders on the ground. We also understand the only evident resolution loss is due to quantization in the encryption process. So don't believe those who would badmouth such additionally secure systems. They do, indeed, work with adequate or superior results.

TELETEXT DECODERS

Although the Federal Communications Commission has yet to set standards for Teletext broadcasts, this service has been telecast experimentally and broadcasts are even now available in many cities both over-the-air and through satellite transmissions.

Such signals are formed by a clock run-in burst, a framing code, a "preamble," and approximate ASCII. The burst permits phase-locked loop circuits to sync lock, framing codes identify where the various bits belong, and the preamble tells the pages and rows being transmitted. Each page has 24 rows, and every row has 40 characters. Any location may be blank, or show an alphanumeric or graphic character. Encoding takes place on two or more scan lines in each TV signal frame. A teletext transmission computer converts all information to a digital code via a typewriter data terminal.

The Teletext decoder in your personal TV receiver then separates this digital information from the remainder of the regulator television signal, decodes whatever is required, and displays all information on the TV screen. Zenith is now working on such a decoder which will output RGB (red, blue, and green) baseband directly into connections of a television monitor. In this way, better color saturation and resolutions are obtained, producing excellent graphics and alphanumerics.

A simplified block diagram of the Teletext decoder is illustrated in Fig. 9-3. Incoming signals are picked up usually at the receiver i-f's as composite video, data search in the vertical blanking interval takes place, and levels are found and adjusted, then the data is isolated. In Data Acquisition, digital signals are detected and rearranged in usable sequence for memory and display circuits. After this, a page of data is stored in Page Memory, and then the Display Generator stores the characteristics of letters, numbers, and symbols in a read-only memory (ROM). Upon commands from the microprocessor control and space command transmitter, all the information is routed to and displayed on the RGB monitor's cathode-ray tube as wideband, interference-free letters, numbers, and symbols of all descriptions.

COMPANDING

While not always a major heading, mention of the word "compander" not many years ago would have almost mandated a jail

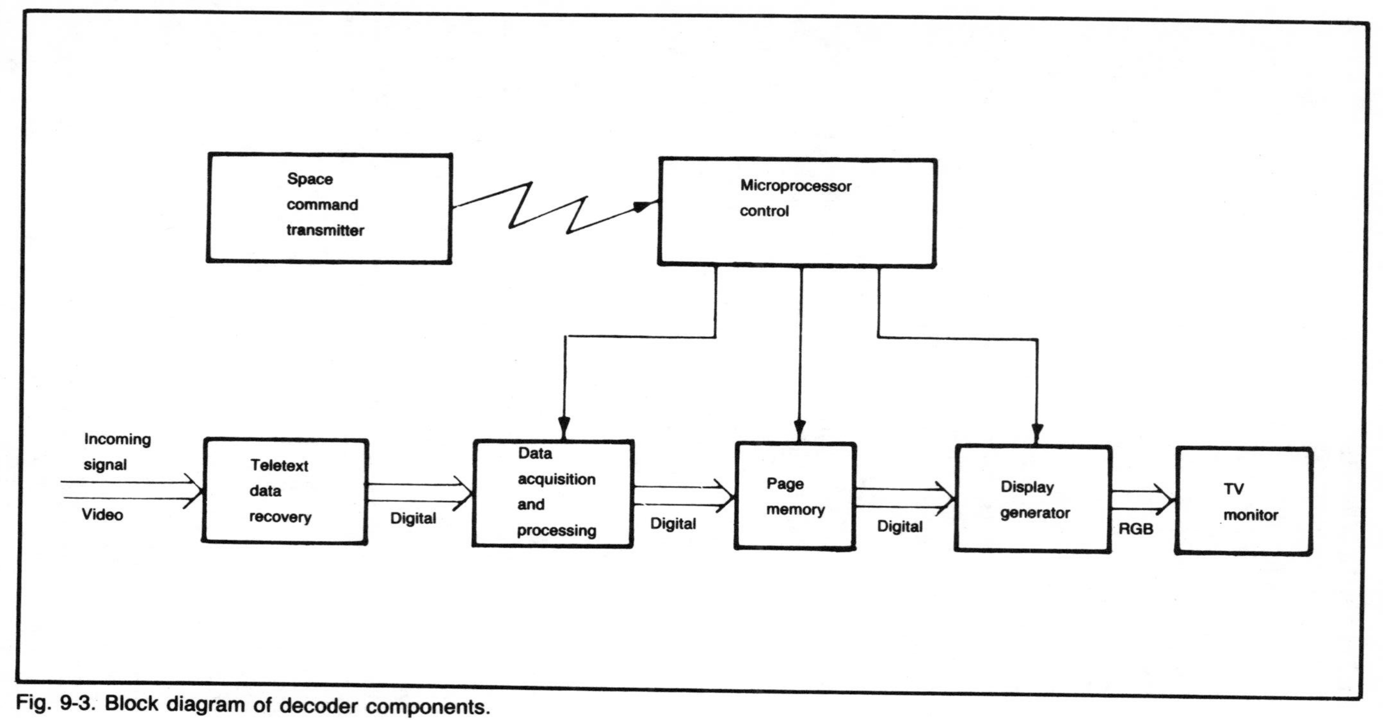

Fig. 9-3. Block diagram of decoder components.

term for breach of national security. Today, however, its use in lower frequency intelligence is becoming more and more important because of limited transponder capacity. American Satellite was the first prime satellite carrier to use FDM/FM syllabic companding in 1978, and since then MCI and Intelsat have approved usage with others scheduled to follow. RCA Americom, for instance, using 36 MHz transponders, found that 1,092 standard message channels could be compared to handle 2,074 channels without quality loss. Compandors, such as Coastcom's Model 937 VF, are designed to double normal output (2:1), and there are others with companding ratios of at least 3:1 already in service.

These compandors consist of compressors and expanders so that input information may be condensed and that of the output amplified. The compandor actually reduces the dynamic range of, say, voice inputs, and the expander restores such signals to their original measurements (Fig. 9-4). In FM, the deviation decreases, and so more channels may be added to any FDM baseband. Then at the receive end, all is restored. Along the way, signal to-noise

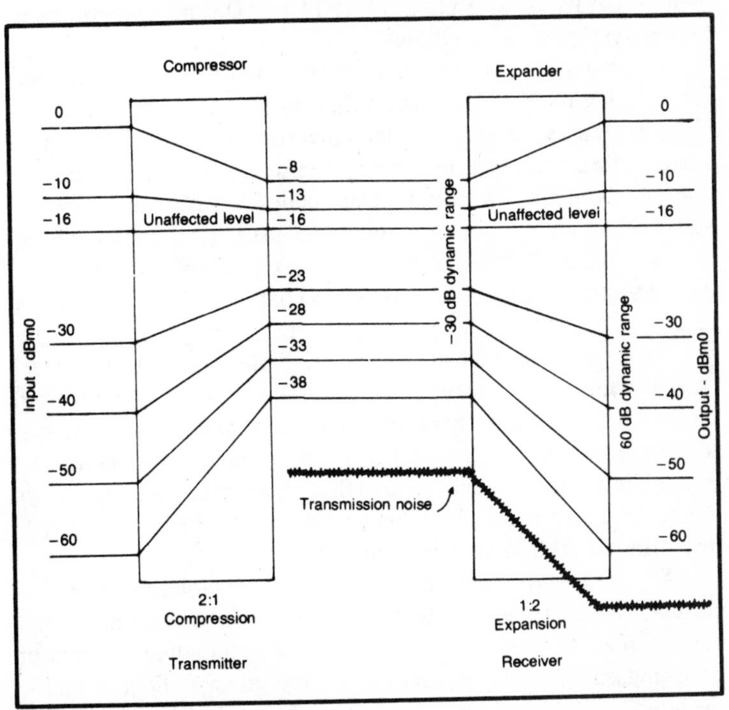

Fig. 9-4. Compression and expansion "companding" is now coming of age (courtesy Communications News).

ratios are improved as much as 25 to 27 dB in 3:1 compandors, and as much as 15 to 17 dB in 2:1 systems. Data rates up to 4.8 kilobits/second seem to be unaffected by companding, and even 9.6 kilobits/second have been successfully attempted without undue effects in some circumstances. These compression/expansion devices can be added to existing FDM/FM systems without disturbing existing equipment.

Z-TAC™

Zenith's Z-TAC has been designed specifically for CATV, is monitor-ready, text-ready and will interface with individual billing systems. Based on American Television and Communication's (ATC) synchronization suppression and active video inversion (SSAVI), it offers 20 program categories, full head-end addressability, Zenith's full-featured remote control with dual cable switching options, sequential and random-channel numbering, channel oriented and program tagged plus being head-end controlled. Recently tested in space applications, Z-TAC is said to perform equally well on CATV, subscription TV (STV), MDS (microwave distribution systems), or satellites.

Video may be inverted and the horizontal sync pulse IRE offset, offering four modes of operation: standard video; sync offset, video normal; sync normal, video inverted; sync offset, video inverted. Horizontal sync levels may be used as static, dynamic random, or dynamic algorithmic. One mode is continuously used in static selection, dynamic random switches between modes in no set pattern, and dynamic algorithmic measures the video and chooses operating parameters such as channel power reduction or enhanced scrambling. In power reduction, horizontal sync is offset into the normal video area, measuring the average picture level (APL), which is ordinarily lower than either normal or inverted video. Choosing correctly reduces average channel power and system performance is improved through more linear range operations. Scrambling enhancement requires that offset sync is almost always below APL, and so high level luma fields are inverted, burying sync and offering additional scrambling (Fig. 9-5).

Audio has three basic options termed selectable, clear, or suppressed. Here, audio may be transmitted as is, or scrambled and sent with a "barker" channel that can carry an advertisement or other message, or may be channeled through an audio jack on the decoder for a second language output or similar purpose. In the clear, audio isn't scrambled but sound is muted in the decoder-off

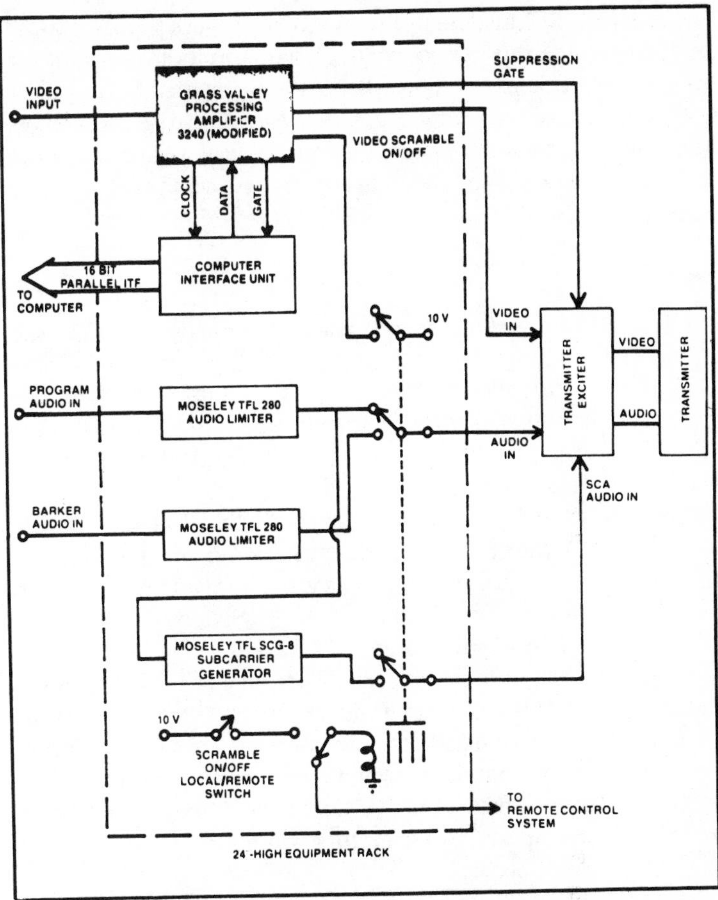

Fig. 9-5. SSAVI encoder designed to operate with Harris or RCA exciters (courtesy American TV and Communications Corp.).

mode. Suppressed audio mutes sound when the picture is scrambled and is done within the decoder and does not involve transmission (head end) portions.

Tiering options permit customers a varied selection of programs included in four channel banks, each of which contains five program categories. Such banks may be changed from the head-end upon demand. Decoders contain both a market code and subscriber number. The market code won't permit the decoder's move from one market to another, while subscribers need only be addressed when changing program categories. Lost or stolen decoders in non-pay status are not addressed and therefore there's no picture or sound.

AZ-TEXT™ module is also available to work with Z-TAC™ and supply one-way text service. Signal formats may be chosen from British Teletext, Antiope, Telidon, the proposed North American standard, or other alphanumeric systems. AZ-ALERT™ option will also sound such alarms as fire, medical, police, etc. when Z-TAC is operating and tuned to any channel containing required data.

SSAVI

If some of the above sounds vaguely familiar, it is, since the origin of the system began with ATC of Englewood, Colorado. SSAVI it is, too, especially with certain Zenith circuits in the decoder unit, reducing costs. Sync-pulse suppression confuses the television receiver in its sweeps that puts erratic vertical tears in the picture, resulting in a totally distorted pattern and swirling colors. Blacks and whites are transposed as well. Audio may be passed through the decoders as either scrambled or plain language, in addition to the "barker" signal as a separate audio output into another speaker or even as simulated stereo. Sixteen-ohm speakers may be driven without extra amplification.

Certain markets are digitized in a permanent PROM memory in the decoder. This is compared with any market name that's transmitted and if the two match, then the decoder does its job. Otherwise, reception is unauthorized and pictures continue scrambled. PROMs, of course, may be substituted in the decoder. Two different STV (subscription TV) operators will not normally be given the same market codes.

Tiering and Storage

In tiering, any decoder may receive any combination of five programming levels, or all of them. Therefore, different options may be sold such as sports, certain movies, plays, or special events. STV packages are tailored to address most customers needs or fancies.

Program codes and on/off commands are stored in a battery-run memory within each decoder. If the decoder is plugged into a power outlet, the rechargeable battery trickle charges constantly. Unplugged, the charged battery maintains memory information for some three weeks. Battery life expectancy amounts to between 7 and 10 years. Of course the decoder operates normally anyway when plugged into ac.

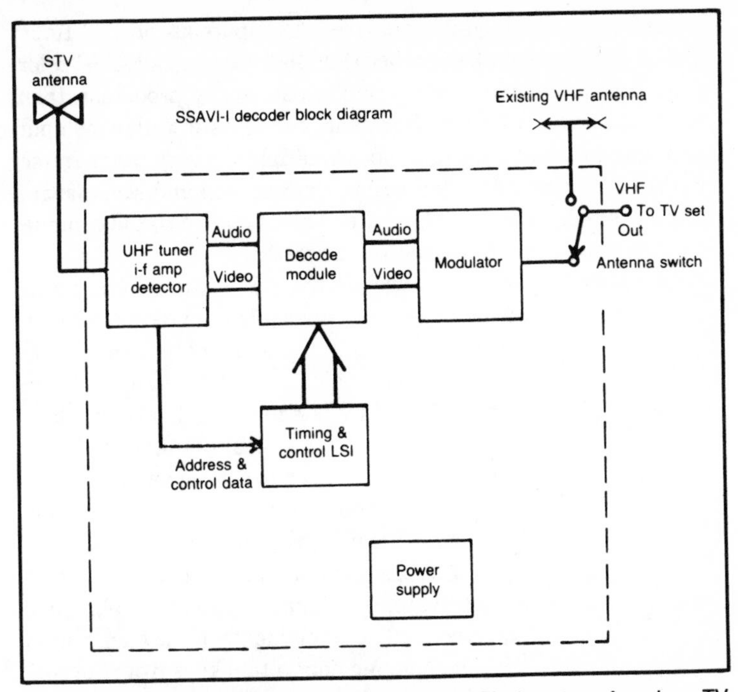

Fig. 9-6. Block diagram of SSAVI decoder with I/O's (courtesy American TV and Communications Corp.).

SSAVI decoders (Fig. 9-6) contain Zenith System 3 receiver/tuner units and will respond to TV signals just the same as first class television receivers. If additional signal pickup is required, the SSAVI decoder can furnish power to a preamplifier stage located somewhere along the antenna lead-in. The decoder also has an antenna switch built in that isolates the unit from delivering spurious radiation back to the U/V antenna—an FCC requirement.

VIDEO TELECONFERENCING SYSTEM (VTS)

In late 1982, we found that VTS has progressed much further than expected through the industrious efforts of Compression Labs Inc. of San Jose California. By digitizing both audio and video and encrypting them at a transmission rate of 1.544 megabits/second, video up to 2.7 MHz and audio between 50 Hz and 5 kHz can be transmitted and received via satellite or terrestrial transmissions, with the video in relatively full color and motion. The system also offers channel time-sharing with graphics, interface to RS-232 and RS-449 data ports, in addition to both local and remote diagnostic troubleshooting and system analysis. All inputs are in real time.

The Compression Labs codec (Fig. 9-7) picks up a NTSC color signal from a camera, routes it through an analog processor, then an A/D converter, filters by H/V sync, and transfers all to an input frame memory and then to a block multiplex. Every other frame from the TV camera is then cosine transformed and scene adaptively encoded at the 1.5 Mbps rate. This is passed to the communications controller for digital transmission.

The receiving terminal picks up the 1.544 Mbps signal, runs it through a scene adaptive decoder, inversely cosine transformed, then block demultiplexed. The output frame memory then permits double frame display to maintain the usual 30 frames/second conventional vertical rate. Followed by horizontal and vertical filtering, the digital bit stream is transformed to analog amplitude levels by the D/A converter and NTSC-processed to a designated TV monitor. Along the way, of course, audio is separated in transmission and recombined in reception before it reaches the TV monitor.

Operating within the Data Encryption Standard of the National Bureau of Standards, a private key algorithm using a 56-bit code word encrypts audio, video, and other elements of the transmitted signal, with user option of changing encryption keys from the control terminal. Audio coding results from a delta modulation algorithm running at 47 kilobits/sec. Voice signals are then synchronized with video by delaying audio for appropriate lip sync. In TV graphics, the video signal from a motion TV camera is interrupted and the graphics display is switched into the communications channel. Several frames of motion video are then "preempted" so that a portion of the graphics may be transmitted during each of these intervals. By this method, highly detailed still images are easily sent and recovered. System diagnostics are possible through a 1200-baud modem.

TVRO SCRAMBLING

By the first quarter of 1986, newly-introduced TVRO scrambling has a devastating effect on the satellite home earth station industry sending it into a sales tailspin that amounted to between 35 and 50 percent of what it was in 1985. Moviemakers and the cable industry argued that "free air" viewing did *not* include the divine right of any and all satellite receivers to pick up programs regardless of origination. Especially vexed were Home Box Office, Cinemax, The Movie Channel, Showtime, CNN, ESPN, Nickelodeon, and Arts & Entertainment who have already begun mask-

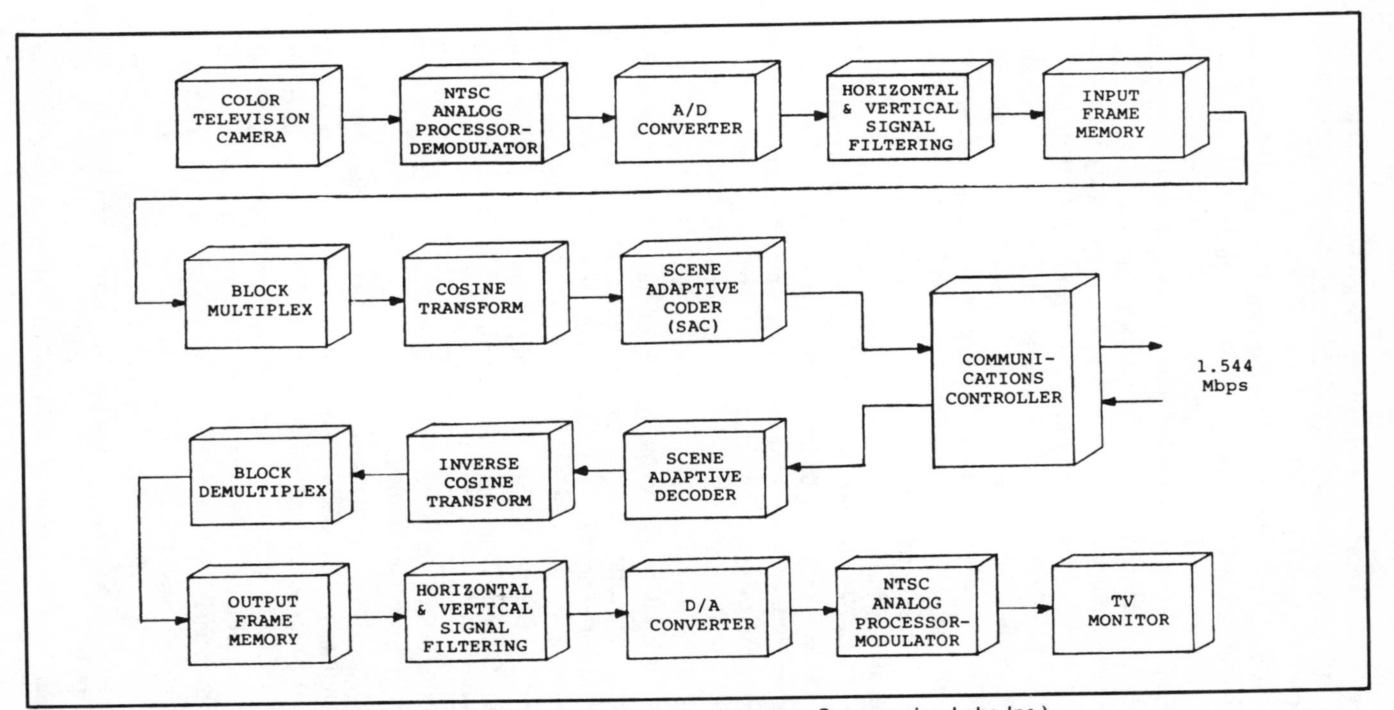

Fig. 9-7. Flow diagram of the CLI VTS 1.5-Megabit full duplex system (courtesy Compression Labs Inc.).

ing signals. These, according to best advice are to be followed soon by others, constituting a total of 21 channels that will "go dark" with the coming of 1987.

Despite the introduction of legislation in the U.S. Congress, intervention and lawsuits by SPACE, and the sudden appearance of so-called codebreaker black boxes, a two-year moratorium on scrambling has not occurred, and the few qualified TVRO retail sales-service organizations are slowly turning to service-only outlets—although these are estimated to constitute only 5-10 percent of the total TVRO industry. Unfortunately, TVRO sellers—other than those in remote areas—have been unable to convince new buyers that 21 channels out of a possible 200 don't amount to a great deal and that satellite sound and pictures are *always* better than local telecasts. Possibly they are not aware of this factor themselves. In addition, the current price of a popular descrambler amounts to $395 plus monthly service program rates, and this would discourage most potential customers who have to pay between $2,000 and $3,500 for a decent satellite system in the first place. Furthermore, adequately programming some of these descramblers is not a job for an adolescent or unsophisticated adult.

Two late entries in the scrambling scramble seem to be making immediate headway but, apparently, will not go head-to-head, since one will be purely commercial and the other CATV and consumer. The first is a MAC (multiplexed analog component) system imported from Great Britain and under vigorous development in Canada and the U.S. by Scientific Atlanta. The second originates from a Bureau of Standards algorithm and is known as VideoCipher® I and VideoCipher® II, with VideoCipher® III on hold. We are told that Private Satellite Network and other similar companies will be using the former, while the VideoCiphers are now becoming well established with cable television and the consumer, although there remains some dealer/subscriber resistance even during 1986. CBS is said to favor VideoCipher I, which has more secure video scrambling, while ABC, Home Box Office, the Movie Channel, and others are actively considering or actually using VideoCipher II. By the end of 1986, some are saying, 21 C-band channels will have been scrambled—a condition that is already evident on Galaxy 1 and ANIK-D 1.

What the future holds in the way of secure programs is, at the moment, anybody's guess, but between January and April of 1985, the TVRO sales industry was off between 35 and 50 percent over immediately prior months, and was just picking up again in June.

This has given considerable comfort to the cable and movie moguls who don't like "backyard pirates" in their hunting preserves, but a severe downer for TVRO manufacturers and retailers who couldn't quite learn to handle the situation without public alarm, Congressional consternation, and several lawsuits.

Considering that any worthwhile TVRO system can access at least a dozen C-band satellites with 12 to 24 transponders each—many of which are broadcasting video—our personal opinion is there is little cause for apoplexy. As the scrambling issue unfolds further before Congress and the "black box" (illegal) entrepreneurs, however, there could be some changes, one way or the other. Regardless, the 1986 sales dip is probably a mere "glitch" in satellite earth station progress, and advanced electronics, stronger (remaining) retailers, servicers, and manufacturers should soon have the summer and fall boom of 30,000 to 40,000 systems per month moving again before this publication is in print.

The MACs

According to W.G. Stallard, of the U.K.'s Independent Broadcasting Authority, there are actually *three* multiplexed analog component MACs, all with somewhat different formats involving digital sound modulation and/or time-division multiplexing. The three are described briefly as follows:

☐ A-MAC: contains one or more FM or digitally-modulated subcarriers added to video *before* FM modulation.

☐ B-MAC: the time-division multiplexing of digital sound into the line blanking period, with the composite signal then FM modulating the carrier. See Fig. 9-8.

☐ C-MAC: during line blanking period sound digitally modulates carrier and video FM modulates carrier during line *scan*.

As you can see, these are considerably different systems and are individually useful to some existing systems and not in others. Since C-MAC does not need a color subcarrier and only uses a small amount of preemphasis, this version appeals to the British who are much in favor of Teletext transmissions. Further, as each television line begins, there are 194 bits, with eight devoted to sync so that video line, frame and color sequences may be timed, in addition to a further trigger for time division multiplex control. Remaining bits are then multiplied by lines/second for sound and data that is stored when received and read out at a continuous rate for con-

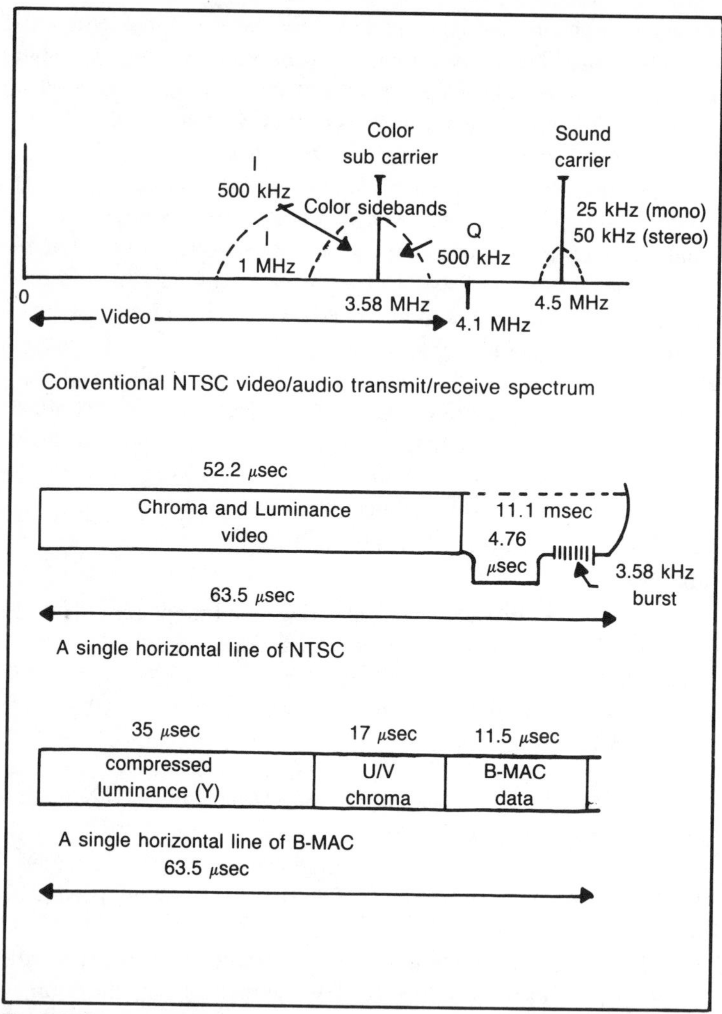

Fig. 9-8. Comparisons between standard NTSC analog and B-MAC multiplexed systems.

tinuity. Synchronous transmissions then occur at the "internationally agreed" sampling rate of 32 kHz.

B-MAC seems more suitable to our own NTSC North American color system since it involves time division multiplexing and will also contribute to wider display screens and greater video bandwidths. While not necessarily scrambled in the sense of enciphering, it will require specific decoders and red, blue, and green (RGB) input receivers to handle the transmissions; therefore, its singular business and commercial suitability. For instead of using the usual 4:3 CRT aspect ratio, some plans are afoot for 16:9 aspect ratios. Final formats, however, seem to be proprietary just now, so we'll proceed with what has been made public already.

First, in all time-multiplexed schemes, chroma for satellite transmissions is transmitted line-by-line and sequentially. Secondly, the Advanced Television Systems Cmte. (U.S.A.) has adopted the wide screen aspect ratio of 16:9 and CCIR will consider such ratio for high definition television. Note that this ratio is simply the square of the present 4:3 and could be sync-clocked for conventional display. Thirdly, a sampling frequency of 14.318 MHz could provide luma bandwidths of 6.4 MHz as opposed to a 3.58 MHz frequency which would increase set bandpass to 4.8 MHz compared with comb-filter equipped conventional receivers whose maximum response is 4.2 MHz.

Based on the French sequential color transmission system and, combined with time division multiplexing, only one signal is transmitted during any interval, removing luma-chroma intermodulation possibilities and the color subcarrier. Chroma could then be transmitted during horizontal blanking and luminance during normal, active line scan, along with certain compression and expansion. Because the horizontal blanking interval amounts to 11.1 μs, the remaining space following chroma would consist of such items as sync and clamping, with 52.4 μs now devoted totally to luminance—all neatly compressed within the allowable 63.5 μs NTSC standard horizontal line. European transmissions would occur in the YUV format that stands for luminance and color difference signals and must take into account both the phase alternation line (PAL) German system—much like the U.S., except that it alternates chroma phase every other line and operates at 50 Hz. The sequential with memory French SECAM system transmits a U color signal on one line and a V color signal on another. These are combined with a delay line and an electronic switch. SECAM is also a 50 Hz system.

Our own I and Q system results from suppressed carrier chroma, folded onto the video transmission and recovered by a 3.58 MHz sync generator in the receiver that is stabilized by transmitter color burst of some eight cycles on the back porch of the horizontal sync pulse during each of the 525 vertical scan lines. Because of the 50 Hz rate, the Europeans operate at 625 scan lines and have slightly wider video bandwidths. All three world-wide systems work on a constant luminance principle where color is not appreciably affected by the luminance content in any picture.

M/A-COM's VideoCipher® Satellite Scrambling Systems

Reputedly to be used by broadcast companies ABC and CBS, along with a number of the private program and film makers, VideoCipher® Scrambling Systems, developed by Linkabit™, a M/A-COM subsidiary, is decidedly the consumer scrambler/descrambler program security electronics for the immediate future.

Originated from a Data Encryption Standard (DES) algorithm evolved by the National Bureau of Standards, individual system addresses have a 56-bit key for each descrambler which amounts to $72,057,590 \times 10^9$ permutations, and is thought to be next to impossible to break. Major attention, however, is devoted to audio, and some classified-illegal electronics have already surfaced which are said to restore compressed sync and inverted and slightly messaged video. Audio, however, remains more or less plain noise, undecipherable for any period, at least until now.

Latest reports say there are at least four versions, numbered as you might expect, VideoCipher® I, II, III, and IV. The latter is a version of VideoCipher® II, while III is no longer used. In each, video V/H sync has been removed, the 3.58 MHz color burst centered at some nonstandard line position, and audio transmits as a *pair* of digital signals or channels during 11.1 μs horizontal blanking intervals in the 525 U.S. NTSC line scan sequence. While VideoCipher I supports somewhat more intense video scrambling, our prime attention is devoted to VideoCipher® II, which is the industry's mainstay just now.

VideoCipher II® descramblers evolve from custom VSLI (very large scale integration) integrated circuits applicable to DBS, broadcast, CATV, or SMATV operations. In 1986, more than 15,000 cable head-end descramblers are now installed and operating successfully, with many more to come.

In transmit, an uplink scrambler accepts NTSC video, two audio inputs, and an auxiliary data signals as channel control computers generate consumer authorization for addressing and control. Other subcarriers are then added as needed and transmitted over either C or Ku spectrums to the satellite and then its receiving locations on earth.

The authorized descrambler produces 4.2 MHz baseband video (for a monitor), two stereo audio outputs, and a channel 3/4 remodulated AM/FM carrier for standard TV. There is also an auxiliary data output for downloading software, device control, and other applications up to 88 kilobits/second.

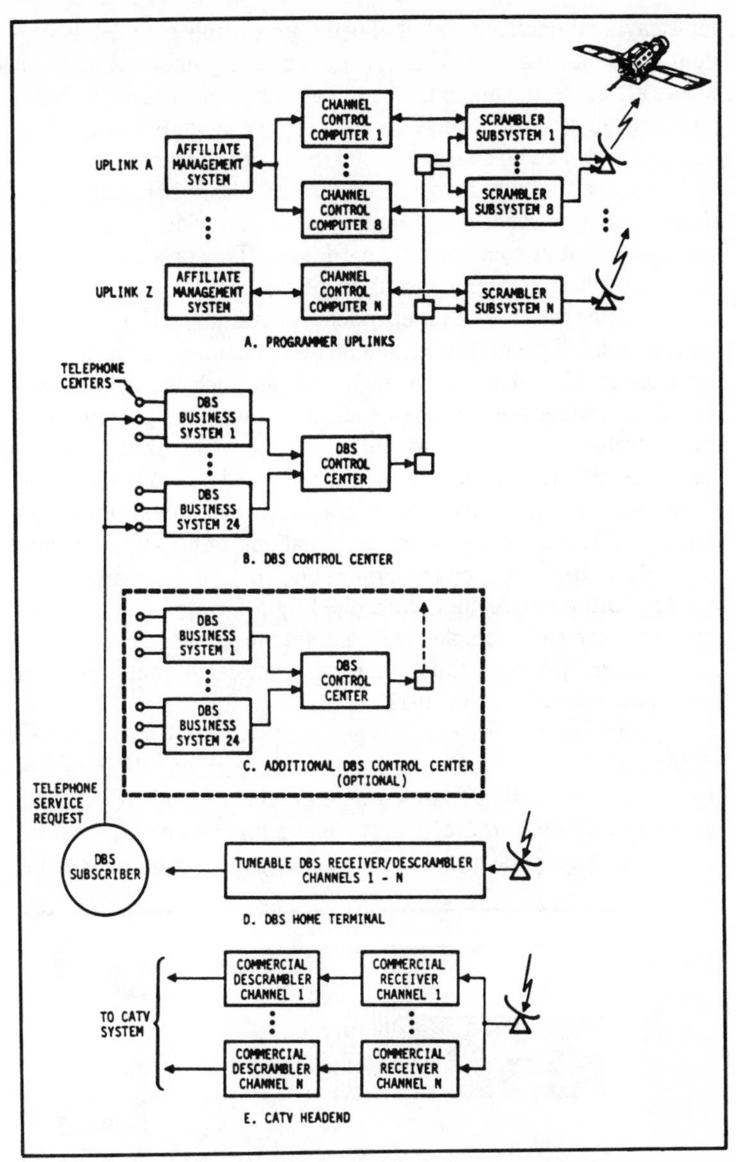

Fig. 9-9. The simpler but more commonly used VideoCipher II® CATV/DBS block diagram (courtesy M/A-COM).

In scrambling, there is filtering and then digitizing of the two audio channels at a digital audio disc rate. Binary samples are subsequently added in random sequence generated by the National Bureau of Standards DES algorithm, combined with error coding bits, interleaving them for transmit. These bits are now completely random and require a specific DES key for descrambling. Single bit errors are corrected and burst errors are random and independent, ensuring high quality, noise-free audio. See Figs. 9-9 and 9-10.

These two audio channels, combined with addressing and control bits, are digitally transmitted instead of the horizontal sync pulse, with all normal sync removed, inverting video, and centering color burst at some nonstandard level. This process removes all audio subcarriers and results in better C/N.

Because there are 56 independent programming tiers, one tier may include all programs on one or more channels, or it may select one specific program on a single channel, such as pay-per-view. The descrambler then compares its tiering with the program key and operates or not, and may not be modified to permit unauthorized programming. Subscribers may then read messages from the scrambled satellite channels, such as authorization, password, selections, coming attractions, financial obligations, or nonauthorization.

VideoCipher I®, on the other hand, has video scrambled in a time-shuffling algorithm, controlled by a keystream which encrypts bit-by-bit and completely scrambles the picture. On-site selectors compare program inputs and descrambler outputs for both audio and video. See Fig. 9-11.

System performance depends on precise back-to-back audio and video specifications, with video luminance weighted S/N at no less than 57 dB and audio dynamic range greater than 85 dB. Coding improves audio and control conditions in high noise environments. A/D and D/A conversion, scrambling, gain control, filtering, clamp-

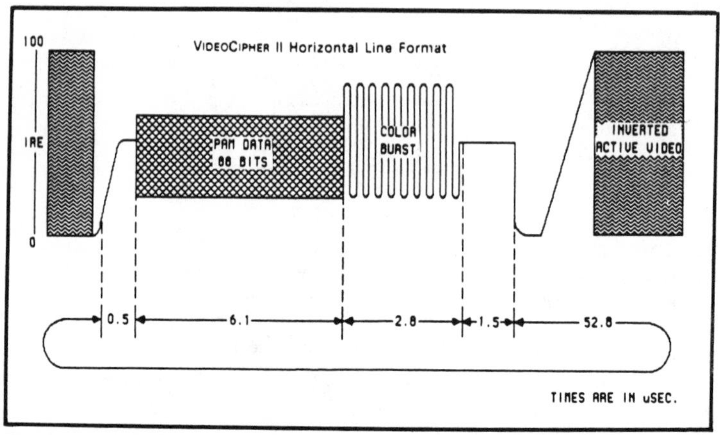

Fig. 9-10. A single horizontal line in VideoCipher II® (courtesy M/A-COM).

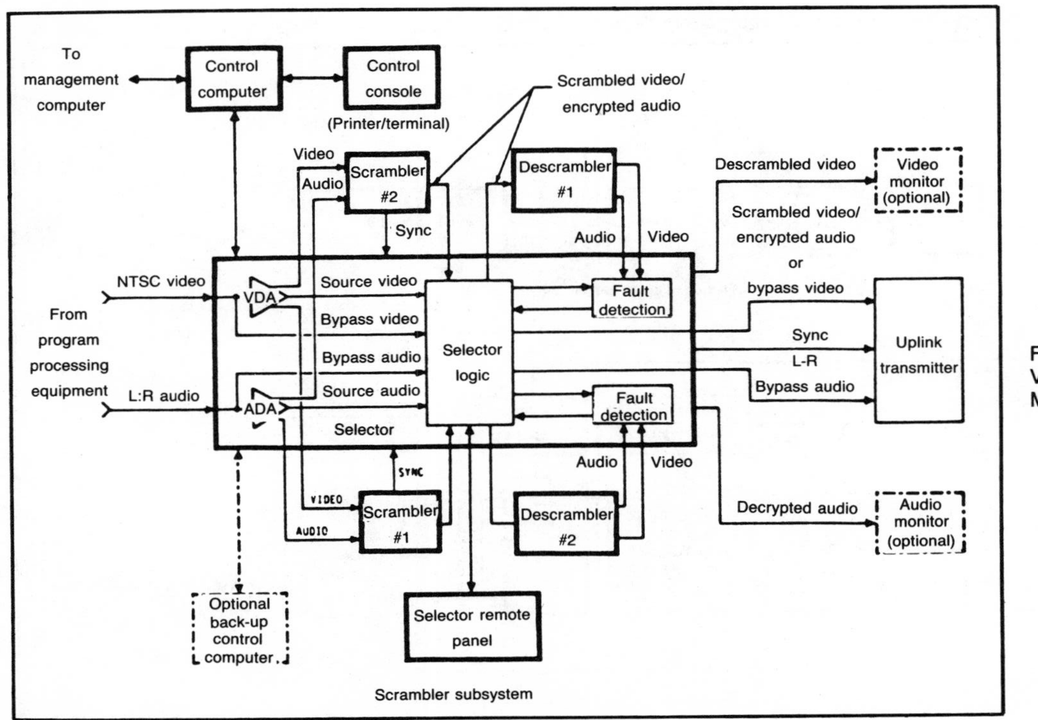

Fig. 9-11. The more complex VideoCipher I® system (courtesy M/A-COM).

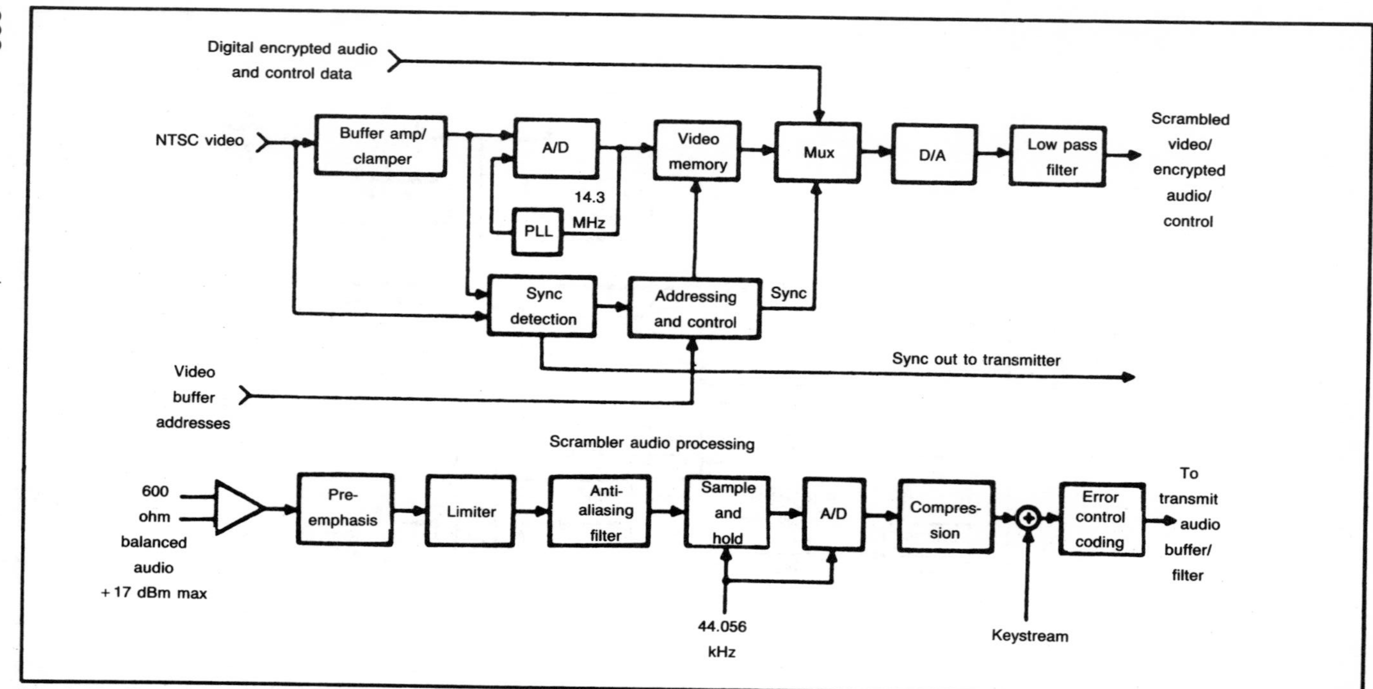

Fig. 9-12. Simplified block diagrams of video/audio scramblers in VideoCipher I® (courtesy M/A-COM).

ing, and phase-lock are all used in the video processing. The active video on each line is A/D converted at 4 × color burst and stored in "multi-line" memory. Digital video lines are then broken up into vari-size pieces, and then read out randomly. They are then further processed and this video is then D/A reconverted for transmission. In Fig. 9-12 you can see the buffer amplifier/clamper, a sync detector, analog-to-digital conversion in addition to phase-locked loop at a sampling rate of 14.3 MHz. Addresses and control are now stored with video in the memory, and encrypted audio is added, along with sync in the multiplexer. Both video and audio are now returned to analog baseband by the D/A and transmitted as encrypted.

Chapter 10

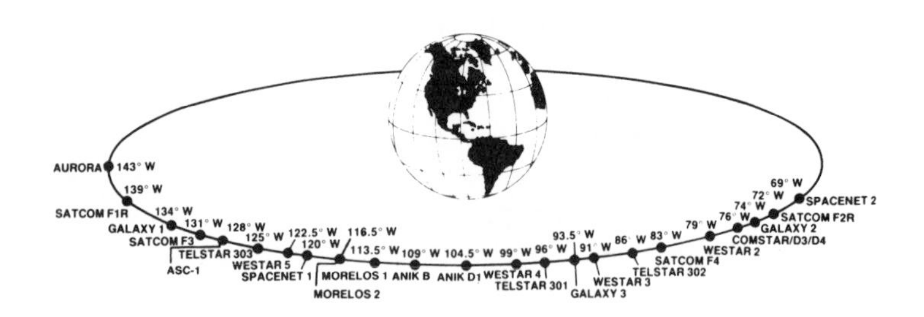

The CATV Satellite Picture

ONCE CALLED "COMMUNITY ANTENNA TELEVISION" IN THE early days of its existence, CATV is now a rapidly growing and overwhelmingly potent force in today's world of consumer and business communications. In 1986 National Cable TV Association in Washington, D.C. estimated that there are today over 6500 CATV operating systems having 20,000 earth receiving stations, serving some 36 million subscribers, with an annual growth rate of 200,000 subscriber homes per month.

If those statistics take your breath away, remember that most new cable franchises are required to install an institutional "B" cable, especially designed for business (and possibly educational) transactions and instructions that are sure to come at a later date. This means that rapid and accurate interchange of information is essential to business needs nationally as well as internationally, and satellite and cable media are prime carriers for the purpose. Teletext and developing fastprint systems are expected to play a major part as press distribution systems, market quotations, shopping news, and business transactions develop. The "B" cable, of course, will encourage additional development of these processes and two-way channels will also permit customers to talk back. As technical CATV people are well aware, broadband cable with excellent S/N ratios, substantial signal linearities, and protected delivery systems, offer almost unparalleled means of video, voice, and data traffic delivery.

Head ends—the processing portion of the CATV signal receiver—will soon be operating in two modes—either analog at the conventional 4 GHz C-band downlink, or digital in the 12 GHz Ku-band downlink. Satellite Business Systems (SBS) has already begun operations in the latter group of frequencies where microwave terrestrial frequency problems have not yet arisen. Analog systems, on the other hand, remain in C-band for the moment, but will be augmented by 12 GHz downlinks in the K-band by at least 1986. Advantages of digital transmissions are initial message privacy signal manipulations, as well as error corrected bit rate of only one in 10^{10}. Up to 40 megabits/sec can be transmitted via satellite which will, in time, probably make possible full color motion for such services as Teletext and live video conferencing. At the moment, only limited motion with certain color is possible but considerable progress is being made in bandwidth compression and companding that can overcome much of the 90-megabit/sec estimated bit rate for non-degraded color motion. By the end of 1985, it is expected that SBS will have almost 100 earth stations in operation throughout the U.S. Because of the higher frequencies involved, only 5.5- and 7.7-meter transceiving dishes are required to handle continental traffic. Perhaps some cable facilities will follow in due time from other corporate consortiums, seven of which have already made applications to the FCC for future service in Ku. So far, cable is doing quite well with analog at C-band. Later, there could be activity in the K-bands after DBS and others have had their experiences. What are called "digital head ends," however, are another matter and equipment such as digital multiplexers, modems, data translators, etc., can be made available for those having a need. Mid-split and high-split amplifiers usually cover bandwidths from 5 to 108 MHz upstream to 174 and 300 MHz downstream, and 5 to 174 MHz upstream and 234 to 440 MHz downstream, as the vernacular goes. There is also a sub-split of 5-30 MHz and 54 to 300 MHz.

PAY CABLE

Although most sports, public affairs, and news remains in the "free" category of satellite-broadcast programs, *Pay Cable*—a natural outgrowth of over-the-air Pay TV—is even now making considerable headway. Coupled with Cable Security it appears to constitute quite an attractive package for many homeowners who have premium requirements for such services.

In 1972, Home Box Office began its program with just over

300 subscribers in Wilkes Barre, Pennsylvania, and in just 10 short years more than 14 million subscribers are now receiving Pay Cable in 10 major programming operations. According to the National Cable Television Association, "for an average monthly fee of $9," cable subscribers may view prime movies, special sports events, variety specials, and even music that's particularly suited to cable TV. New movies are said to be available to Pay Cable distributors as early as 8 or 9 months after first release. Today, more than 570 cable markets offer combination pay services to the public which seems to rapidly be picking up the options. Two-way cable is also making inroads into the market in large metro systems. How about ordering Broadway theater tickets, scanning an enticing restaurant menu, shop in the stores, and answer opinion polls from your easy chair by just pressing a remote-control button?

This "talk back" feature is also making possible home security that's almost like having private doctors, firemen, and the police on duty in the house at all times. Cable is now offering protection from fire, medical problems, and intrusion by way of constant computer scans from the head end that immediately alerts police and fire fighters when any of these emergencies occur. Then, by simply pushing a button, fast medical care may be summoned within minutes through this same computer arrangement. Is it any wonder that the astronomical growth rate of CATV continues? For your information, the top 10 cable systems operations are run by: American TV & Communications Corp., Teleprompter Corp., Tele-Communications Inc., Warner Amex Cable Communications, Storer Cable Communications, Times Mirror Cable Television, Viacom International Inc., UA/Columbia Cablevision Inc., and United Cable TV Corp. Three of these, by the way are located in Colorado; New York has two.

SECURITY SYSTEMS

Security systems are available both for homes and business and are connected to a service computer in the cable company's business offices. Menus and display formats are available to the operator, with any alarm taking priority over everything else. Keyboards and printouts serving as records are included in the equipment. A block diagram outlining the various functions at the head end and at the subscriber's location is shown in Fig. 10-1. All security systems may not have this much elaborate equipment, but the files, keyboard, printer, and service computer are all basics for full service. Note that on the subscriber's end there are three panic but-

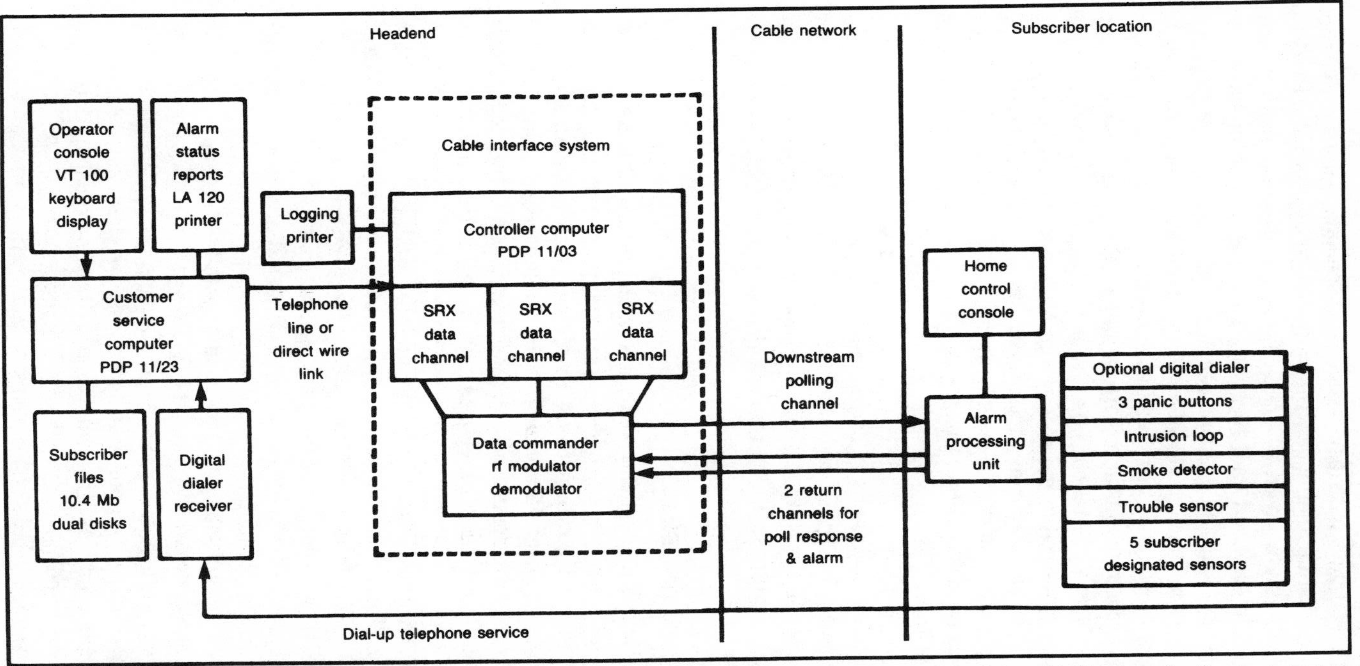

Fig. 10-1. A security system block diagram illustrating headend and subscriber services (courtesy of Jerrold Division of, General Instrument Corp.).

tons and five subscriber designated sensors. There is also an optional dial-up telephone wire connected (if desired) into the digital dialer and receiver. The cable interface, as you can see, is also located in the head end, receiving two channels from the alarm processing unit and returning one.

Sometimes the cable interface may not be located at the head end and, instead, is installed in the network's business office and subsequent communications are by either cable drop/trunk or microwave. Here, an FM modulator accepts polling information from the SRX-11 transmitter and sends it on to the upconverter at the head end. There they are converted to 88 to 108 MHz FM and distributed throughout the system, eventually received and demodulated by the receive SRX-11 units at the business office and cable interface (Fig. 10-2).

Descriptions of cable television systems have been done many times before, so we thought you might like to go directly into measurement conditions that always must be carefully controlled for good signal distribution. We complete the chapter with a further description of Pay TV and some of the trends involved in the early 1980s.

CATV MEASUREMENTS

Television broadcast and cable video processing are, by far, the most complex signals in all of consumer or commercial rf communications. Known as vestigial-sideband transmission, since one sideband has been filtered and virtually suppressed, these are all combinations of pulse, phase, frequency, and amplitude signals fitted into a 6 MHz channel bandwidth with 4 MHz composite video. There are vertical, horizontal, and chroma burst sync, amplitude-modulated video, frequency-modulated sound, and chroma phase and amplitude which must all be coordinated and frequency locked within a single transmitted and received envelope. All this, of course, then requires FM, chroma, and luminance detection, finally being separated and displayed as coordinated components of a single video/audio signal.

If there are also microwave and/or satellite transmissions, you must then consider effects of the uplinks, downlinks, transmitters, receivers, transponders, LNAs and all the rest of the electronics and mechanical components that constitute the whole. As usual, the whole is only as strong and effective as its principal parts. You lose sound and the picture is worth little; cancel picture, then few listen to sound; but drop horizontal and vertical sync, and subse-

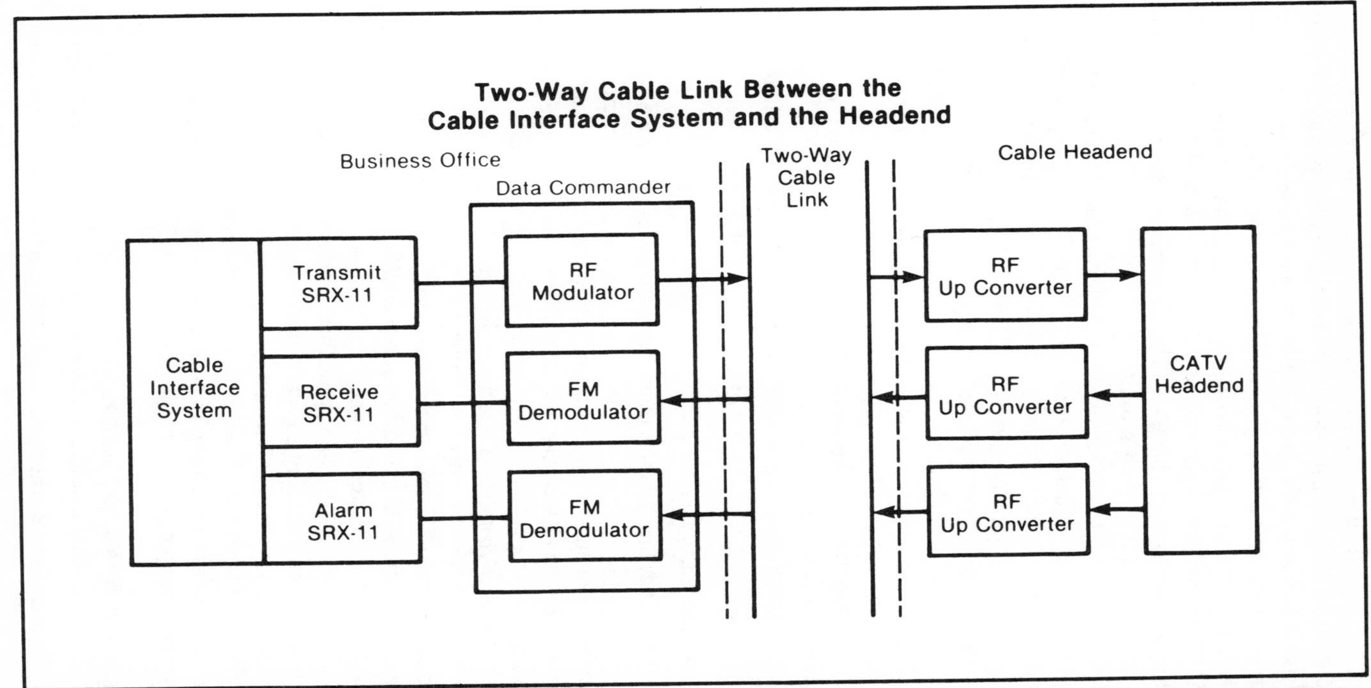

Fig. 10-2. An alternate cable link between business office and CATV headend (courtesy of Jerrold Division, General Instrument Corp.).

quent picture flicker drives everyone away. Of course, overmodulation buzz, intermittent color, and phase changes producing purple people have their effects too; so never let it be said that all problems are major. Trunk amplifiers, feeder lines, taps, drops, and distribution amplifiers contribute their collective problems as well. Therefore, anyone who doesn't share the view that CATV is an octopus proposition, should really rearrange his mental and metaphysical outlook. It's tough!

More difficult yet, however, is the hiring and training of competent technicians to maintain an elaborate CATV setup, and keep it operating even within gross specifications. Here, we'll try and help with some valuable aid in the form of suggestions from prime industry sources Tektronix and Hewlett-Packard. Fortunately, they've been through the pitfalls before and we'll heed their admonitions and add a few instrumentation thoughts and measurements of our own. Such attention to specifications aren't necessarily difficult, but without proper equipment they're impossible. This is why twenty to fifty thousand dollars for a good spectrum analyzer offers nothing less than good insurance. But first, let's blaze the electronics trail with some explanatory definitions.

□ *Headends* pick up broadcast, satellite, and microwave signals from the "air," heterodyne them up or down for convenient handling and include some or all on one or more channels for commercial or consumer consumption. They may also originate programming such as weather and local events as part of the general service.

□ *Trunk Amplifiers* are low distortion, non-noisy, medium gain devices which compensate inevitable cable losses with both auto slope and AGC (automatic gain control). In the same box as Trunk Amplifiers, are also *Bridger Amplifiers,* splitting outgoing signals into several feeder lines.

□ *Distribution Amplifiers* follow the bridgers, and amplify signal levels, compensating for cable and tap losses.

□ *Taps* split feeder line signals for subscribers, and *drops* are the direct cable feeds between taps and the consumer, which are required to offer measurements of some − 6 to + 14 dBmV for snow-free images.

Spectrum Analyzer Readings

Now, if you look at the simplified diagram in Fig. 10-3, all this can be gathered together and assimilated before hard-core mea-

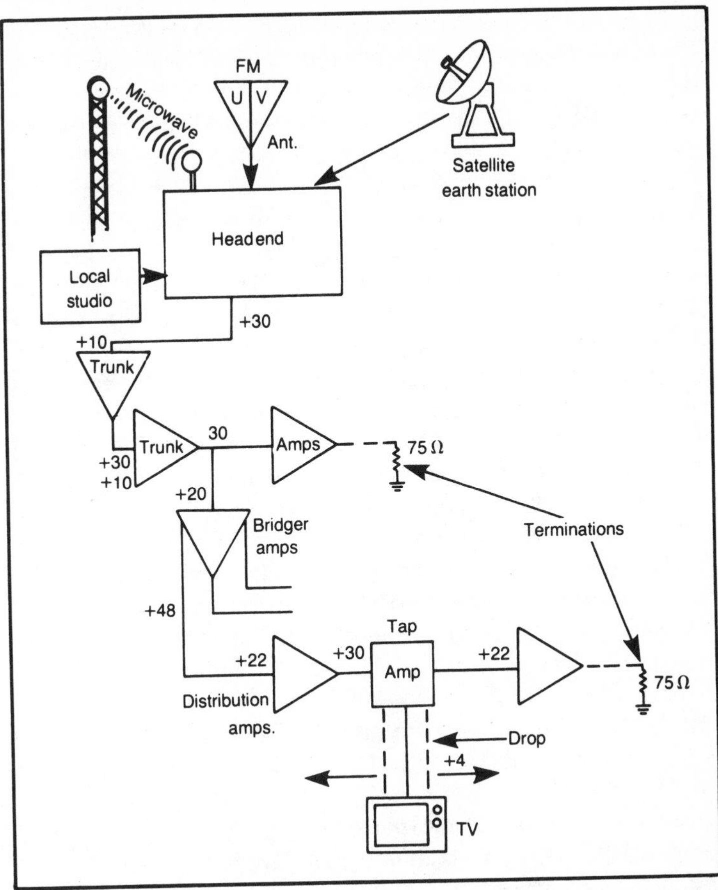

Fig. 10-3. CATV system from headend to subscriber TV. All signals specified in dBmV.

surements begin. Note distances and successive (general) dB levels for various amplifier inputs and outputs as suggested by HP and practiced by many of the CATV installations. A personal check verified these figures which hold true in most instances with only minor variations. Longer distances and different amplifiers, of course, could produce changes which each manufacturer will, undoubtedly, specify. We'll supply more information as the chapter unfolds. In the meantime, let's look at the usual methods of analyzing and troubleshooting individual boxes as well as the network.

As you would suspect, they're called signal and network analysis and differ only in signal source input. The former takes routine intelligence from the trunk and measures it to see if there is both sufficient output and lack of distortion. The latter bases its results on a known signal input/output (a signal generator, for instance) to see if the results are changed enough to warrant repairs or replacement. You may also sweep the entire system from its originating head end, measuring the effects downstream at the connecting amplifiers and feeder lines. Of course field strength meters, frequency counters, and marker generators all help the cause, too. For our purposes, however, we'll rely on ordinary oscilloscopes and spectrum analyzers to make points along the way since broadcast and CATV signals are relatively synonymous—used here as being of similar meaning.

Since a spectrum analyzer both sweeps and detects what it sees from hertz to GHz, there's no difficulty in looking at a pair of television broadcast stations simultaneously at 300 kHz resolution and a frequency rate of 2 MHz/div. All levels begin at −30 dBm, and are then evaluated at 10 dB down per division (Fig. 10-4). But, we had to match the 50 ohms impedance of the analyzer to the 75-ohm coaxial cable of our input with a low-loss impedance matcher. So to all of these readings, you *add* 5.72 dB as loss compensation. On the left of each of the two signals you see the video carrier being amplitude modulated, then the so-called suppressed 3.58 MHz chroma subcarrier, followed by the FM audio carrier, 4.5 MHz separated from video. In our particular area, these channels have to be 4 and 5 because 3 and 4 or 6 and 7 are not adjacent, and also that the 7L12 Tek analyzer has its midpoint setting at less than 80 MHz. Observe that the audio carrier for the first station and the taller video carrier for the second are a full 6 MHz apart. Since these are Federal requirements, CATV must observe the same channel spacings and produce carriers in identified locations accordingly.

Reminding you that an addition of 5.72 dB has to be *added* to each reading algebraically, the first video carrier on the left would appear at a true level of −56 dB + 5.72, or −50.28 dB. Its allegedly suppressed subcarrier then measures some −76 dB + 5.72, or −70.28, and the audio carrier becomes −62 dB + 5.72, or −56.28 dB. The same reasoning applies to the next *higher* frequency channel video, subcarrier, and audio carriers to the right. With well-defined readouts on the picture tube, itself, such measurements-aren't difficult at all. Try channel 5 on your own, it's good practice. Then if you want to translate all this into either microvolts/millivolts or picowatts/nanowatts/milliwatts, the accompanying Tektronix chart does this for you quite easily (Fig. 10-5). As an example −70.28 dBm (an absolute measurement) becomes about 52

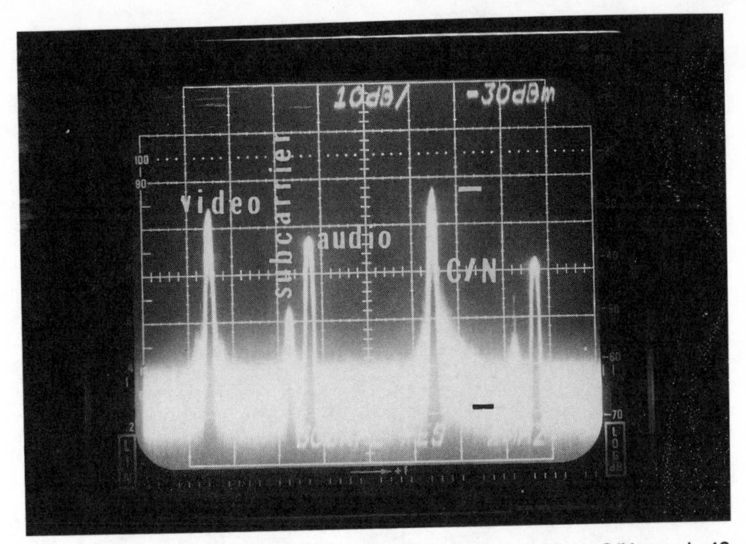

Fig. 10-4. CATV carriers should be similar to those broadcast. C/N equals 48 dB, −15.5 dB, or 32.5 dB.

picowatts or about 58 microvolts. All this is so since we're now working strictly at 50 ohms input impedance, having already compensated for the impedance transition.

Carrier-to-Noise Measurements, however, are another matter entirely, no easy off-the-scale readings like we've tackled so far. The reason is that the characteristics of your own spectrum analyzer must be taken into account for a true reading, and this, sometimes, requires a little sidewinding arithmetic.

We're investigating a 4 MHz video signal and, in this instance, looking at it with a resolution bandwidth of 300 kHz. In addition, it's also necessary to add in 2.5 dB for log amplifier and detector correction. Therefore, you have the following corrective interpolation:

$$\begin{aligned} C/N &= 10 \log (4\ \text{MHz}/.3\ \text{MHz}) + 2.5\ \text{dB} \\ &= 10 \log 13.3 + 2.5\ \text{dB} \\ &= 13.8\ \text{dB}\ \textit{correction} \end{aligned}$$

This subtracted from the *apparent* S/N of the number two video carrier, which is 48 dB, and you have a corrected factor of:

$$\textit{True C/N} = 48\ \text{dB} - 13.8, \text{ or } \textit{34.2 dB}$$

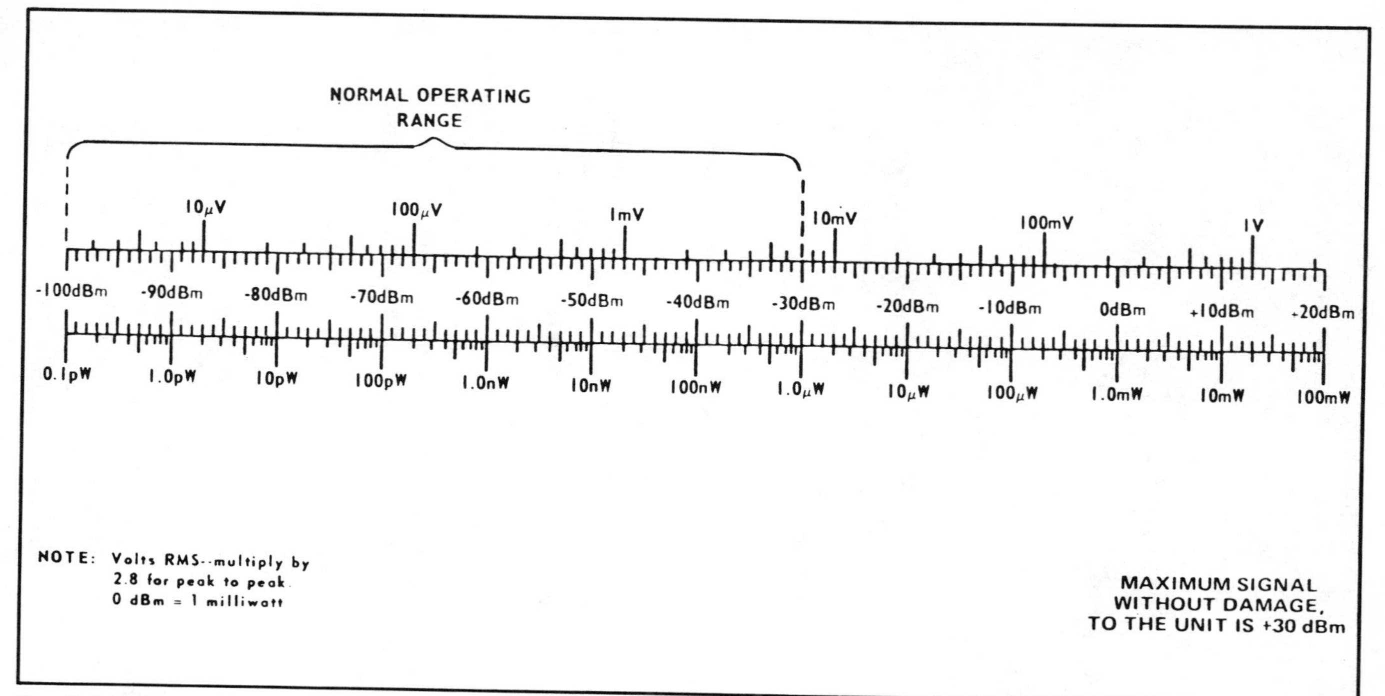

Fig. 10-5. dB/Volts/Watts conversion chart at 50 ohms (courtesy Tektronix).

Naturally, more restrictive resolutions will change the figures accordingly. Observe, by the way, the considerable difference in raw S/N with signal under modulation (which is really carrier-to-noise) and actual C/N. Too many measures, unfortunately, make this obviously inaccurate mistake.

Another useful equation supplied by M/A-COM is the video S/N relative to carrier level:

$$Vs/n = 126.5 + RCL - NF$$

Where RCL is the received carrier level and NF, the receiver noise figure, which is typically 8 to 12 dB. This is, of course, for satellite receive computations, and includes EIA emphasis and color weighting. Here, the carrier is FM and not AM and its level is in dB, somewhere between −79 and −35 dBm. So if your CATV operation is using satellite downlinks, this signal/noise representation should come in rather handily.

S/N

We know you'd like to take this same CATV or off-the-air voltage and put it through a demodulator and immediately measure S/N. But, sorry to say, this can't be easily done because of varying signal amplitudes and program content. Therefore, a color bar generator is required with a cw color bar output to do the job. In this instance, we used swept chroma (Fig. 10-6) at 3.08, 3.58, and 4.08 MHz (reading from left to right), respectively. As you can readily see, the fall-off in bandpass between 3 and 4 MHz is considerable, so be forewarned when you purchase either a video monitor or an ordinary television receiver for business or personal use. They may not be what they're cracked up to be, even with the highly-touted comb filter.

If you have any doubt that luma bandpass also affects chroma (Fig. 10-7), look at the swept chroma waveforms at the video detector and cathode-ray tube of this same receiver. Viewed with an ordinary oscilloscope, the roll off in both instances, while not terrible, is substantially more than it should be, especially at the CRT. Although not quite the 10 dB drop exhibited by the spectrum analyzer, this is certainly a 6 dB down proposition, as opposed to a more substantial and reasonable reading of 3 dB. However, had this been an ordinary 3.58 MHz chroma trap receiver, color reproduction could and would have been substantially less. This

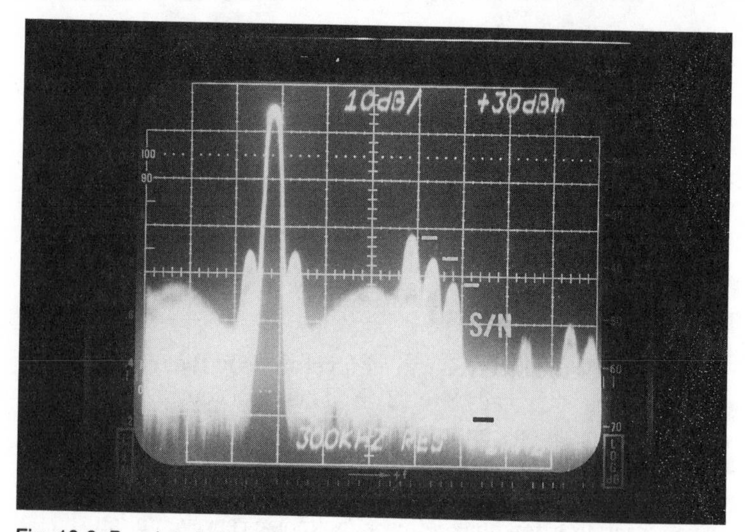

Fig. 10-6. Baseband of standard 3.08 to 4.08 MHz chroma at CRT cathodes. S/N here at 3 MHz is 39 dB, but only 29 dB at 4 MHz.

is not a bad point to remember when, cash in hand, you make the fateful transaction.

Would the response have been better through a broadband monitor without the limiting effects of tuner and intermediate frequency amplifiers? In a good monitor, yes; in a cheap one, not necessarily,

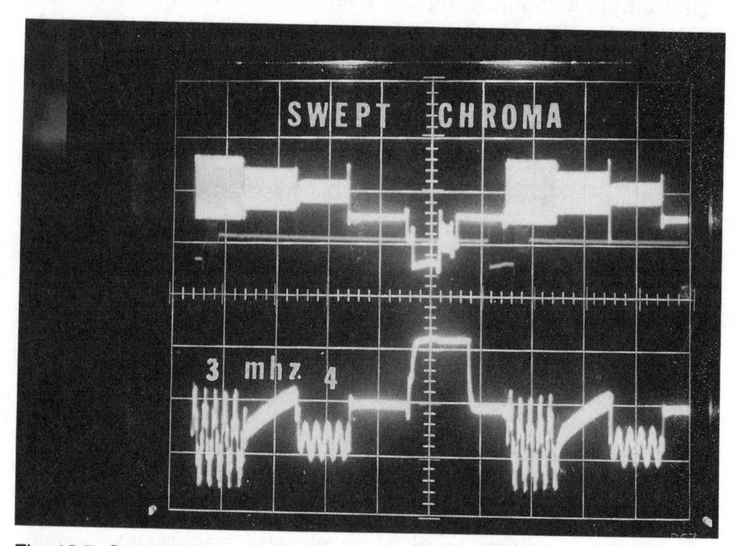

Fig. 10-7. Swept chroma at the video detector and CRT terminals shows roll-off at both test points.

since this was a comb filter set to begin with and does have a 4 MHz bandpass at the cathode-ray tube, as Fig. 10-8 dutifully shows. Note the dropoff in frequency between the starting multiburst of 0.75 and 4.08 MHz, especially in the lower waveform at the CRT. As you can see, luma processing and chroma processing are tricky businesses. Unless there are thorough tests you'll never know.

One such test you'll rarely see from any source is the gated rainbow vector (Fig. 10-9). This takes a very clean color bar generator, a means to invert the scope traces because of modern cathode CRT connections, and little phase shift between the scope's two vertical amplifiers. What the 10-petal display represents is the vector sum and difference voltages shown in the two R-Y and B-Y waveforms above. From this combination alone you can readily discern correct or incorrect front-end tuning, chroma bandpass response, angles of demodulation, and effective chroma range. As you can see in this pattern, which should form a perfect 300-degree arc—30 degrees between each petal—all isn't too bad but certain conditions aren't too good, either. The first petal is not well formed, the first three are compressed together, and a clean 300-degree arc is by no means apparent. That's just for starters and, were this a television book, we could tell you considerably more, not all of it bad. Actually, this is a pretty good pattern in comparison with many foreign and domestic receivers currently on the market. Does any-

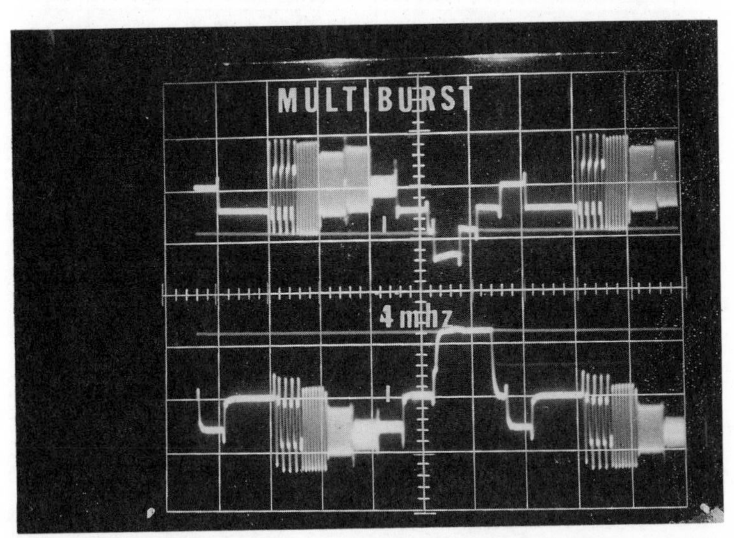

Fig. 10-8. Multiburst at video detector and CRT verifies 4 MHz luminance bandpass.

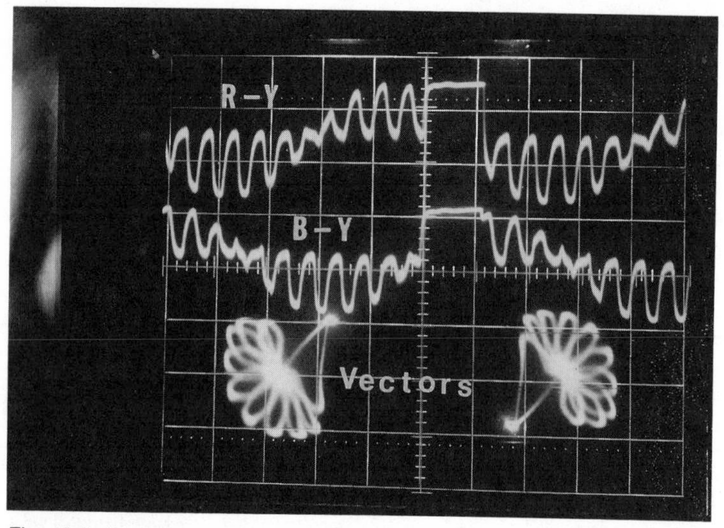

Fig. 10-9. A very clean color bar generator and vectorscope show passable vector patterns.

one *now* produce a perfect gated rainbow vector? Not that we have seen, especially with non-VIR (vertical interval color reference) "idiot" buttons that are supposed to correct for changing broadcast flesh-tones during camera switches, tape, or film projections. NTSC waveforms for television receivers, by the way, are virtually useless because of variable luma, chroma, and picture controls changing preset levels. A clean gated rainbow pattern permits these adjustments and still tells most of the chroma story. Amen! Incidentally, if you'd like to change 50 ohms dBm to 75 ohms dBmV, simply *add* 54.5 dB to each level measured. For instance, −30 dBm equals +24.5 dBmV (−30 + 54.5). To change dBmV (75 ohms) to dBm (75 ohms), subtract 48.8 dB (−10 − 48.8 = −58.8 dBm). We might also remind you that the difference between two levels such as −40 and −10 in any sort of value, is just read as 30 dB since it is not an absolute measurement of one value or another.

Distortions

Let's return briefly to the rf portion and part of its spectrum to look at intermodulation and harmonic distortions. CATV recognizes that spurious beats "less than 36 to 46 dB down" are often visible in the television picture.

Either interference within those measurements can be much too objectionable for picture enjoyment and normally must be found quickly and overcome either by filtering, antenna tuning, or somehow physically or electronically removing the undesirable source.

First, we have an excellent example of *intermodulation distortion*, which is the modulation of two or more signals by one another, producing results equal to the sum and differences of integral multiples of the original signals. Figure 10-10 shows the effects of intermod between the video carrier and its nonsuppressed 3.58 chroma subcarrier. Since the analyzer's center frequency is tuned to about 190 MHz, this unwanted voltage appears close to 192 MHz and can easily be identified by your spectrum analyzer's marker or a separate signal introduced from a signal generator through a splitter.

Next, let's try an example of *crossmodulation*. This is where some sort of interference modulates the main signal—in this instance, the video carrier. Shifted just to the right of this carrier, you see the interfering signal climbing up its skirts. Since amplitude difference is only about 26 dB (Fig. 10-11), you can well imagine what's happening to the picture as it weaves and bobbles with additional disturbance. This also illustrates how easily intermodulation distortion may become crossmodulation distortion. As both illustrations show, there's little difference between the two except a shift in frequency, and there are such things as drift and changing conditions. Sum and differences can quickly become fundamentals, as crossmod aptly suggests when interference and primary signals combine.

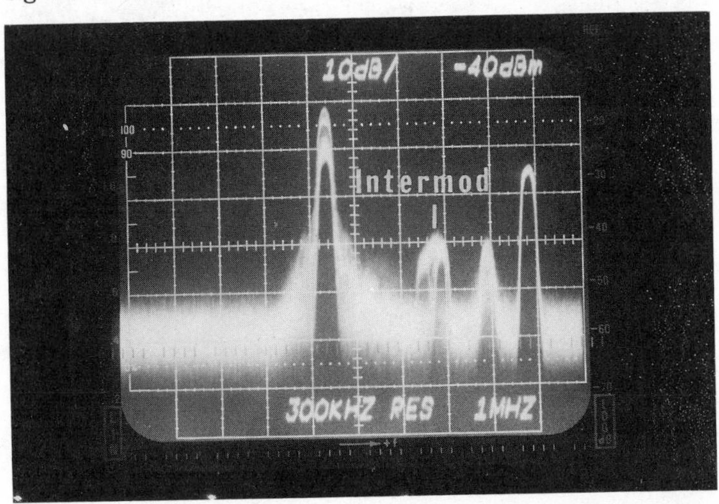

Fig. 10-10. Intermodulation distortion results from product of at least two signals.

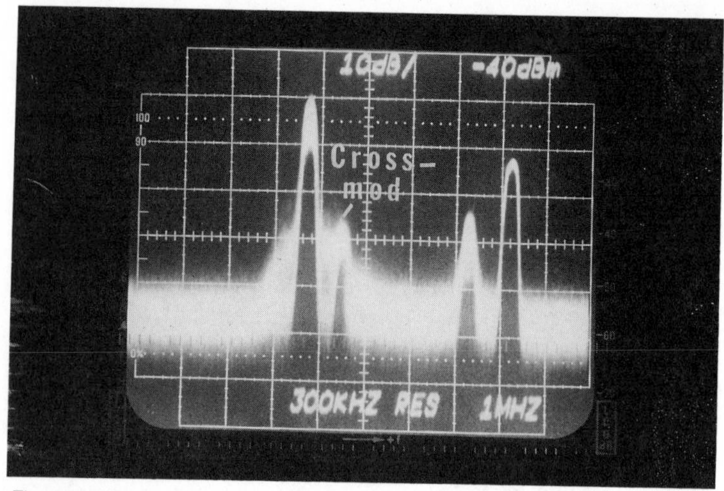

Fig. 10-11. Crossmodulation interferes directly with the video carrier—our main signal.

Other interferences in the CATV system may also be acquired by the head end, enter through the various "sealed" amplifiers, or originate from strong local TV broadcasts leaking into the cable itself. Should the latter appear on co-channel, a secondary image (ghost) will often *lead* the CATV transmission, producing annoying primary and shadow mirror-image pictures. Ordinary *reflective* images that manage to seep through in strength produce *lagging*, indistinct secondaries, indicating their origin may or may not be co-channel, depending on ghost outline. Other distortions on channels two or four, for instance, resulting in diagonal fine lines, often are the result of Citizens Band Radio harmonic interference originating about 27 MHz.

Inside the CATV system, leakage, hum, amplitude modulation, internal cross modulation, poor grounds, bad couplings, slumping amplifiers, frayed or broken cables, illegal tapoffs, and vandalism are all part of the many problems that arise at one time or another. To aid you in one necessary statistic, here's how to calculate low frequency problems such as hum using both dc and ac levels.

$$\% \text{ hum} = \text{ac p-p/dc operating voltage} \times 2 \times 100$$
$$\% \text{ hum} = 1 \text{ V}/50 \text{ V } (2) \times 100 = 1\%$$

Another equation involving *Field Strength* is also helpful during certain measurements. If E is the radiated field strength in volts,

H, the magnetic field strength in webers/meter^2, then space impedance equals 377 ohms:

$$E/H = 377 \text{ ohms}$$

You already know the impedance of free space, therefore either of the other two will complete an equation for the third.

CATV EQUIPMENT AND SYSTEMS

As we move further into CATV systems, you might be able to make good use of several system calculation aids worked out by C-COR Electronics, Inc. of State College, Pennsylvania which should be of considerable help in power and voltage anomalies as well as identifying the various CATV channels already in use.

Beginning with the sub-vhf (5 - 54 MHz) and continuing through the hyper band (300 - 400 MHz), this chart (Fig. 10-12) gives all of the current CATV channel designations, including video, chroma, and audio carriers operating in the various systems throughout the United States. All frequencies are in megahertz, and channel numbers are those established for common use. Notice there are 60 listed with possibly as many as twice that number eventually to be placed in service.

With the power and cross mod charts, one may calculate noise and signal problems throughout any system, based on manufacturers' specifications of their various trunk, bridger, and distribution amplifiers. Measurements at any point, of course, will verify such conditions or indicate further problems possibly due to leakage, amplifier slumps, component failures, or external signals downstream. Where triple beat noise (A + B − C) enters the picture and produces third order distortions that may number in the hundreds or thousands, an illustration is also supplied by C-COR (Fig. 10-13) detailing "incoherent" beats as a function of noise power or voltage degradation in cascaded amplifiers in terms of decibels (dB). Examples 1 and 2 shown in the figure illustrate both noise and beats as graduated numbers between 1 and 1,000 versus dB measurements from 0 to 40. Again, these examples will help in determining satisfactory operation of CATV systems in good condition, but local test equipment should be used to find both general and specific faults, if such exist. As you can presume, the foregoing aids are extremely valuable in laying out or "proving" your system on paper before it becomes fully operational. After that, oscilloscopes, spectrum analyzers, power meters, and field strength

Channel	Freq. range (MHz)	Video (MHz)	Chroma (MHz)	Audio (MHz)
SUB VHF 5-54 MHz				
T7	5.75-11.75	7	10.58	11.5
T8	11.75-17.75	13	16.58	17.5
T9	17.75-23.75	19	22.58	23.5
T10	23.75-29.75	25	28.58	29.5
T11	29.75-35.75	31	34.58	35.5
T12	35.75-41.75	37	40.58	41.5
T13	41.75-47.75	43	46.58	47.5
T14	47.75-53.75	49	52.58	53.5
LOW VHF 54-88 MHz				
2	54-60	55.25	58.83	59.75
3	60-66	61.25	64.83	65.75
4 Pilot*	66-72	67.25	70.83	71.75
5	76-82	77.25	80.83	81.75
6	82-88	83.25	86.83	87.75
FM 88-108 MHz				
SPECIAL 108-120 MHz				
A''	108-114	109.25	112.83	113.75
A'	114-120	115.25	118.83	119.75
MID 120-174 MHz				
A	120-126	121.25	124.83	125.75
B	126-132	127.25	130.83	131.75
C	132-138	133.25	136.83	137.75
D	138-144	139.25	142.83	143.75
E	144-150	145.25	148.83	149.75
F	150-156	151.25	154.83	155.75
G	156-162	157.25	160.83	161.75
H	162-168	163.25	166.83	167.75
I	168-174	169.25	172.83	173.75
HIGH VHF 174-216 MHz				
7	174-180	175.25	178.83	179.75
8	180-186	181.25	184.83	185.75
9	186-192	187.25	190.83	191.75
10	192-198	193.25	196.83	197.75
11	198-204	199.25	202.83	203.75
12	204-210	205.25	208.83	209.75
13	210-216	211.25	214.83	215.75
SUPER 216-300 MHz				
J*	216-222	217.25	220.83	221.75
K	222-228	223.25	226.83	227.75
L	228-234	229.25	232.83	233.75
M	234-240	235.25	238.83	239.75
N*	240-246	241.25	244.83	245.75
O	246-252	247.25	250.83	251.75
P	252-258	253.25	256.83	257.75
Q	258-264	259.25	262.83	263.75
R	264-270	265.25	268.83	269.75
S*	270-276	271.25	274.83	275.75
T	276-282	277.25	280.83	281.75
U	282-288	283.25	286.83	287.75
V	288-294	289.25	292.83	293.75
W	294-300	295.25	298.83	299.75
X*		301.25		
HYPER 300-400 MHz				
AA*	300-306	301.25	304.83	305.75
BB	306-312	307.25	310.83	311.75
CC	312-318	313.25	316.83	317.75
DD	318-324	319.25	322.83	323.75
EE	324-330	325.25	328.83	329.75
FF*	330-336	331.25	334.83	335.75
GG	336-342	337.25	340.83	341.75
HH	342-348	343.25	346.83	347.75
II	348-354	349.29	352.83	353.75
JJ	354-360	355.25	358.83	359.75
KK	360-366	361.25	364.83	365.75
LL	366-372	367.25	370.83	371.75
MM	372-378	373.25	376.83	377.75
NN*	378-384	379.25	382.83	383.75
OO	384-390	385.25	388.83	389.75
PP	390-396	391.25	394.83	395.75
QQ	396-400	397.25	400.83	401.75

* Normal pilot frequencies.

Fig. 10-12. An up-to-date list of CATV channels available throughout the U.S. (courtesy of C-COR Electronics Inc.).

meters must come to the rescue if and when downstream and upstream conditions that are out of tolerance persist.

Headends

No longer a matter of simple up/downconversion of limited signals, the head end now must permit a variety of inputs and drive an increasing number of outputs, especially in metro areas where multi-service requires a great deal more than nominal consumer support. This is especially true of scrambled satellite information that's either here or on the way, as well as one and two-way data services, auto program substitution (to guard against unannounced signal failures), and channel emergency warnings. All this means good surface wave filters to cleanly reject adjacent channel interference, many levels for channel coding, excellent spurious signal reduction, and extra wide bandwidths for truer signal processing and full-channel service. Better U/V demodulation is also required, special data channels, security overrides (where the community is so equipped), baseband matrix switching, status monitors, and satellite receivers for those very important channels not available through microwave, broadcast, or radio/studio reception.

A good example of one of these modulators comes from the Jerrold Division of General Instrument Corp (Fig. 10-14). At 450 MHz, it delivers a full 67 channels of service, handles scrambling and emergency override, program source switching, and full feature i-fs for loop-throughs and AGC. All this, in addition to many front panel control functions and a slide-out drawer for signal leveling without interrupting on-going service. Along with the processor and satellite receiver, the modulator can *transmit* previously scrambled signals which most of its competition cannot, according to Jerrold.

Mainstations

Packed with redundant amplifiers, backup power supply, surge protectors, status monitors, both 1-way and 2-way configurations, feeder disconnects, modular design, and A-B cables for home service and institutions, Jerrold supplies highly flexible mainstations. Each amplifier station can be fitted with two amplifiers, permitting one to remain in service (automatically) if the other fails; and if there is status monitoring, any such failure will be noted at the headend and allow convenient repairs. All of which, of course, is done by plug-in modules that can be removed and replaced at will.

SYSTEM PERFORMANCE CALCULATION AIDS

Voltage Addition

TWO SIGNALS OR CROSS-MOD. COMPONENTS, COMBINED ON A VOLTAGE BASIS.
GIVEN: dB DIFFERENCE BETWEEN THE TWO LEVELS
FIND: dB TO ADD TO HIGHER LEVEL TO GET TOTAL LEVEL IN dB.

dB difference	0.0	0.1	0.2	0.3	0.4	0.5	0.6	0.7	0.8	0.9
0	6.02	5.97	5.92	5.87	5.82	5.77	5.73	5.68	5.63	5.58
1	5.53	5.49	5.44	5.39	5.35	5.30	5.26	5.21	5.17	5.12
2	5.03	5.03	4.99	4.94	4.90	4.86	4.82	4.78	4.73	4.69
3	4.65	4.61	4.57	4.53	4.49	4.45	4.41	4.37	4.33	4.29
4	4.25	4.21	4.17	4.13	4.10	4.06	4.02	3.98	3.95	3.91
5	3.88	3.84	3.80	3.77	3.73	3.70	3.66	3.63	3.60	3.56
6	3.53	3.50	3.46	0.43	3.40	3.36	3.33	3.30	3.27	3.24
7	3.21	3.18	3.15	3.12	3.09	3.06	3.03	3.00	2.97	2.94
8	2.91	2.88	2.86	2.83	2.80	2.77	2.74	2.72	2.69	2.68
9	2.64	2.61	2.59	2.56	2.53	2.51	2.48	2.46	2.44	2.41
10	2.39	2.36	2.34	2.32	2.29	2.27	2.25	2.22	2.20	2.18
11	2.16	2.13	2.11	2.09	2.07	2.05	2.03.	2.01	1.99	1.97
12	1.95	1.93	1.91	1.89	1.87	1.85	1.83	1.81	1.79	1.77
13	1.75	1.74	1.72	1.70	1.68	1.67	1.65	1.63	1.61	1.60
14	1.58	1.56	1.55	1.53	1.51	1.50	1.48	1.47	1.45	1.44
15	1.42	1.41	1.39	1.38	1.36	1.35	1.33	1.32	1.31	1.29
16	1.28	1.26	1.25	1.24	1.22	1.21	1.20	1 19	1.17	1.16
17	1.15	1.14	1.12	1.11	1.10	1.09	1.08	1.06	1.05	1.04
18	1.03	1.02	1.01	1.00	0.99	0.98	0.96	0.95	0.94	0.93
19	0.92	0.91	0.90	0.89	0.88	0.87	0.86	0.86	0.85	0.84
20	0.83	0.82	0.81	0.80	0.79	0.78	0.77	0.77	0.76	0.75
21	0.74	0.73	0.73	0.72	0.71	0.70	0.69	0.69	0.68	0.67
22	0.66	0.66	0.65	0.64	0.64	0.63	0.62	0.61	0.61	0.60
23	0.59	0.59	0.58	0.57	0.57	0.56	0.56	0.55	0.54	0.54
24	0.53	0.53	0.52	0.51	0.51	0.50	0.50	0.49	0.49	0.48
25	0.48	0.47	0.46	0.46	0.45	0.45	0.44	0.44	0.43	0.43
26	0.42	0.42	0.42	0.41	0.41	0.40	0.40	0.39	0.39	0.38
27	0.38	0.38	0.37	0.37	0.36	0.36	0.35	0.35	0.35	0.34
28	0.34	0.34	0.33	0.33	0.32	0.32	0.32	0.31	0.31	0.31
29	0.30	0.30	0.30	0.29	0.29	0.29	0.28	0.28	0.28	0.27
30	0.27	0.27	0.26	0.26	0.26	0.26	0.25	0.25	0.25	0.24
31	0.24	0.24	0.24	0.23	0.23	0.23	0.23	0.22	0.22	0.22
32	0.22	0.21	0.21	0.21	0.21	0.20	0.20	0.20	0.20	0.19
33	0.19	0.19	0.19	0.19	0.18	0.18	0.18	0.18	0.18	0.17
34	0.17	0.17	0.17	0.17	0.16	0.16	0.16	0.16	0.16	0.15
35	0.15	0.15	0.15	0.15	0.15	0.14	0.14	0.14	0.14	0.14
36	0.14	0.14	0.13	0.13	0.13	0.13	0.13	0.13	0.12	0.12
37	0.12	0.12	0.12	0.12	0.12	0.12	0.11	0.11	0.11	0.11
38	0.11	0.11	0.11	0.10	0.10	0.10	0.10	0.10	0.10	0.10
39	0.10	0.10	0.09	0.09	0.09	0.09	0.09	0.09	0.09	0.09
40	0.09	0.09	0.08	0.08	0.08	0.08	0.08	0.08	0.08	0.08

Power Addition

TWO SIGNALS OR NOISE LEVELS COMBINED ON A POWER BASIS.
GIVEN: dB DIFFERENCE BETWEEN THE TWO LEVELS
FIND: dB TO ADD TO THE HIGHER LEVEL TO GET TOTAL LEVEL IN dB

dB difference	0.0	0.1	0.2	0.3	0.4	0.5	0.6	0.7	0.8	0.9
0	3.01	2.96	2.91	2.86	2.82	2.77	2.72	2.67	2.63	2.58
1	2.54	2.50	2.45	2.41	2.37	2.33	2.28	2.24	2.20	2.16
2	2.13	2.09	2.05	2.01	1.97	1.94	1.90	1.87	1.83	1.80
3	1.76	1.73	1.70	1.67	1.64	1.60	1.57	1.54	1.51	1.48
4	1.46	1.43	1.40	1.37	1.35	1.32	1.29	1.27	1.24	1.22
5	1.19	1.17	1.15	1.12	1.10	1.08	1.06	1.04	1.01	0.99
6	0.97	0.95	0.93	0.91	0.90	0.88	0.86	0.84	0.82	0.81
7	0.79	0.77	0.76	0.74	0.72	0.71	0.70	0.68	0.67	0.65
8	0.64	0.62	0.61	0.60	0.59	0.57	0.56	0.55	0.54	0.53
9	0.51	0.50	0.49	0.48	0.47	0.46	0.45	0.44	0.43	0.42
10	0.41	0.40	0.39	0.39	0.38	0.37	0.36	0.35	0.35	0.34
11	0.33	0.32	0.31	0.31	0.30	0.30	0.29	0.28	0.28	0.27
12	0.26	0.26	0.25	0.25	0.24	0.24	0.23	0.23	0.22	0.22
13	0.21	0.21	0.20	0.20	0.19	0.19	0.19	0.18	0.18	0.17
14	0.17	0.16	0.16	0.16	0.15	0.15	0.15	0.15	0.14	0.14
15	0.13	0.13	0.13	0.13	0.12	0.12	0.12	0.12	0.11	0.11
16	0.10	0.10	0.10	0.10	0.10	0.09	0.09	0.09	0.09	0.09
17	0.08	0.08	0.08	0.08	0.08	0.08	0.07	0.07	0.07	0.07
18	0.07	0.07	0.07	0.06	0.06	0.06	0.06	0.06	0.06	0.06
19	0.05	0.05	0.05	0.05	0.05	0.05	0.05	0.05	0.05	0.04
20	0.04	0.04	0.04	0.04	0.04	0.04	0.04	0.04	0.04	0.03

C-COR®
ELECTRONICS INC

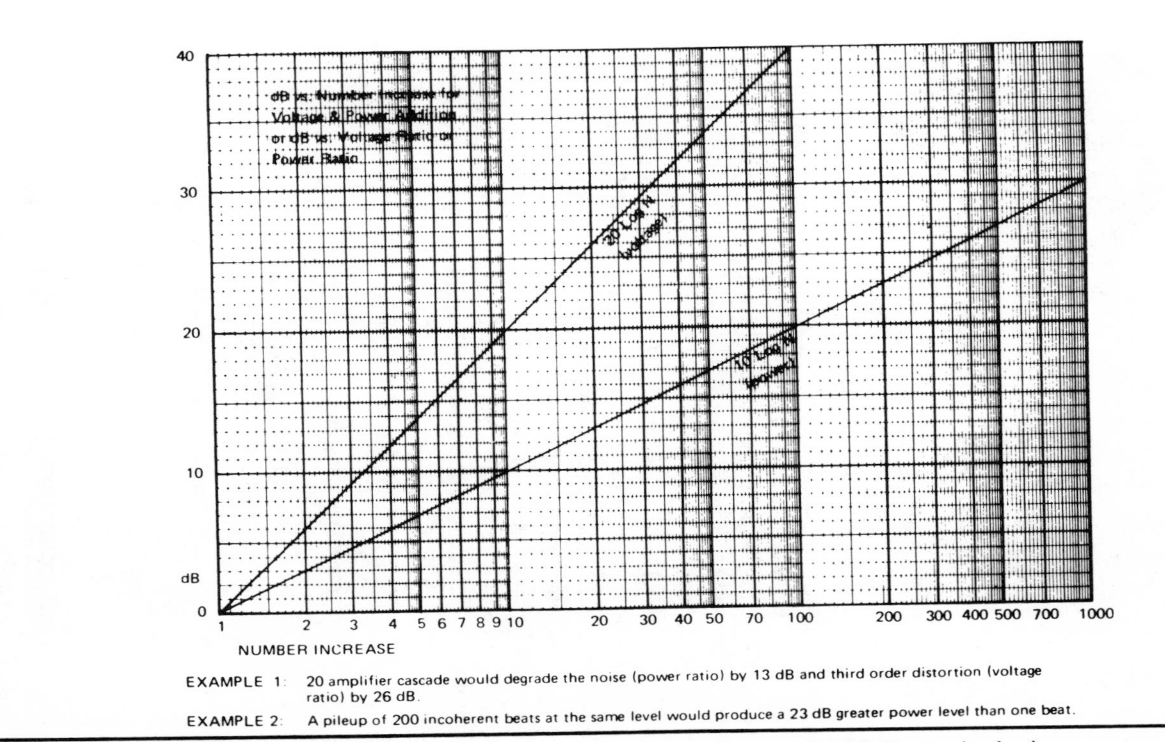

EXAMPLE 1: 20 amplifier cascade would degrade the noise (power ratio) by 13 dB and third order distortion (voltage ratio) by 26 dB.

EXAMPLE 2: A pileup of 200 incoherent beats at the same level would produce a 23 dB greater power level than one beat.

Fig. 10-13. System calculation aids in terms of both voltage and power (courtesy of C-COR Electronics Inc.).

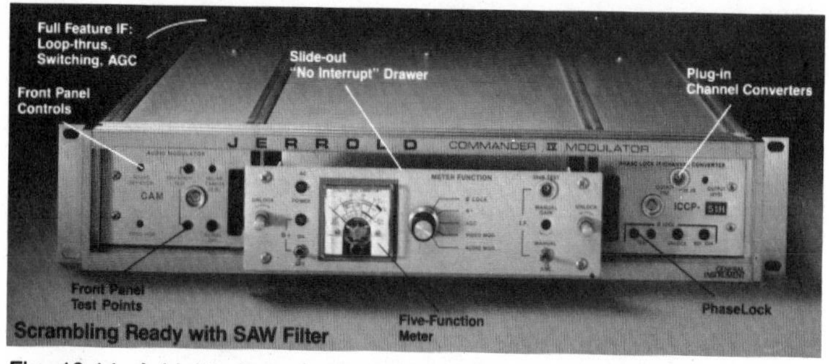

Fig. 10-14. A highly functional modulator that will also handle scrambling (courtesy of Jerrold Division, General Instrument Corp.).

External interfering signals may also be localized in the headend via feeder disconnections so the problem line can be identified. In addition, the Starline JN series permits modular buildup so that economical systems may be installed initially, and more complex functions added as requirements or operating capital increase. One of these JN Mainstations is shown in Fig. 10-15. Note the modular construction and observe that protective housings can be opened in "both directions."

Most CATV manufacturers and Jerrold make less expensive and less flexible amplifiers of various descriptions, of course, to fulfill different electrical and economic conditions. But, it is these modular, two-way, somewhat more expensive units that will serve both public and financial interests eventually, certainly in the multi-thousand subscriber category. News, weather, financial markets, shopping, education, and job opportunities will become prime programming and income considerations with the coming of the late

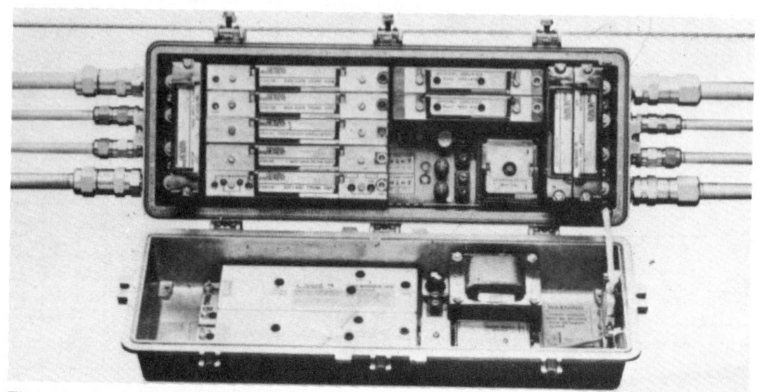

Fig. 10-15. The JN mainstation. It's totally modular (courtesy of Jerrold Division, General Instrument Corp.).

1980s and 1990s. Combine all this with the huge advantage of even more extensive satellite entertainment and other fare, and cable can offer substantial competition to any and all other electronic media. A vast home entertainment and business market is about to become a reality, affecting every segment of our daily and nightly lives! Once again, the home will become the center of all family activities, just as it was before the turn of the century—but for vastly different reasons, including a remarkably informed existence for those who will look and listen.

Trunk Stations

C-COR offers T-50X through T-52-X trunk stations which have 22 dB, 26 dB, and 32 dB respective spacings with forward band-passes between 54 and 400 MHz that vary with each series (Fig. 10-16). These feature slope-compensatived auto level controls, plug-in equalizers, reverse amplifier and plug-in bridger amplifier modules, plus splitters, directional couplers, striplines, and filters for four distribution outputs. There's also powered housing, plug-in gas discharge tubes at rf ports, and fuseable ac power distribution,

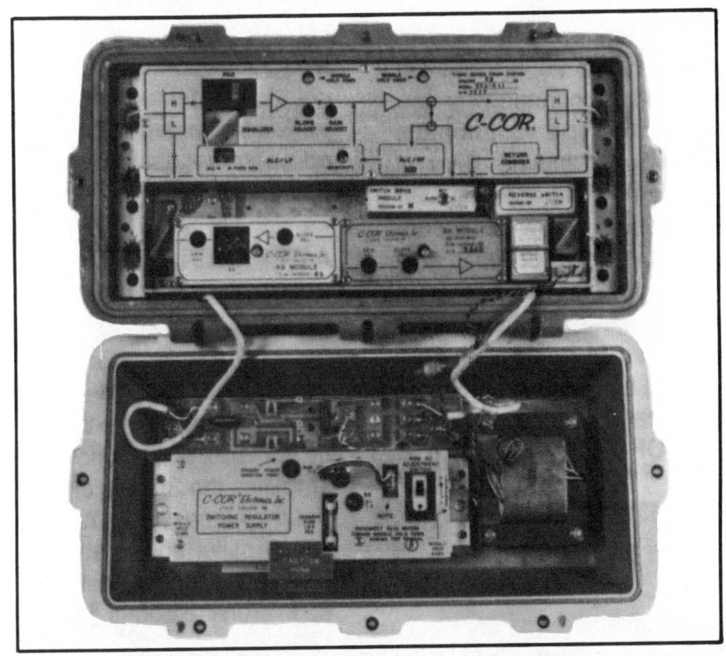

Fig. 10-16. C-COR's sophisticated trunk station amplifier with good MTBF stats (courtesy of C-COR Electronics, Inc.).

in addition to switching regulated power supplies. External ac and test-point connectors complete the description, which calls for mean time between failures of over 400,000 hours.

Such specifications not only indicate exceedingly high reliability as well as state-of-the-art features and advanced development for those who care to install excellent equipment. Two-way operation is available, of course, where diplex filters and reverse amplifier modules are included. C-COR claims the coolest amplifiers in the industry—just an example of what's available to those with specific needs.

Phase Lock

Phase-locked carrier systems are uniquely suited to handle large cable systems expanding to 50 or more channels, according to Jerrold's Michael Jeffers. For modest cable expansion in the 300 MHz region, supported by hybrid integrated circuits and larger cables, this works well up to about 35 channels. But, with greater demands, some cross modulation, and especially triple beat, new techniques must be used to supply adequate video throughout these enlarged systems.

Consequently, Mr. Jeffers suggests that the third order term in the power series expansion, resulting in three separate sideband components and a carrier component from three TV channels, may be handled by carrier phase lock to minimize triple beat. This drives 35- or 52-channel systems 5 dB higher "in level" with phase lock than otherwise in a noncoherent system. When phase lock is applied, sideband components become symmetrical about the carrier and any remaining distortion shows a cross modulation and is much less likely to become as annoying as triple beat.

There is also a frame synchronizer available at considerable cost that delays the TV signal so that all sync pulses may be coordinated with that of the transmitter. When sync-locked, all video may then be modulated on separate carriers and transmitted accordingly. Sync regeneration, of course, takes place at baseband only and not at rf since the delay provides time for proper sync reinsertion. You might recall that vertical sync is transmitted at a 59.94 Hz color rate, there are 262.5 lines per field, two fields per frame, and 30 frames per second. The horizontal rate, of course, occurs at 15,734 Hz/sec and amounts to a single sync pulse rather than the 6-6-6-series of equalizing and vertical pulses required for vertical sync.

Monitoring Trunk Amplifiers

Another of Jerrold's innovative ideas comes from Donald Groff in a paper entitled "An Amplifier Status Monitoring and Control System," which we will paraphrase to get its worthwhile contents in the book.

A transceiver positioned within a trunk amplifier receives a headend control signal that is 30-Hz frequency-shift keyed at 53.75 MHz and Manchester coded for ready detection at a 19.2 kHz rate. There are 16 bits, 9 of which are used for address, 5 for independent commands, plus start and stop bits. All are transmitted asynchronously, with the 9-bit address permitting 2^9 possible states and allowing all stations to be addressed within 0.5 second.

The bridger output picks off the control signal at the amplifier, a heterodyne receiver with crystal-controlled local oscillator converts it to a 10.7 MHz i-f. Recovered information now becomes digital and is further processed by CMOS logic which checks for the proper address. If modular-stored address and transmitted addresses coincide, then ensuing information is logically stored. Thereafter, only one station may reply on the upstream line at a time as diode switches permit reverse feed intelligence to continue into the reverse trunk.

Manually, addresses and commands are "dialed in" with thumbwheel and toggle switches. During scan, all 512 addresses are transmitted sequentially, with the memory mode allowing different commands to proceed to different addresses. Addresses and status information are displayed by front panel LEDs. Reverse video transmissions are also under development.

PAY TV

Pay TV is the subject of another interesting paper by Jerrold's Thomas O'Brien, Jr. His ideas are to partition addresses, control and authorization information into groups that are transmitted sequentially to approximate the EIA format. This is an 11-bit word, including start and stop bits, 8 data bits, and a parity bit located just before the stop bit.

Each word is then received, its address compared, and necessary information stored for later use. When groups have been recognized and accepted, all terminals would be commanded to take the same actions during each sequence. LSI logic permits the designing of complex converter/descrambler microcomputer types. In this instance, an FM receiver demodulates transmitted intelligence and

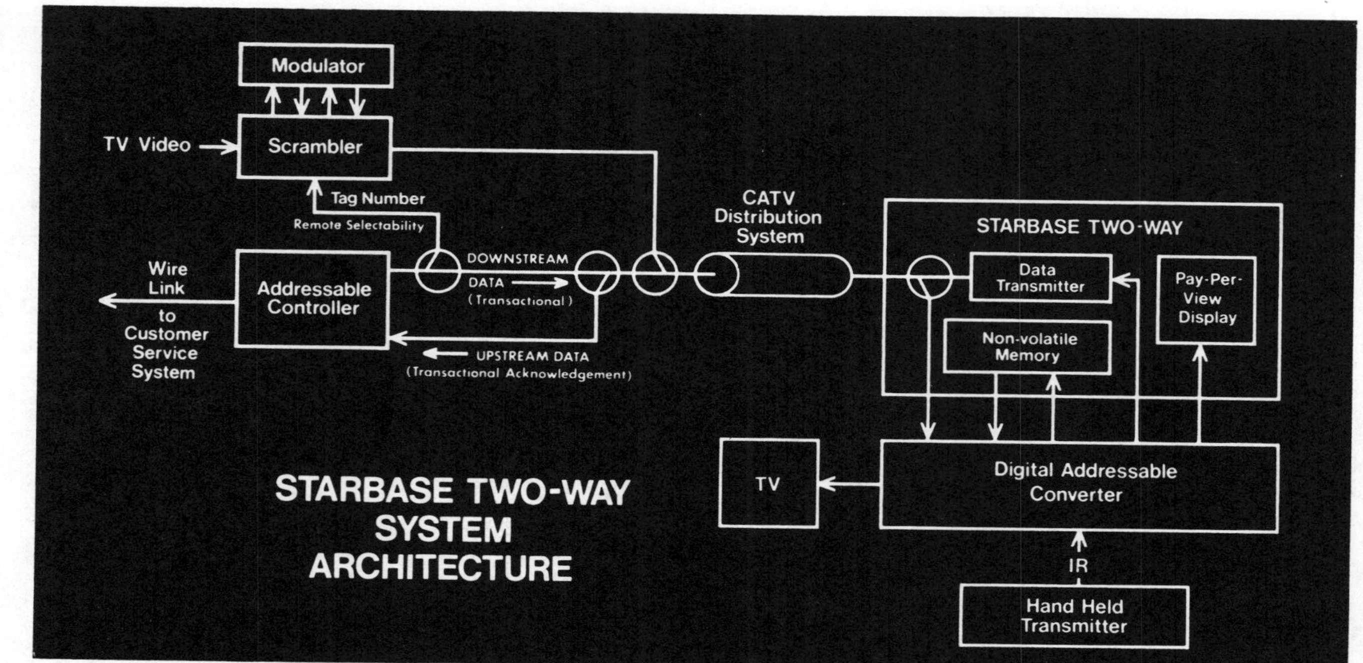

Fig. 10-17. Jerrold's customer program-select block diagram of its new STARBASE™ pay TV system (courtesy of Jerrold Division of General Instrument Corp.).

converts it to a serial bit stream. A decoder offers an 8-bit parallel data stream to the microcomputer. Meanwhile, an AM receiver demodulates timing signals for composite sync. The microcomputer then controls both converter and descrambler circuits. Finally, a programmable read-only memory stores the terminal's logic address for use in positive recognition.

The address for pay TV can access two million subscriber terminals, handle eight cable systems at 256 thousand terminals per system, and provide eight levels of update for 128 thousand subscribers in four minutes. Timing and 128 program levels are embedded in the TV signal and data transmission is narrow band FM.

At the headend, a microcomputer "stores, formats, and transmits" all address, control and other information to the rf data modulator in the smaller systems. As such systems increase in number, more modules can be added, including customer service operations, and even billing.

The Jerrold "pay-per-view" systems have both STARVUE™ and STARBASE™ systems for downstream 106.5 MHz systems and single upstream channels operating between 8.6 and 10.1 MHz for the former and 18 to 30 MHz for the latter. A two-way system

Fig. 10-18. The STARCOM™ 450 handles 66 channels (courtesy of Jerrold Division of General Instrument Corp.).

architecture drawing is illustrated in Fig. 10-17. As you can see, subscribers may make program selections virtually up to the final moments before programs change. Through a hand-held transmitter, the appropriate channel is selected, a given code keyed, and the digital addressable converter jogs the non-volatile memory, transmits to the pay-view display, and passes a signal to the data transmitter on its way upstream. The CATV distribution system moves these commands and upstream "requests" find their way both to the scrambler and the addressable controller in almost microseconds. A unique coded entry authorization eliminates the possibility of unintended pay program selections, according to Jerrold. If the upstream channel does not function as well as it should, the addressable controller may select another channel operating in the same group of 18-30 MHz frequencies.

A photo of Jerrold's STARCOM™ 450, which handles 66 channels through its digital settop converter and remote control is shown in Fig. 10-18. Built around a General Instrument microprocessor, this large-channel addressable unit provides maximum flexibility for major CATV systems. The remote unit is keyed much like the better remote television controls.

Index

Edited by Roland S. Phelps

Satellite Communications—2nd Edition

Stan Prentiss

An up-to-the-minute overview of satellite technology, equipment, and services.

"Packed between its covers is a veritable one-source encyclopedia of information ranging from satellites in orbit and under construction to types of services offered to home reception systems to security and scrambling devices."
—Modern Electronics

"Contains a treasure of useful information on how to install one's own earth station, data on the satellites presently in orbit and those under construction . . ."
—Satellite Dish Magazine

All those who found the first bestselling edition of *Satellite Communications* useful will be even more impressed with this revised, updated, and expanded edition! It provides a comprehensive, easy-to-follow look at satellite technology as it exists today, with special emphasis on television receive-only earth stations (TVROs) and signals to them from geosynchronous orbiting satellites.

In this edition Stan Prentiss, a former NASA satellite engineer, gives detailed explanations of many of the latest developments in satellite communications such as: Video Cipher® I, II, and B-MAC scrambling techniques . . . the new RCA Ku-band communications and video satellites . . . the impact of the Space Shuttle disaster on the satellite TV business as a whole . . . updated information on Western Union's Integrated Landline-Satellite Wideband Transmission Network . . . new contour maps of the U.S. that show the effective radiated power of some of the satellites at various locations . . . plus an updated chart of all satellites in the U.S. Domestic Satellite System, including the name of each satellite, its orbital position, its frequency, date launched, and number of transponders.

Also highlighted are up-to-date nomograph charts to aid you in locating any satellite from your own specific longitude and latitude using only a straightedge ruler. You'll also get complete information on TVRO dishes, feeds, receivers, and special space noise/loss/EIRP, as well as the prime means of uplink transmissions such as TCMA, and more.

Stan Prentiss is a former NASA satellite engineer and author of over 40 books on electronics and TV topics. He is a member of IEEE.

TAB TAB BOOKS Inc.
Blue Ridge Summit, Pa. 17214

Send for FREE TAB Catalog describing over 1200 current titles in print.

FPT > $16.95

ISBN 0-8306-2792-8

PRICES HIGHER IN CANADA

1660-0287